AMERICAN MATHEMATICAL SOCIETY TRANSLATIONS

Series 2

Volume 101

SIX PAPERS IN ANALYSIS

by

B. M. Levitan

V. A. Marčenko

B. L. Roždestvenskiĭ

Published by the

AMERICAN MATHEMATICAL SOCIETY

Providence, Rhode Island

1973

Library of Congress Catalog Number A51-5559
International Standard Book Number 0-8218-3051-1

AMS (MOS) subject classifications (1970).
Primary 34A10, 34A25, 34B25, 34C20, 35B30,
35L45, 35L60, 40E05, 42A76, 42A84, 47E05.
Secondary 34B25, 35F15, 35L65, 35P20,
40A05, 40E05, 41A50, 45D05, 47A70.

Library of Congress Cataloging in Publication Data

CIP

Levitan, Boris Moiseevich.
Six papers in analysis.

(American Mathematical Society. Translations, series 2, v. 101)
CONTENTS: Marcenko, V. A. Some questions in the theory of one-dimensional linear differential operators of the second order, (I-II).--Levitan, B. M. Remark on a theorem of V. A. Marcenko. On the asymptotic behavior of the spectral function of a self-adjoint second order differential equation. On a special Tauberian theorem. [etc.]
1. Differential equations--Addresses, essays, lectures. 2. Differential equations, Partial--Addresses, essays, lectures. 3. Differential operators--Addresses, essays, lectures. I. Marchenko, Vladimir Aleksandrovich. II. Rozhdestvenskiĭ, Boris Leonidovich. III. Title. IV. Series.
QA3.A572 ser. 2, vol. 101 [QA371] 515'.3'08 73-15614
ISBN 0-8218-3051-1

Printed in the United States of America

TABLE OF CONTENTS

Amer. Math. Soc. Transl.
(2) Vol. 101, 1973

SOME QUESTIONS IN THE THEORY OF ONE-DIMENSIONAL LINEAR DIFFERENTIAL OPERATORS OF THE SECOND ORDER. I

V. A. MARČENKO

CONTENTS

---

* Translation of Trudy Moskov. Mat. Obšč. 1 (1952), 327–420.
AMS (*MOS*) *subject classifications* (1970). Primary 34B25, 47E05; Secondary 35F15, 45D05, 47A70, 40A05.

INTRODUCTION

By a one-dimensional linear differential operator of the second order we mean an operator L of the form

$$L[u] = \frac{d^2}{dx^2} u(x) - q(x) \cdot u(x), \tag{1}$$

defined on an interval $(-a, a)$ or $[0, a)$, where $a \leqslant \infty$. The function $q(x)$ is assumed to be real and integrable on each subinterval $(-b, b)$ (respectively $[0, b)$) for arbitrary $b < a$.

The foundations of the theory of operators of this type were laid by Bernoulli, d'Alembert, and Euler over two centuries ago. Almost simultaneously d'Alembert and Bernoulli proposed distinct methods for integrating the equations of motion of a vibrating string. Out of the ideas which underlay Bernoulli's method and which were developed further by Fourier, Sturm, and Liouville arose the spectral theory of linear differential operators. This theory is applicable to boundary value problems of mathematical physics. Eigenfunctions are a central feature of the theory.

The fundamental theorem on completeness of the system of eigenfunctions of a boundary value problem on a finite interval was first proved rigorously by Steklov in 1896 [1]. Further developments of the theory occurred in conjunction with developments in the theory of integral equations; the proof of the fundamental theorem was simplified, and the theorem was generalized to the case of an infinite interval by Weyl [2].

Thus the fundamental propositions and methods in the spectral theory of one-dimensional linear differential operators of the second order were established with satisfactory completeness and generality by 1910.

Additional accounts of the theory can be found in Weyl [2], Titchmarsh [3], and Levitan [4].

The present work is devoted to the development and systematic application of a new device—transformation operators—to the task of solving problems in the theory of differential operators of the form (1).

This device is closely related to the theory of generalized shift operators which was developed by Delsarte [8], [9], Levitan [4], [5], and Povzner [10] on the basis of the following considerations.

Suppose that in the differential operators

$$L_1[u] = \frac{d^2}{dx^2} u(x) - q_1(x) u(x), \quad L_2[u] = \frac{d^2}{dx^2} u(x) - q_2(x) u(x)$$

the functions $q_1(x)$ and $q_2(x)$ are even. We consider the solution $u(x, y; f)$ of the partial differential equation

$$\frac{\partial^2}{\partial x^2} u(x, y) - q_1(x) u(x, y) = \frac{\partial^2}{\partial y^2} u(x, y) - q_2(y) u(x, y) \tag{2}$$

with initial conditions

$$u(x, 0) = f(x), \quad u'_y(x, 0) = 0,$$

where $f(x)$ is an arbitrary twice-differentiable function. The solution $u(x, y; f)$ can be considered formally as the result of applying an additive homogeneous operator T^y (where y is a parameter) to $f(x)$:

$$u(x, y; f) = T_x^y [f(x)].$$

The operators T^y of Delsarte are called generalized shift operators.

If, following Povzner, we apply the Riemann method for integrating (2), the solution $u(x, y; f)$ can be written in the form

$$u(x, y; f) = \frac{1}{2}[f(x+y) + f(x-y)] + \int_{x-y}^{x+y} T(x, y, t) f(t)\, dt, \tag{3}$$

where $T(x, y, t)$ is the function obtained from the Riemann function by changing coordinates from the characteristics of equation (2) to standard coordinates. This allows us to extend generalized shift operators T^y in a natural way to nondifferentiable functions $f(x)$ by means of the formula

$$T_x^y [f(x)] = \frac{1}{2}[f(x+y) + f(x-y)] + \int_{x-y}^{x+y} T(x, y, t) f(t)\, dt. \tag{4}$$

We denote by $\omega^{(1)}(\lambda, x)$ $(\omega^{(2)}(\lambda, y))$ the solution of the equation

$$\underset{x}{L_1}[u] + \lambda u = 0 \qquad (\underset{y}{L_2}[u] + \lambda u = 0)$$

with initial conditions $u(0) = 1$ and $u'(0) = 0$.

From the definition of generalized shift operators we obtain the important relation

$$T_x^y [\omega^{(1)}(\lambda, x)] = \omega^{(1)}(\lambda, x)\, \omega^{(2)}(\lambda, y),$$

from which for $x = 0$ it follows that

$$T_x^y[\omega^{(1)}(\lambda, x)]_{x=0} = \omega^{(2)}(\lambda, x).$$

By virtue of (4) and the fact that $\omega^{(1)}(\lambda, x)$ is even we can write the preceding equation in the form

$$\omega^{(2)}(\lambda, y) = \omega^{(1)}(\lambda, y) + \int_{-y}^{y} T(x, y, t)\,\omega^{(1)}(\lambda, t)\,dt.$$

Consequently the operator W defined by the formula

$$W[\varphi(x)] = \varphi(x) + \int_{-x}^{x} T(0, x, t)\,\varphi(t)\,dt,$$

transforms the solution $\omega^{(1)}(\lambda, x)$ of the equation $L_1[u] + \lambda u = 0$ into the solution $\omega^{(2)}(\lambda, x)$ of the equation $L_2[u] + \lambda u = 0$:

$$W[\omega^{(1)}(\lambda, x)] = \omega^{(2)}(\lambda, x).$$

The operator W is called a transformation operator.

In the works of Delsarte, Levitan, and Povzner we find that generalized shift operators T^y are applied with the goal of finding generalized Taylor formulas for almost periodic functions and positive definite functions. Also Levitan and Povzner showed that generalized shift operators can be used in establishing Weyl's inversion formula [2]. In these investigations the authors make frequent use of transformation operators, but these operators play a lesser role than that of generalized shift operators.

By contrast, in the present work it is transformation operators which play the leading role. And it seems to us that the present work demonstrates the fact that for many problems in the theory of one-dimensional linear differential operators of the second order the apparatus of transformation operators is a very natural and powerful instrument of investigation. We now briefly outline the contents of the present work.

Transformation operators are introduced in Chapter 1. Here we depart from the scheme mentioned above in which the partial differential equation (2) plays a basic role. This is done for two reasons.

First, our method can easily be generalized to the case in which $q(x)$ is singular at 0.

Second, in the scheme mentioned above we have to deal with the Riemann functions of equation (2), and this often introduces serious difficulties.

Let L be a differential operator of the form (1). We let $\omega_h(\lambda, x)$ denote the solution of the equation

$$L[u] + \lambda u = 0 \tag{5}$$

with initial conditions

$$u(0) = 1, \quad u'(0) = h, \quad \text{if} \quad h \neq \infty;$$

$$u(0) = 0, \quad u'(0) = 1, \quad \text{if} \quad h = \infty.$$

The fundamental results of Chapter 1 are contained in the following theorems.

THEOREM 1.2.1. *For each pair of operators* L_1 *and* L_2 *of the form* (1) *given on an interval* $(-a, a)$, $a \leq \infty$, *and for an arbitrary real number* $\theta \neq \infty$ *there exists a transformation operator* $W = W_{\{L_1 L_2 \theta\}}$ *on the set of all functions* $f(x)$ *which are integrable in each subinterval* $(-b, b)$, $b < a$, *defined by the formula*

$$W[f] = f(x) + \int_{-|x|}^{|x|} K(x, t) f(t)\, dt,$$

such that

$$W[\omega_h^{(1)}(\lambda, x)] = \omega_{h+\theta}^{(2)}(\lambda, x),$$

where $\omega^{(1)}(\lambda, x)$ *and* $\omega^{(2)}(\lambda, x)$ *are the solutions of equations* (5) *for the operators* L_1 *and* L_2 *respectively. The kernel* $K(x, t)$ *is real and bounded in each square* $-b \leq x \leq b$, $-b \leq t \leq b$, $b < a$.

THEOREM 1.2.2. *For each pair of operators* L_1 *and* L_2 *of the form* (1) *given on an interval* $[0, a)$, $a \leq \infty$, *and for arbitrary real numbers* h_1 *and* h_2 *which are both finite or both infinite there exists a transformation operator* $V = V_{\{L_1 L_2 h_1 h_2\}}$ *on the set of all functions* $f(x)$ *which are integrable in each subinterval* $[0, b)$, $b < a$, *defined by the formula*

$$V[f] = f(x) + \int_0^x K(x, t) f(t)\, dt,$$

such that

$$V[\omega_{h_1}^{(1)}(\lambda, x)] = \omega_{h_2}^{(2)}(\lambda, x),$$

where $\omega_{h_1}^{(1)}(\lambda, x)$ *and* $\omega_{h_2}^{(2)}(\lambda, x)$ *are the solutions of equation* (5) *for the operators* L_1 *and* L_2 *respectively. The kernel* $K(x, t)$ *is real and bounded in each square* $0 \leq x \leq b$, $0 \leq t \leq b$, $b < a$.

The proofs of these theorems depend on a few simple properties of classical Fourier integrals and integral equations of Volterra type. Along the way we obtain bounds for $K(x, t)$.

In Chapter 2 we apply transformation operators to some questions in spectral analysis of differential operators of the form (1). We remark that in these very important (especially for physics) applications we use very simple properties of transformation operators which follow immediately from their form.

In spectral theory it is shown that a boundary value problem on the interval $[0, a)$, $a \leqslant \infty$, for an operator L of the form (1) is determined by a boundary condition at 0

$$y(0)\cos\alpha + y'(0)\sin\alpha = 0$$

and possibly another condition at the point a. To each boundary value problem there corresponds a nondecreasing function $\rho(\lambda)$ $(-\infty < \lambda < \infty)$ which generates an isometric transformation of $L^2[0, a)$ onto $L^2_{\{\rho\}}(-\infty, \infty)$ according to the formulas

$$\begin{aligned} F(\lambda) &= \int_0^a f(x)\,\omega_h(\lambda, x)\,dx \quad (h = -\cot\alpha), \\ f(x) &= \int_{-\infty}^{\infty} F(\lambda)\,\omega_h(\lambda, x)\,d\rho(\lambda), \end{aligned} \tag{6}$$

where the integrals converge in the metrics of $L^2_{\{\rho\}}(-\infty, \infty)$ and $L^2[0, a)$ respectively. By the spectrum of a boundary value problem we mean the set of points of increase of the function $\rho(\lambda)$.

The importance of investigating $\rho(\lambda)$ is clear. But until now this has only been done in the simple case where the interval $[0, a)$ is finite and $q(x)$ is integrable on all of $[0, a)$.

Transformation operators allow us to study the behavior of $\rho(\lambda)$ for boundary value problems on $[0, \infty)$ for arbitrary operators of the form (1). This is done in §2 of Chapter 2, where we prove the following theorem.

THEOREM 2.2.1. *As* $\lambda \to +\infty$ *we have the following asymptotic formulas for* $\rho(\lambda)$:

$$\begin{aligned} \rho(\lambda) &= \frac{2}{\pi}\sqrt{\lambda} + o(\sqrt{\lambda}), && \textit{if} \quad \cot\alpha \neq \infty; \\ \rho(\lambda) &= \frac{2}{3\pi}\sqrt{\lambda^3} + o(\sqrt{\lambda^3}), && \textit{if} \quad \cot\alpha = \infty. \end{aligned}$$

We should mention that Lemmas 2.2.1 and 2.2.5 give more precise descriptions of the behavior of $\rho(\lambda)$.

In classical spectral theory of differential operators a typical problem is to find some spectral properties of a given operator L.

At the present time physics is presenting us with new problems in which we must reconstruct a differential operator from some of its spectral properties. We will refer to such problems as inverse problems of spectral analysis.

It is clear that in order to solve this kind of problem we must first of all find out which spectral properties uniquely determine an operator L of the form (1) (that is, determine the function $q(x)$). We devote §3 of Chapter 2 to this question.

Our most general result is contained in the following theorem.

THEOREM 2.3.1. *Suppose that operators* L_1 *and* L_2 *of the form* (1)

$$L_1[u] = \frac{d^2}{dx^2} u(x) - q_1(x) u(x), \quad L_2[u] = \frac{d^2}{dx^2} u(x) - q_2(x) u(x)$$

are given on $[0, a)$, $a \leqslant \infty$. *If we consider all possible boundary value problems for* L_1 *and* L_2 *we obtain two sets* $R(L_1)$ *and* $R(L_2)$ *of resolution functions* $\rho(\lambda)$ *from which to construct the inversion formulas* (6).

If there exists a pair of functions $\rho_1(\lambda) \in R(L_1)$ *and* $\rho_2(\lambda) \in R(L_2)$ *such that*

$$\rho_1(\lambda) = c\rho_2(\lambda) \qquad (c \text{ is a constant})$$

then $q_1(x) = q_2(x)$ *almost everywhere on* $[0, a)$.

In other words, the resolution function $\rho(\lambda)$ for a boundary value problem uniquely determines the corresponding operator L and the boundary conditions.

By using this theorem we can show that in some special cases the operator L is uniquely determined by some spectral properties in which $\rho(\lambda)$ does not play an explicit role.

Thus for operators L with discrete spectra we have

THEOREM 2.3.2. *Suppose that on* $[0, a)$, $a \leqslant \infty$, *we are given operators* L_1 *and* L_2 *of the form* (1):

$$L_1[u] = \frac{d^2}{dx^2} u(x) - q_1(x) u(x), \quad L_2[u] = \frac{d^2}{dx^2} u(x) - q_2(x) u(x)$$

which have discrete spectra. We consider the boundary value problems:

(α_1) *and* (β_1) *for* L_1 *with boundary conditions at* 0

$$y(0)\cos\alpha_1 + y'(0)\sin\alpha_1 = 0, \qquad (\alpha_1)$$
$$y(0)\cos\beta_1 + y'(0)\sin\beta_1 = 0 \qquad (\beta_1)$$

and the same condition at a *if this condition is needed.*

(α_2) *and* (β_2) *for* L_2 *with boundary conditions at* 0

$$y(0)\cos\alpha_2 + y'(0)\sin\alpha_2 = 0, \qquad (\alpha_2)$$
$$y(0)\cos\beta_2 + y'(0)\sin\beta_2 = 0 \qquad (\beta_2)$$

and the same condition at a *if this condition is needed.*

If $\cot(\alpha_1 - \beta_1) \neq \infty$, $\cot(\alpha_2 - \beta_2) \neq \infty$, *and the spectrum of problem*

(α_1) *coincides with the spectrum of problem* (α_2), *and the spectrum of problem* (β_1) *coincides with the spectrum of problem* (β_2), *then* $q_1(x) = q_2(x)$ *almost everywhere on* $[0, a)$.

In other words, if an operator L has a discrete spectrum, then the spectra of two boundary value problems for this operator with different boundary conditions at one end (and identical conditions repeated at the other end if necessary) uniquely determine the operator as well as the boundary conditions.

The simplest example of an operator L of the form (1) with a discrete spectrum is an operator given on a finite interval $[0, a)$ where the function $q(x)$ is integrable on all of $[0, a)$. The problem of reconstructing such an operator from the spectra of its boundary value problems is known in the literature as the inverse Sturm-Liouville problem. It was first posed by Ambarzumian [13] in connection with some physics problems. A thorough investigation of the uniqueness of the solution of the inverse Sturm-Liouville problem was first presented by Borg [14]. We observe, however, that even in this special case Theorem 2.3.2 is stronger than the corresponding theorem of Borg.

In various physics problems we encounter operators L of the form (1) given on $[0, \infty)$ and satisfying

$$\int_0^\infty (1+x)\,|q(x)|\,dx < \infty.$$

It is known that boundary value problems for such operators are determined by giving a boundary condition at zero, and the problem has a finite number of negative eigenvalues. Furthermore, for a real number s the solution $\omega_h(s^2, x)$ is described by the asymptotic formula

$$\omega_h(s^2, x) = |M(s)| \sin(sx + \delta(s)) + o(1) \quad (x \to \infty),$$

where $\delta(s) = \arg M(s)$. The function $\delta(s)$ is called the limiting phase of the solution $\omega_h(s^2, x)$. It plays an important role in quantum scattering theory, and the question of whether L is uniquely determined by the limiting phase is of great interest from the point of view of physics.

Levinson [15] proved that the limiting phase uniquely determines an operator if the corresponding boundary value problem has no negative eigenvalues. On the other hand, Bargmann [16] showed in some simple cases that the limiting phase together with the locations of all negative eigenvalues does not uniquely determine an operator.

In connection with these examples Bargmann presented the following generalization of Levinson's theorem.

THEOREM 2.3.3. *Let operators* L_1 *and* L_2 *of the form* (1)

$$L_1[u] = \frac{d^2}{dx^2} u(x) - q_1(x) u(x), \qquad L_2[u] = \frac{d^2}{dx^2} u(x) - q_2(x) u(x),$$

be given on $[0, \infty)$ *and satisfy*

$$\int_0^\infty (1+x) \, | q_1(x) | \, dx < \infty, \qquad \int_0^\infty (1+x) \, | q_2(x) | \, dx < \infty.$$

For some α *let the following conditions be satisfied:*

1) *The boundary value problems for* L_1 *and* L_2 *with the boundary condition*

$$y(0) \cos \alpha + y'(0) \sin \alpha = 0 \qquad (h = -\cot \alpha)$$

have the same negative eigenvalues $\lambda_1, \cdots, \lambda_n$.

2) *The norms of the eigenfunctions* $\omega_h^{(1)}(\lambda_k, x)$ *of the boundary value problem for* L_1 *are equal to the norms of the eigenfunctions* $\omega_h^{(2)}(\lambda_k, x)$ *of the boundary value problem for* L_2:

$$\int_0^\infty | \omega_h^{(1)}(\lambda_k, x) |^2 \, dx = \int_0^\infty | \omega_h^{(2)}(\lambda_k, x) |^2 \, dx \qquad (k = 1, 2, \ldots, n).$$

3) *For real numbers* $s \neq 0$ *the limiting phases* $\delta^{(1)}(s)$ *of the solutions* $\omega_h^{(1)}(s^2, x)$ *equal the limiting phases* $\delta^{(2)}(s)$ *of the solutions* $\omega_h^{(2)}(s^2, x)$; *that is,* $\delta^{(1)}(s) = \delta^{(2)}(s)$.

Then $q_1(x) = q_2(x)$ *almost everywhere on* $[0, \infty)$.

The Supplement to Chapter 2 is devoted to solving the inverse Sturm-Liouville problem; that is, to constructing an algorithm for reconstructing an unknown operator L of the form (1) from the spectra of two boundary value problems.

The idea underlying the solution consists of the following. Let $K(x, t)$ be the kernel of the transformation operator $V = V_{\{D^2 L o h\}}$ which takes $\cos \sqrt{\lambda}\, x$ to the solution $\omega_h(\lambda, x)$ of $L[u] + \lambda u = 0$, where L is the operator we are seeking. If we know the spectra of two boundary value problems, it is not difficult to find the function $F(x, y)$ which is related to $K(x, y)$ by the formula

$$F(x, y) = K(x, y) + \int_x^a K(z, x) K(z, y) \, dz \qquad (y \leqslant x).$$

This equation can be solved for $K(x, y)$ by the method of successive approximations. If we know $K(x, y)$, then we can find $\omega_h(\lambda, x)$ by means of the formula

$$\omega_h(\lambda, x) = \cos \sqrt{\lambda}\, x + \int_0^x K(x, t) \cos \sqrt{\lambda}\, t \, dt,$$

and then

$$q(x) = \frac{\omega_h''(0, x)}{\omega_h(0, x)}.$$

By the same method we can reconstruct $q(x)$ from the limiting phase $\delta(s)$ if we know that $q(x) = 0$ outside some finite interval.

Another method of solving the inverse Sturm-Liouville problem was presented by Kreĭn in [18].(1)

Chapter 3 contains some applications of transformation operators to the theory of generalized shifts. These applications are based on the following theorem.

THEOREM 3.1.1. *Let operators* L_1 *and* L_2 *of the form* (1) *be given on* $(-a, a)$, $a \leqslant \infty$. *Let* $W = W_{\{L_2 L_1 \theta\}}$ *be the transformation operator introduced in Theorem* 1.2.1. *Then*

$$L_1 \equiv W L_2 W^{-1}, \quad L_1^{-1} \equiv W L_2^{-1} W^{-1},$$

where L^{-1} *denotes the operator which gives the solution of* $L[u] = f(x)$ *for zero initial conditions.*

An analogous assertion (Theorem 3.1.1′) is true for operators given on $[0, a)$, $a \leqslant \infty$.

Generalized shift operators are introduced in §2 of Chapter 3 through the following definition.

DEFINITION 3.2.2 (3.2.1). By the *generalized shift operator of the second (first) kind associated with the differential operators* L_1 *and* L_2 of the form (1) we mean the operator T_x^y (R_x^y) which is defined on the set of all functions $f(x)$ integrable in each finite interval of the real line as follows:

$$T_x^y[f] = \underset{x}{W_1} \underset{y}{W_2} S_x^y [W_1^{-1}[f]]$$

$$\left(R_x^y[f] = \underset{x}{W_1} \underset{y}{W_2} U_x^y [W_1^{-1}[f]]\right),$$

where(2)

$$W_1 = W_{\{D^2 L_1 \theta_1\}}, \quad W_2 = W_{\{D^2 L_2 \theta_2\}},$$

$$S_x^y[\varphi] = \frac{1}{2}[\varphi(x+y) + \varphi(x-y)] \quad \left(U_x^y[\varphi] = \varphi(x-y)\right).$$

---

(1) After the present work was written, the paper [24] of Gel′fand and Levitan appeared, which presented a complete solution of the problem of reconstructing a differential operator from $\rho(\lambda)$. But we decided to retain the Supplement to Chapter 2, since some of its details are still interesting.

(2) D always denotes the differential operator $D^2[u] \equiv d^2u(x)/dx^2$, which is the simplest operator of the form (1).

If $f(x)$ is twice-differentiable, then from Theorem 3.1.1 we see that $u(x, y; f) = T_x^y[f]$ is the solution of the partial differential equation $\underset{x}{L}_1[u] = \underset{y}{L}_2[u]$ with initial conditions $u(x, 0) = f(x)$ and $u'_y(x, 0) = \theta_2 f(x)$. Thus generalized shift operators of the second kind are the same as Delsarte operators.

To illustrate the possible generalizations of classical harmonic analysis which can be obtained by replacing the usual shift operator $U_x^y[f] = f(x-y)$ by a generalized shift operator R_x^y of the first kind, we consider generalized positive definite functions and we prove an assertion concerning them (Theorem 3.2.1) which is analogous to Bochner's theorem.

In §3 of Chapter 3 we investigate the Taylor-Delsarte formula associated with a generalized shift operator T_x^y of the second kind. In particular, we give (Theorem 3.3.2) necessary and sufficient conditions which a function must satisfy in order to be expandable in a Maclaurin-Delsarte series which converges on the whole real line.

In the applications of transformation operators considered in Chapters 2 and 3 we use their simplest properties (in fact, only their form). We proceed on the assumption that the applications will not deal with questions concerning uniform metrics on the whole real line.

For applications of transformation operators to questions which are inseparably connected with a uniform metric on the whole real line or on a half-line (in the first case, applications to the theory of generalized almost periodic functions) we must first study the properties of these transformation operators which are connected with the metric. It is very important to study the structure of the operators $V_{\{D^2 L 0 h\}}$, $V_{\{D^2 L \infty\infty\}}$ and their inverses $V_{\{L D^2 h 0\}}$, $V_{\{L D^2 \infty\infty\}}$ (recall that $D^2 \equiv d^2/dx^2$).

We shall devote Chapter 4 (Part II) to such questions.

In that fourth chapter we will place upon L the additional restriction

$$\int_0^\infty (1+x^2)\,|\,q(x)\,|\,dx < \infty, \tag{7}$$

since in the general case the problem makes no sense.

The structure of the operators $V_{\{D^2 L 0 h\}}$ is relatively easy to investigate. Their characteristic properties were first described by Povzner [10] for the example $V_{\{D^2 L 0 0\}}$. A generalization of Povzner's results is contained in the following theorem.

THEOREM 4.1.1. *If an operator* L *of the form* (1) *satisfies* (7), *then*

1) $V_{\{D^2 L 0 h\}}[f] = f(x) + c\int_0^x f(t)\,dt + \int_0^x K_h(x,t)\,f(t)\,dt,$

where $c = \lim_{x\to\infty} \omega_h'(0, x)$, *and*

2) $$\sup_{0\leqslant x<\infty} \int_0^x |K_h(x, t)|\, dt < \infty.$$

It is significantly harder to investigate the inverse operator $V_{\{LD^2h0\}}$, whose properties depend heavily on the behavior of the solutions $\omega_h(s^2, x)$ for small values of s. Therefore we will first study in detail (§2 of Chapter 4) the properties of $\omega_h(s^2, x)$. The investigation of $V_{\{LD^2h0\}}$ then takes place in §3 of Chapter 4 and climaxes with the following theorem.

THEOREM 4.3.1. *If an operator* L *of the form* (1) *satisfies* (7), *then*

1) $$V_{\{LD^2h\,0\}}[f] = f(x) + \int_0^\infty \sum_{k=1}^n \frac{1}{\alpha_k} \cosh x\sqrt{|\lambda_k|}\cdot \omega_h(\lambda_k, t) f(t)\, dt + \int_0^\infty H(x, t) f(t)\, dt,$$

2) $$\sup_{0\leqslant x<\infty} \int_0^\infty |H(x, t)|\, dt < \infty,$$

where $\lambda_1 < \lambda_2 < \cdots < \lambda_n < 0$ *are the eigenvalues of the boundary value problem for* L *with boundary condition* $y'(0) - hy(0) = 0$. *Formula* 1) *holds for all* $f(x)$ *such that* $\int_0^\infty e^{-t\sqrt{|\lambda_n|}}|f(t)|\, dt < \infty$; *in particular, for all bounded functions on* $[0, \infty)$.

Similar theorems are proved for the operators $V_{\{D^2L\infty\infty\}}$, $V_{\{LD^2\infty\infty\}}$.

In §4 of Chapter 4 we look at two applications of the above theorem. One of these relates to a generalization of Bernstein's inequality, and the other is concerned with the structure of commutative normed rings associated with L, which were introduced by Povzner [10] and later studied by Agranovič [20].

The last (fifth) chapter is devoted to a generalization of almost periodic functions. A formal generalization of almost periodic functions, connected with the generalized shift operator T_x^y, was first introduced by Delsarte [9].

We recall the definition given by Delsarte. Suppose that in the differential operators L_1 and L_2 of the form (1) the functions $q_1(x)$ and $q_2(x)$ are even. We consider the generalized shift operator T_x^y of the second kind associated with L_1 and L_2, defined by the formula

$$T_x^y[f] = \underset{x}{W_1}\underset{y}{W_2} S_x^y[W_1^{-1}[f]],$$

where $W_1 = W_{\{D^2L_10\}}$, $W_2 = W_{\{D^2L_20\}}$ (see Definition 3.2.2). (If $f(x)$ is twice differentiable, then $u(x, y; f) = T_x^y[f]$ is the solution of the partial differential equation $\underset{x}{L_1}[u] = \underset{y}{L_2}[u]$ with initial conditions $u(x, 0) = f(x)$ and $u_y'(x, 0) = 0$.)

DEFINITION 5.1.1. A continuous function $F(x)$ is said to be $\{L_1, L_2\}$-*almost periodic* if the family of functions $T_x^y[F]$ (where y is a parameter) is compact in the sense of uniform convergence on the whole real line.

The fundamental problem in the theory of $\{L_1, L_2\}$-almost periodic functions is to find their constructive characteristics. No complete solution of this problem has yet been found for any pair of operators L_1, L_2 of the form (1). Levitan (see [5] and [6]) considered the problem under the assumption that the functions $q_1(x)$ and $q_2(x)$ in L_1 and L_2 satisfied

$$\int_{-\infty}^{\infty} (1+x^2)\,|q_1(x)|\,dx < \infty, \qquad \int_{-\infty}^{\infty} (1+x^2)\,|q_2(x)|\,dx < \infty. \tag{8}$$

Inequality (8) enabled Levitan to come close to solving the fundamental problem of the theory of $\{L_1, L_2\}$-almost periodic functions. But he did not completely succeed in overcoming all of the technical difficulties, and as a result he did not find the complete solution.

Our investigation of transformation operators in Chapter 4 enables us to solve completely the fundamental problem of the theory of $\{L_1, L_2\}$-almost periodic functions if we assume inequality (8).

Since $q_1(x)$ and $q_2(x)$ are even, we can seek a solution of the fundamental problem separately for even and odd $\{L_1, L_2\}$-almost periodic functions.

Let Λ_0 be the set of all values of the parameter λ for which even solutions $\omega_0(\lambda, x)$ of the equations $L_1[u] + \lambda u = 0$ are $\{L_1, L_2\}$-almost periodic functions. For given operators L_1 and L_2 the set Λ_0 can be characterized as follows: it always contains all positive numbers and sometimes a finite number of nonpositive numbers.

The solution of the fundamental problem for even $\{L_1, L_2\}$-almost periodic functions is contained in the following theorem.

THEOREM 5.4.1. *A function $F(x)$ is an even $\{L_1, L_2\}$-almost periodic function if and only if for each $\epsilon > 0$ there exists a function $\sigma(x)$ of the form*

$$\sigma(x) = \sum_{k=1}^{N} a_k \omega_0(\lambda_k, x) \qquad (\lambda_k \in \Lambda_0)$$

(a linear combination of a finite number of even $\{L_1, L_2\}$-almost periodic solutions of $L_1[u] + \lambda u = 0$) such that

$$\sup_{\substack{-\infty<x<\infty \\ -\infty<y<\infty}} |T_x^y[F(x) - \sigma(x)]| < \varepsilon.$$

An analogous theorem is true for odd $\{L_1, L_2\}$-almost periodic functions. Some of the results of this paper have been announced in [25]–[28].

CHAPTER 1

TRANSFORMATION OPERATORS

§1. Auxiliary properties

Let L denote a linear differential operator of the second order, given on an interval $(-a, a)$, $a \leqslant \infty$, of the form

$$L[u] = \frac{d^2}{dx^2} u(x) - q(x) u(x), \tag{1.1.1}$$

where $q(x)$ is a real function which is integrable on each subinterval $(-b, b)$ for arbitrary $b < a$.

Let $\omega_h(\lambda, x)$ denote the solution of the equation

$$L[u] + \lambda u = 0 \tag{1.1.2}$$

(λ is an arbitrary complex number) with initial conditions

$$u(0) = 1, \quad u'(0) = h, \quad \text{if} \quad h \neq \infty;$$
$$u(0) = 0, \quad u'(0) = 1, \quad \text{if} \quad h = \infty.$$

The functions $\omega_0(\lambda, x)$ and $\omega_\infty(\lambda, x)$ form a fundamental system of solutions of equation (1.1.2). Therefore in terms of these functions we can express the function $\omega_h(\lambda, x)$:

$$\omega_h(\lambda, x) = \omega_0(\lambda, x) + h\omega_\infty(\lambda, x)$$

as well as all other solutions of (1.1.2); among these we will frequently encounter the following solutions for which we introduce special notation:

$$c(\lambda, x) = \omega_0(\lambda, x), \quad s(\lambda, x) = \sqrt{\lambda}\, \omega_\infty(\lambda, x),$$
$$e(\lambda, x) = c(\lambda, x) + is(\lambda, x),$$
$$X^0(x) = c(0, x), \quad X^1(x) = \omega_\infty(0, x).$$

We note that when $q(x) \equiv 0$ we simply obtain

$$c(\lambda, x) = \cos\sqrt{\lambda}\, x, \ s(\lambda, x) = \sin\sqrt{\lambda}\, x, \ X^0(x) = 1, \quad X^1(x) = x.$$

Later on, as we introduce various bounds, we will often switch from $q(x)$ to $|q(x)|$. Hereafter we will denote this change by writing $\sim$ over the expression in which the change is made. For example,

$$\widetilde{L}[u] = \frac{d^2}{dx^2} u(x) - |q(x)|\, u(x),$$

and $\widetilde{\omega}_h(\lambda, x)$ is the solution of the equation $\widetilde{L}[u] + \lambda u = 0$ with initial conditions $u(0) = 1$ and $u'(0) = h$, and so forth.

It is known that solving (1.1.2) is equivalent to solving the following Volterra integral equation:

$$u(x) = a \cos \sqrt{\lambda} x + b \frac{\sin \sqrt{\lambda} x}{\sqrt{\lambda}} + \int_0^x \frac{\sin \sqrt{\lambda}(x-t)}{\sqrt{\lambda}} q(t) u(t)\, dt,$$

where $a = u(0)$ and $b = u'(0)$.

In particular, for the functions $e(\lambda^2, x)$ and $\omega_\infty(\lambda^2, x)$ we obtain

$$e(\lambda^2, x) = e^{i\lambda x} + \int_0^x \frac{\sin \lambda (x-t)}{\lambda} q(t) e(\lambda^2, t)\, dt, \qquad (1.1.3)$$

$$\omega_\infty(\lambda^2, x) = \frac{\sin \lambda x}{\lambda} + \int_0^x \frac{\sin \lambda (x-t)}{\lambda} q(t) \omega_\infty(\lambda^2, t)\, dt. \qquad (1.1.3')$$

From the general theory of Volterra equations it follows that the preceding equations can be solved by iteration if $q(x)$ is bounded. In that case we have

$$e(\lambda^2, x) = e^{i\lambda x} + \varphi_1(\lambda, x) + \varphi_2(\lambda, x) + \dots, \qquad (1.1.4)$$

$$\omega_\infty(\lambda^2, x) = \frac{\sin \lambda x}{\lambda} + \psi_1(\lambda, x) + \psi_2(\lambda, x) + \dots, \qquad (1.1.4')$$

where

$$\varphi_n(\lambda, x) = \int_0^x d\tau_1 \dots \int_0^{\tau_{n-1}} d\tau_n q(\tau_1) \dots q(\tau_n) \frac{\sin \lambda (x-\tau_1)}{\lambda} \dots \frac{\sin \lambda (\tau_{n-1}-\tau_n)}{\lambda} e^{i\lambda \tau_n}, \qquad (1.1.5)$$

$$\psi_n(\lambda, x) = \int_0^x d\tau_1 \dots \int_0^{\tau_{n-1}} d\tau_n q(\tau_1) \dots q(\tau_n) \frac{\sin \lambda (x-\tau_1)}{\lambda} \dots \frac{\sin \lambda (\tau_{n-1}-\tau_n)}{\lambda} \frac{\sin \lambda \tau_n}{\lambda}. \qquad (1.1.5')$$

We do not assume that $q(x)$ is bounded, but only that is integrable on each subinterval $(-b, b)$, $b < a$. Thus we cannot just quote the theory of Volterra equations. But, as we will show later, the series (1.1.4) and (1.1.4′) under our assumptions will converge uniformly in each subinterval $(-b, b)$, $b < a$, and therefore they in fact represent solutions of the corresponding Volterra equations.

The following expressions for the functions $\phi_n(\lambda, x)$ and $\psi_n(\lambda, x)$, which are more suitable than (1.1.5) and (1.1.5′), will play an important role. Here we make use of Fourier transforms.

Let h be an arbitrary nonzero real number. Let $h(y)$ be the function equal to $1/2|h|$ in the interval $(-|h|, |h|)$ and equal to zero elsewhere.

We have

$$\int_{-\infty}^{\infty} h(y) e^{-i\lambda y}\, dy = \frac{1}{2|h|} \int_{-|h|}^{|h|} e^{-i\lambda y}\, dy = \frac{\sin \lambda |h|}{\lambda |h|} = \frac{\sin \lambda h}{\lambda h},$$

and using the theorem on Fourier integrals of convolutions we find that

$$\int_{-\infty}^{\infty} h_1 * h_2 * \cdots * h_n(y)\, e^{-i\lambda y}\, dy = \frac{\sin \lambda h_1}{\lambda h_1} \frac{\sin \lambda h_2}{\lambda h_2} \cdots \frac{\sin \lambda h_n}{\lambda h_n},$$

where

$$h_1 * h_2 * \cdots * h_n(y) = \int_{-\infty}^{\infty} h_1 * h_2 * \cdots * h_{n-1}(y - t)\, h_n(t)\, dt.$$

Since the functions $h_k(y)$ are nonnegative, even, and equal to zero outside the intervals $(-|h_k|, |h_k|)$, the preceding formula implies that $h_1 * \cdots * h_n(y)$ is also nonnegative, even, and equal to zero for $|y| \geqslant |h_1| + \cdots + |h_n|$. Let

$$H_n = \sup_{-\infty < x < \infty} h_1 * h_2 * \cdots * h_n(y).$$

Then

$$H_n \leqslant \int_{-\infty}^{\infty} H_{n-1} h_n(t)\, dt = H_{n-1};$$

that is,

$$H_n \leqslant H_{n-1} \leqslant \cdots \leqslant H_1 = \frac{1}{2\,|h_1|}$$

and

$$0 \leqslant h_1 * h_2 * \cdots * h_n(y) \leqslant \frac{1}{2\,|h_1|} \qquad (-\infty < y < \infty).$$

Furthermore,

$$\int_{-\infty}^{\infty} h_1 * h_2 * \cdots * h_n(-y + b)\, e^{i\lambda y}\, dy = -\int_{\infty}^{-\infty} h_1 * h_2 * \cdots * h_n(t)\, e^{-i\lambda (t-b)}\, dt$$

$$= e^{i\lambda b} \int_{-\infty}^{\infty} h_1 * h_2 * \cdots * h_n(t)\, e^{-i\lambda t}\, dt = \frac{\sin \lambda h_1}{\lambda h_1} \frac{\sin \lambda h_2}{\lambda h_2} \cdots \frac{\sin \lambda h_n}{\lambda h_n} e^{i\lambda b}.$$

Therefore, setting $\mathscr{E}(y) = h_1 * \cdots * h_n(-y + b)$, we have

$$\frac{\sin \lambda h_1}{\lambda h_1} \frac{\sin \lambda h_2}{\lambda h_2} \cdots \frac{\sin \lambda h_n}{\lambda h_n} e^{i\lambda b} = \int_{-\infty}^{\infty} \mathscr{E}(y)\, e^{i\lambda y}\, dy,$$

and, according to what we said before,

$$\mathscr{E}(y) = 0, \quad \text{if} \quad |y| \geqslant |h_1| + |h_2| + \cdots + |h_n| + |b|,$$

$$0 \leqslant \mathscr{E}(y) \leqslant \frac{1}{2\,|h_1|} \qquad (-\infty < y < \infty),$$

$$\mathscr{E}(y) = \mathscr{E}(-y), \qquad \text{if} \quad b = 0.$$

In particular,

$$\frac{\sin\lambda(x-\tau_1)}{\lambda(x-\tau_1)}\frac{\sin\lambda(\tau_1-\tau_2)}{\lambda(\tau_1-\tau_2)}\cdots\frac{\sin\lambda(\tau_{n-1}-\tau_n)}{\lambda(\tau_{n-1}-\tau_n)}e^{i\lambda\tau_n}$$

$$=\int_{-\infty}^{\infty}\mathcal{E}_1(\tau_1,\tau_2,\ldots,\tau_n;x,y)e^{i\lambda y}\,dy, \tag{1.1.6}$$

$$\frac{\sin\lambda(x-\tau_1)}{\lambda(x-\tau_1)}\frac{\sin\lambda(\tau_1-\tau_2)}{\lambda(\tau_1-\tau_2)}\cdots\frac{\sin\lambda(\tau_{n-1}-\tau_n)}{\lambda(\tau_{n-1}-\tau_n)}\frac{\sin\lambda\tau_n}{\lambda\tau_n}$$

$$=\int_{-\infty}^{\infty}\mathcal{E}_2(\tau_1,\tau_2,\ldots,\tau_n;x,y)e^{i\lambda y}\,dy, \tag{1.1.6'}$$

where

$$0\leqslant\mathcal{E}_1(\tau_1,\tau_2,\ldots,\tau_n;x,y)\leqslant\frac{1}{2|x-\tau_1|},$$

$$0\leqslant\mathcal{E}_2(\tau_1,\tau_2,\ldots,\tau_n;x,y)\leqslant\frac{1}{2|x-\tau_1|},$$

$$\mathcal{E}_1(\tau_1,\tau_2,\ldots,\tau_n;x,y)=\mathcal{E}_2(\tau_1,\tau_2,\ldots,\tau_n;x,y)=0\quad\text{for } |y|\geqslant|x|,$$

$$\mathcal{E}_2(\tau_1,\tau_2,\ldots,\tau_n;x,y)=\mathcal{E}_2(\tau_1,\tau_2,\ldots,\tau_n,x,-y)$$

(we assume that the numbers $x,\tau_1,\cdots,\tau_n$ are of the same sign and $|x|>|\tau_1|>\cdots>|\tau_n|$.

We introduce the notation

$$\mathfrak{A}_n(x,y)=\int_0^x d\tau_1\ldots\int_0^{\tau_{n-1}}d\tau_n$$
$$\times\, q(\tau_1)\ldots q(\tau_n)(x-\tau_1)\ldots(\tau_{n-1}-\tau_n)\tau_n\mathcal{E}_2(\tau_1,\ldots,\tau_n;x,y), \tag{1.1.7}$$

$$\mathfrak{B}_n(x,y)=\int_0^x d\tau_1\ldots\int_0^{\tau_{n-1}}d\tau_n$$
$$\times\, q(\tau_1)\ldots q(\tau_n)(x-\tau_1)\ldots(\tau_{n-1}-\tau_n)\mathcal{E}_1(\tau_1,\ldots,\tau_n;x,y), \tag{1.1.7'}$$

and from (1.1.5), (1.1.6) and (1.1.5′), (1.1.6′) we obtain

$$\varphi_n(\lambda,x)=\int_{-\infty}^{\infty}\mathfrak{A}_n(x,y)e^{i\lambda y}\,dy=\int_{-|x|}^{|x|}\mathfrak{A}_n(x,y)e^{i\lambda y}\,dy,$$

$$\psi_n(\lambda,x)=\int_{-\infty}^{\infty}\mathfrak{B}_n(x,y)e^{i\lambda y}\,dy=\int_{-|x|}^{|x|}\mathfrak{B}_n(x,y)e^{i\lambda y}\,dy.$$

(Here the integration is actually performed on the finite interval $(-|x|;|x|)$, since outside this interval the functions $\mathfrak{A}_n(x,y)$ and $\mathfrak{B}_n(x,y)$ equal zero along with $\mathcal{E}_1(\tau_1,\cdots,\tau_n;x,y)$ and $\mathcal{E}_2(\tau_1,\cdots,\tau_n;x,y)$.)

These formulas give us the representations of $\phi_n(\lambda,x)$ and $\psi_n(\lambda,x)$ that we need. We note that the kernel $\mathfrak{B}_n(x,y)$ is even in y since the functions $\mathcal{E}_2(\tau_1,\cdots,\tau_n;x,y)$ have this property.

Now the series (1.1.4) and (1.1.4′) can be written in the form

$$e^{i\lambda x}+\sum_{n=1}^{\infty}\varphi_n(\lambda, x)=e^{i\lambda x}+\sum_{n=1}^{\infty}\int_{-|x|}^{|x|}\mathfrak{A}_n(x, y)\,e^{i\lambda y}\,dy,$$

$$\frac{\sin\lambda x}{\lambda}+\sum_{n=1}^{\infty}\psi_n(\lambda, x)=\frac{x}{2|x|}\int_{-|x|}^{|x|}e^{i\lambda y}\,dy+\sum_{n=1}^{\infty}\int_{-|x|}^{|x|}\mathfrak{B}_n(x, y)\,e^{i\lambda y}\,dy.$$

Therefore we can prove uniform convergence of the series (1.1.4) and (1.1.4′) for $|x|\leqslant b<a$ by showing that the series

$$\mathfrak{A}_1(x, y)+\mathfrak{A}_2(x, y)+\dots+\mathfrak{A}_n(x, y)+\dots, \tag{1.1.8}$$

$$\frac{x}{2|x|}+\mathfrak{B}_1(x, y)+\mathfrak{B}_2(x, y)+\cdots+\mathfrak{B}_n(x, y)+\cdots \tag{1.1.8'}$$

converge uniformly for $|y|\leqslant|x|\leqslant b<a$.

From (1.1.7) and (1.1.7′) we have

$$|\mathfrak{A}_n(x, y)|\leqslant\int_0^x d\tau_1\dots\int_0^{\tau_{n-1}}d\tau_n\,|q(\tau_1)\dots q(\tau_n)|$$

$$\times(x-\tau_1)\dots(\tau_{n-1}-\tau_n)\,\mathcal{E}_1(\tau_1,\dots,\tau_n;\,x, y)=\widetilde{\mathfrak{A}}_n(x, y),$$

$$|\mathfrak{B}_n(x, y)|\leqslant\int_0^x d\tau_1\dots\int_0^{\tau_{n-1}}d\tau_n\,|q(\tau_1)\dots q(\tau_n)(x-\tau_1)|$$

$$\times(\tau_1-\tau_2)\dots(\tau_{n-1}-\tau_n)\,\tau_n\,\mathcal{E}_2(\tau_1,\dots,\tau_n;\,x, y)=\frac{x}{|x|}\widetilde{\mathfrak{B}}_n(x, y),$$

and since $|\tau_i-\tau_{i+1}|\leqslant|\tau_i|$, while $2|x-\tau_1|\,\mathcal{E}_1(\tau_1,\cdots,\tau_n;\,x, y)\leqslant 1$ and $2|x-\tau_1|\,\mathcal{E}_2(\tau_1,\cdots,\tau_n;\,x, y)\leqslant 1$, it follows that

$$\widetilde{\mathfrak{A}}_n(x, y)$$

$$\leqslant\frac{x}{2|x|}\int_0^x|q(\tau_n)|\,d\tau_n\left\{\int_0^x d\tau_1\dots\int_0^{\tau_{n-2}}d\tau_{n-1}\,|q(\tau_1)|\,\tau_1\dots|q(\tau_{n-1})|\,\tau_{n-1}\right\},$$

$$\frac{x}{|x|}\widetilde{\mathfrak{B}}_n(x, y)$$

$$\leqslant\frac{1}{2}\int_0^x|q(\tau_n)|\,\tau_n d\tau_n\left\{\int_0^x d\tau_1\dots\int_0^{\tau_{n-2}}d\tau_{n-1}\,|q(\tau_1)|\,\tau_1\dots|q(\tau_{n-1})|\,\tau_{n-1}\right\}.$$

Hence the series

$$\sum_{n=1}^{\infty}\widetilde{\mathfrak{A}}_n(x, y),\qquad\frac{x}{|x|}\left(\frac{x}{2|x|}+\sum_{n=1}^{\infty}\widetilde{\mathfrak{B}}_n(x, y)\right) \tag{1.1.9}$$

majorize the series

$$\sum^{\infty} |\mathfrak{A}_n(x, y)|, \qquad \frac{1}{2} + \sum_{n=1}^{\infty} |\mathfrak{B}_n(x, y)|$$

and in turn are majorized by the series

$$\frac{x}{2|x|}\int_0^x |q(t)|\,dt$$

$$\times\left\{1 + \int_0^x |q(\tau_1)|\,\tau_1 d\tau_1 + \int_0^x d\tau_1 \int_0^{\tau_1} d\tau_2 \,|q(\tau_1)|\,\tau_1\,|q(\tau_2)|\,\tau_2 + \cdots\right\},$$

$$\frac{1}{2} + \frac{1}{2}\int_0^x |q(t)|\,t dt$$

$$\times\left\{1 + \int_0^x |q(\tau_1)|\,\tau_1 d\tau_1 + \int_0^x d\tau_1 \int_0^{\tau_1} d\tau_2 \,|q(\tau_1)|\,\tau_1\,|q(\tau_2)|\,\tau_2 + \cdots\right\}.$$

If $q(x)$ is continuous, these series automatically converge since the expression in braces is just the solution $z(x)$ obtained by iterating the Volterra equation

$$z(x) = 1 + \int_0^x |q(t)|\,t\,z(t)\,dt$$

or the solution of the equivalent differential equation

$$z'(x) = |q(x)|\,x\,z(x), \quad z(0) = 1;$$

that is,

$$1 + \int_0^x |q(\tau_1)|\,\tau_1 d\tau_1 + \int_0^x d\tau_1 \int_0^{\tau_1} d\tau_2\,|q(\tau_1)|\,\tau_1\,|q(\tau_2)|\,\tau_2 + \cdots = e^{\int_0^x |q(t)|\,tdt}.$$

By approximating $q(x)$ with continuous functions it is easy to show that this equality is preserved for functions which are integrable in each subinterval $(-b, b)$, $b < a$.

Therefore we see that for $|y| \leqslant |x| \leqslant b < a$ the series in (1.1.9) converge uniformly, and their sums $\widetilde{\mathfrak{A}}(x, y)$ and $(x/|x|)\widetilde{\mathfrak{B}}(x, y)$ satisfy the inequalities

$$\widetilde{\mathfrak{A}}(x, y) \leqslant \frac{x}{2|x|}\int_0^x |q(t)|\,dt\, e^{\int_0^x |q(t)|\,tdt},$$

$$\frac{x}{|x|}\widetilde{\mathfrak{B}}(x, y) \leqslant \frac{1}{2}\left(1 + \int_0^x |q(t)|\,tdt\, e^{\int_0^x |q(t)|\,tdt}\right).$$

Hence for $|y| \leqslant |x| \leqslant b < a_1$ the series (1.1.8) and (1.1.8′) converge absolutely and uniformly, and their sums $\mathfrak{A}(x, y)$ and $\mathfrak{B}(x, y)$ satisfy the inequalities

$$|\mathfrak{A}(x, y)| \leqslant \tilde{\mathfrak{A}}(x, y) \leqslant \frac{x}{2|x|}\int_0^x |q(t)|\,dt\, e^{\int_0^x |q(t)|\,t dt},$$

$$|\mathfrak{B}(x, y)| \leqslant \frac{x}{|x|}\tilde{\mathfrak{B}}(x, y) \leqslant \frac{1}{2}\left(1 + \int_0^x |q(t)|\,t dt\, e^{\int_0^x |q(t)|\,t dt}\right).$$

We conclude that the series (1.1.4) and (1.1.4′) converge uniformly for $|x| \leqslant b < a$, and therefore they represent solutions of the Volterra equations (1.1.3) and (1.1.3′), and

$$e(\lambda^2, x) = e^{i\lambda x} + \int_{-|x|}^{|x|} \mathfrak{A}(x, y)\, e^{i\lambda y}\, dy,$$

$$\omega_\infty(\lambda^2, x) = \int_{-|x|}^{|x|} \mathfrak{B}(x, y)\, e^{i\lambda y}\, dy$$

or, in view of the evenness of $\mathfrak{B}(x, y)$ in y,

$$\omega_\infty(\lambda^2, x) = \int_{-|x|}^{|x|} \mathfrak{B}(x, y)\cos \lambda y\, dy.$$

Recalling the definitions of $c(\lambda, x)$ and $s(\lambda, x)$, we have

$$c(\lambda^2, x) = \frac{e(\lambda^2, x) + e(-\lambda^2, x)}{2} = \frac{e^{i\lambda x} + e^{-i\lambda x}}{2} + \int_{-|x|}^{|x|} \mathfrak{A}(x, y)\frac{e^{+i\lambda y} + e^{-i\lambda y}}{2}\, dy$$

$$= \cos \lambda x + \int_{-|x|}^{|x|} \mathfrak{A}(x, y)\cos \lambda y\, dy,$$

$$s(\lambda^2, x) = \frac{e(\lambda^2, x) - e(-\lambda^2, x)}{2i} = \frac{e^{i\lambda x} - e^{-i\lambda x}}{2i} + \int_{-|x|}^{|x|} \mathfrak{A}(x, y)\,\frac{e^{i\lambda y} - e^{-i\lambda y}}{2i}\, dy$$

$$= \sin \lambda x + \int_{-|x|}^{|x|} \mathfrak{A}(x, y)\sin \lambda y\, dy.$$

We collect the preceding results in a lemma.

LEMMA 1.1.1. *For* $|x| < a$ *we have*

$$c(\lambda, x) = \cos \sqrt{\lambda}\, x + \int_{-|x|}^{|x|} \mathfrak{A}(x, y)\cos \sqrt{\lambda}\, y\, dy,$$

$$s(\lambda, x) = \sin \sqrt{\lambda}\, x + \int_{-|x|}^{|x|} \mathfrak{A}(x, y)\sin \sqrt{\lambda}\, y\, dy,$$

$$e(\lambda^2, x) = e^{i\lambda x} + \int_{-|x|}^{|x|} \mathfrak{A}(x, y)\, e^{i\lambda y}\, dy,$$

$$\omega_\infty(\lambda, x) = \int_{-|x|}^{|x|} \mathfrak{B}(x, y) \cos\sqrt{\lambda}\, y\, dy,$$

where the kernels $\mathfrak{A}(x, y)$ *and* $\mathfrak{B}(x, y)$ *are real and satisfy the inequalities*

$$|\mathfrak{A}(x, y)| \leqslant \tilde{\mathfrak{A}}(x, y) \leqslant \frac{x}{2|x|} \int_0^x |q(t)|\, dt\, e^{\int_0^x |q(t)|\, t\, dt},$$

$$|\mathfrak{B}(x, y)| \leqslant \frac{x}{|x|} \tilde{\mathfrak{B}}(x, y) \leqslant \frac{1}{2}\left(1 + \int_0^x |q(t)|\, t\, dt\, e^{\int_0^x |q(t)|\, t\, dt}\right).$$

COROLLARY. *If an operator* L *of the form* (1.1.1) *is given on the whole real line* $-\infty < x < \infty$ *and* $q(x)$ *satisfies the condition*

$$\int_0^\infty (1 + t)|q(t)|\, dt = \alpha < \infty \quad \left(\int_{-\infty}^0 (1 - t)|q(t)|\, dt = \beta < \infty\right),$$

then the kernels $\mathfrak{A}(x, y)$ *and* $\mathfrak{B}(x, y)$ *are uniformly bounded for* $x \geqslant 0$ $(x \leqslant 0)$:

$$|\mathfrak{A}(x, y)| \leqslant \tilde{\mathfrak{A}}(x, y) \leqslant \frac{\alpha}{2} e^{\alpha} \quad \left(|\mathfrak{A}(x, y)| \leqslant \tilde{\mathfrak{A}}(x, y) \leqslant \frac{\beta}{2} e^{\beta}\right),$$

$$|\mathfrak{B}(x, y)| \leqslant \tilde{\mathfrak{B}}(x, y) \leqslant \frac{1}{2}(1 + \alpha e^{\alpha}) \quad \left(|\mathfrak{B}(x, y)| \leqslant -\tilde{\mathfrak{B}}(x, y) \leqslant \frac{1}{2}(1 + \beta e^{\beta})\right).$$

Later (Chapter 4) we will need expansions of $\mathfrak{A}(x, y)$ and $\mathfrak{B}(x, y)$ which display their essential parts explicitly. These expressions can be obtained as follows. By transforming equations (1.1.7) and (1.1.7′) we obtain

$$\mathfrak{A}_n(x, y) = \int_0^x d\tau_1 \cdots \int_0^{\tau_{n-1}} d\tau_n\, q(\tau_1) \cdots q(\tau_n) \frac{(x - \tau_1)}{|x - \tau_1|} (\tau_1 - \tau_2) \cdots (\tau_{n-1} - \tau_n)$$

$$\times \left\{\frac{1}{2} - \left[\frac{1}{2} - |x - \tau_1|\, \mathscr{E}_1(\tau_1 \cdots \tau_n;\ x, y)\right]\right\}$$

$$= \frac{x}{2|x|}\left\{\int_0^x d\tau_1 \cdots \int_0^{\tau_{n-1}} d\tau_n\, q(\tau_1) \cdots q(\tau_n)(\tau_1 - \tau_2) \cdots (\tau_{n-1} - \tau_n) - \int_0^x d\tau_1\right.$$

$$\left.\cdots \int_0^{\tau_{n-1}} d\tau_n q(\tau_1) \cdots q(\tau_n)(\tau_1 - \tau_2) \cdots (\tau_{n-1} - \tau_n)[1 - 2|x - \tau_1|\, \mathscr{E}_1(\tau_1, \ldots, \tau_n; x, y)]\right\},$$

$$\mathfrak{B}_n(x, y) = \int_0^x d\tau_1 \cdots \int_0^{\tau_{n-1}} d\tau_n q(\tau_1) \cdots q(\tau_n) \frac{(x - \tau_1)}{|x - \tau_1|} (\tau_1 - \tau_2) \cdots (\tau_{n-1} - \tau_n)\, \tau_n$$

$$\times \left\{\frac{1}{2} - \left[\frac{1}{2} - |x - \tau_1|\, \mathscr{E}_2(\tau_1, \ldots, \tau_n;\ x, y)\right]\right\}$$

$$= \frac{x}{2|x|}\left\{\int_0^x d\tau_1 \cdots \int_0^{\tau_{n-1}} d\tau_n q(\tau_1)\cdots q(\tau_n)(\tau_1-\tau_2)\cdots(\tau_{n-1}-\tau_n)\tau_n - \int_0^x d\tau_1\right.$$

$$\left.\cdots\int_0^{\tau_{n-1}} d\tau_n q(\tau_1)\cdots q(\tau_n)(\tau_1-\tau_2)\cdots(\tau_{n-1}-\tau_n)\tau_n[1-2|x-\tau_1|\mathcal{E}_2(\tau_1,\ldots,\tau_n;x,y)]\right\},$$

where

$$\int_0^x d\tau_1 \cdots \int_0^{\tau_{n-1}} d\tau_n q(\tau_1)\cdots q(\tau_n)(\tau_1-\tau_2)\cdots(\tau_{n-1}-\tau_n) = \int_0^x q(\tau_1)\phi_{n-1}(0,\tau_1)\,d\tau_1,$$

$$\int_0^x d\tau_1 \cdots \int_0^{\tau_{n-1}} d\tau_n q(\tau_1)\cdots q(\tau_n)(\tau_1-\tau_2)\cdots(\tau_{n-1}-\tau_n)\tau_n = \int_0^x q(\tau_1)\psi_{n-1}(0,\tau_1)\,d\tau_1.$$

Therefore, setting

$$\mathfrak{A}'_n(x,y) = \int_0^x d\tau_1 \cdots \int_0^{\tau_{n-1}} d\tau_n q(\tau_1)\cdots q(\tau_n)(\tau_1-\tau_2)$$
$$\cdots(\tau_{n-1}-\tau_n)[1-2|x-\tau_1|\mathcal{E}_1(\tau_1,\ldots,\tau_n;x,y)], \tag{1.1.10}$$

$$\mathfrak{B}'_n(x,y) = \int_0^x d\tau_1 \cdots \int_0^{\tau_{n-1}} d\tau_n q(\tau_1)\cdots q(\tau_n)(\tau_1-\tau_2)$$
$$\cdots(\tau_{n-1}-\tau_n)\tau_n[1-2|x-\tau_1|\mathcal{E}_2(\tau_1,\ldots,\tau_n;x,y)], \tag{1.1.10$'$}$$

we have

$$\mathfrak{A}_n(x,y) = \frac{x}{2|x|}\left\{\int_0^x q(t)\varphi_{n-1}(0,t)\,dt - \mathfrak{A}'_n(x,y)\right\},$$

$$\mathfrak{B}_n(x,y) = \frac{x}{2|x|}\left\{\int_0^x q(t)\psi_{n-1}(0,t)\,dt - \mathfrak{B}'_n(x,y)\right\}.$$

We observe that for $|t| \leqslant b < a$ the series $\Sigma_{n=1}^{\infty}\phi_{n-1}(0,t)$ and $\Sigma_{n=1}^{\infty}\psi_{n-1}(0,t)$ converge uniformly to the functions $c(0,t) = X^0(t)$ and $\omega_\infty(0,t) = X^1(t)$, and the series $\Sigma_{n=1}^{\infty}\mathfrak{A}_n(x,y)$ and $\Sigma_{n=1}^{\infty}\mathfrak{B}_n(x,y)$ for $|y| \leqslant |x| \leqslant b < a$ converge uniformly to the kernels $\mathfrak{A}(x,y)$ and $\mathfrak{B}(x,y) - x/2|x|$.

Therefore the preceding formulas can be summed for all $n = 1, 2, \cdots$, and we obtain

$$\mathfrak{A}(x,y) = \frac{x}{2|x|}\left\{\int_0^x q(t)X^0(t)\,dt - \mathfrak{A}'(x,y)\right\},$$

$$\mathfrak{B}(x,y) - \frac{x}{2|x|} = \frac{x}{2|x|}\left\{\int_0^x q(t)X^1(t)\,dt - \mathfrak{B}'(x,y)\right\}$$

or

$$\mathfrak{B}(x, y) = \frac{x}{2|x|}\left\{1 + \int_0^x q(t) X^1(t)\, dt - \mathfrak{B}'(x, y)\right\},$$

where

$$\mathfrak{A}'(x, y) = \sum_{n=1}^{\infty} \mathfrak{A}'_n(x, y), \quad \mathfrak{B}'(x, y) = \sum_{n=1}^{\infty} \mathfrak{B}'_n(x, y).$$

We can find bounds for $\mathfrak{A}'(x, y)$ and $\mathfrak{B}'(x, y)$. We have

$$|\mathfrak{A}'(x, y)| \leqslant \sum_{n=1}^{\infty} |\mathfrak{A}'_n(x, y)|, \quad |\mathfrak{B}'(x, y)| \leqslant \sum_{n=1}^{\infty} |\mathfrak{B}'_n(x, y)|.$$

From (1.1.10) and (1.1.10'), noting that

$$1 - 2|x - \tau_1| \mathscr{E}_1(\tau_1, \cdots, \tau_n;\ x, y) \geqslant 0$$

and

$$1 - 2|x - \tau_1| \mathscr{E}_2(\tau_1, \cdots, \tau_n;\ x, y) \geqslant 0,$$

we have

$$|\mathfrak{A}'_n(x, y)| \leqslant \frac{x}{|x|}\int_0^x d\tau_1 \cdots \int_0^{\tau_{n-1}} d\tau_n |q(\tau_1) \cdots q(\tau_n)| (\tau_1 - \tau_2) \cdots (\tau_{n-1} - \tau_n)$$
$$\times [1 - 2|x - \tau_1| \mathscr{E}_1(\tau_1, \cdots, \tau_n;\ x, y)] = \frac{x}{|x|} \widetilde{\mathfrak{A}}'_n(x, y),$$

$$|\mathfrak{B}'_n(x, y)| \leqslant \int_0^x d\tau_1 \cdots \int_0^{\tau_{n-1}} d\tau_n |q(\tau_1) \cdots q(\tau_n)| (\tau_1 - \tau_2) \cdots (\tau_{n-1} - \tau_n) \tau_n$$
$$\times [1 - 2|x - \tau_1| \mathscr{E}_2(\tau_1, \cdots, \tau_n;\ x, y)] = \widetilde{\mathfrak{B}}'_n(x, y).$$

Therefore

$$|\mathfrak{A}'(x, y)| \leqslant \frac{x}{|x|} \sum_{n=1}^{\infty} \widetilde{\mathfrak{A}}'_n(x, y) = \frac{x}{|x|} \widetilde{\mathfrak{A}}'(x, y),$$

$$|\mathfrak{B}'(x, y)| \leqslant \sum_{n=1}^{\infty} \widetilde{\mathfrak{B}}'_n(x, y) = \widetilde{\mathfrak{B}}'(x, y),$$

which changes the bounds for $\mathfrak{A}'(x, y)$ and $\mathfrak{B}'(x, y)$ into bounds for $\widetilde{\mathfrak{A}}'(x, y)$ and $\widetilde{\mathfrak{B}}'(x, y)$.

The functions $\widetilde{X}^0(t)$ and $\widetilde{X}^1(t)$ clearly satisfy the following integral equations:

$$\widetilde{X}^0(t) = 1 + \int_0^t (t-u)|q(u)| \widetilde{X}^0(u)\, du = 1 + t\int_0^t |q(u)| \widetilde{X}^0(u)\, du - \int_0^t u|q(u)| \widetilde{X}^0(u)\, du,$$

$$\widetilde{X}^1(t) = t + \int_0^t (t-u)|q(u)| \widetilde{X}^1(u)\, du = t + t\int_0^t |q(u)| \widetilde{X}^1(u)\, du - \int_0^t u|q(u)| \widetilde{X}^1(u)\, du.$$

On the other hand, from Lemma 1.1.1 we have

$$\tilde{X}^0(x) = 1 + \int_{-|x|}^{|x|} \tilde{\mathfrak{A}}(x, y)\,dy = 1 + \frac{x}{2|x|} \int_{-|x|}^{|x|} \left\{ \int_0^x |q(t)|\, \tilde{X}^0(t)\,dt - \tilde{\mathfrak{A}}'(x, y) \right\} dy$$

$$= 1 + x \int_0^x |q(t)|\, \tilde{X}^0(t)\,dt - \frac{x}{2|x|} \int_{-|x|}^{|x|} \tilde{\mathfrak{A}}'(x, y)\,dy,$$

$$\tilde{X}^1(x) = \int_{-|x|}^{|x|} \tilde{\mathfrak{B}}(x, y)\,dy = \frac{x}{2|x|} \int_{-|x|}^{|x|} \left\{ 1 + \int_0^x |q(t)|\, \tilde{X}^1(t)\,dt - \tilde{\mathfrak{B}}'(x, y) \right\} dy$$

$$= x + x \int_0^x |q(t)|\, \tilde{X}^1(t)\,dt - \frac{x}{2|x|} \int_{-|x|}^{|x|} \tilde{\mathfrak{B}}'(x, y)\,dy.$$

Comparing these equations we see that

$$\frac{1}{2} \int_{-|x|}^{|x|} \frac{x}{|x|} \tilde{\mathfrak{A}}'(x, y)\,dy = \int_0^x t\, |q(t)|\, \tilde{X}^0(t)\,dt,$$

$$\frac{1}{2} \int_{-|x|}^{|x|} \tilde{\mathfrak{B}}'(x, y)\,dy = \frac{x}{|x|} \int_0^x t\, |q(t)|\, \tilde{X}^1(t)\,dt,$$

from which it follows that

$$\frac{1}{2} \int_{-|x|}^{|x|} |\mathfrak{A}'(x, y)|\,dy \leqslant \frac{1}{2} \int_{-|x|}^{|x|} \frac{x}{|x|} \tilde{\mathfrak{A}}'(x, y)\,dy = \int_0^x t\, |q(t)|\, \tilde{X}^0(t)\,dt,$$

$$\frac{1}{2} \int_{-|x|}^{|x|} |\mathfrak{B}'(x, y)|\,dy \leqslant \frac{1}{2} \int_{-|x|}^{|x|} \tilde{\mathfrak{B}}'(x, y)\,dy = \frac{x}{|x|} \int_0^x t\, |q(t)|\, \tilde{X}^1(t)\,dt.$$

Thus we have proved

LEMMA 1.1.2. *The kernels* $\mathfrak{A}(x, y)$ *and* $\mathfrak{B}(x, y)$ *admit the following representations*:

$$\mathfrak{A}(x, y) = \frac{x}{2|x|} \left\{ \int_0^x q(t)\, X^0(t)\,dt - \mathfrak{A}'(x, y) \right\},$$

$$\mathfrak{B}(x, y) = \frac{x}{2|x|} \left\{ 1 + \int_0^x q(t)\, X^1(t)\,dt - \mathfrak{B}'(x, y) \right\},$$

where

$$|\mathfrak{A}'(x, y)| \leqslant \frac{x}{|x|} \tilde{\mathfrak{A}}'(x, y); \qquad |\mathfrak{B}'(x, y)| \leqslant \tilde{\mathfrak{B}}'(x, y),$$

$$\int_{-|x|}^{|x|} |\mathfrak{A}'(x, y)|\,dy \leqslant \int_{-|x|}^{|x|} \frac{x}{|x|} \tilde{\mathfrak{A}}'(x, y)\,dy = 2 \int_0^x t\, |q(t)|\, \tilde{X}^0(t)\,dt,$$

$$\int_{-|x|}^{|x|} |\mathfrak{B}'(x, y)|\, dy \leqslant \int_{-|x|}^{|x|} \tilde{\mathfrak{B}}'(x, y)\, dy = \frac{2x}{|x|} \int_0^x t\, |q(t)|\, \tilde{X}^1(t)\, dt.$$

§2. Definition and elementary properties of transformation operators

We let $\mathfrak{M}^p_{(-a,a)}$ ($\mathfrak{M}^p_{[0,a)}$) denote the set of all functions $f(x)$ defined and pth power integrable ($p \geqslant 1$) on each subinterval $(-b, b)$ (respectively $(0, b)$) for arbitrary $b < a$. Here $(-a, a)$, $a \leqslant \infty$, is the interval on which the differential operator L of the form (1.1.1) is given.

Clearly $f(x)$ belongs to $\mathfrak{M}^p_{(-a,a)}$ ($\mathfrak{M}^p_{[0,a)}$) if and only if the following quantity is finite for each $b < a$:

$$\left(\int_\Delta |f(x)|^p\, dx\right)^{\frac{1}{p}} = \|f\|_b^p \qquad (p \neq \infty),$$

$$\operatorname*{ess\,sup}_{x \in \Delta} |f(x)| = \|f\|_b^\infty \quad (p = \infty), \qquad x \in \Delta,$$

where Δ denotes the subinterval $(-b, b)$ (respectively $(0, b)$). A set with index p contains all of the sets with higher indices, and the set $\mathfrak{M}^\infty_{(-a,a)}$ ($\mathfrak{M}^\infty_{[0,a)}$) contains the set $C_{(-a,a)}$ ($C_{[0,a)}$) which consists of all continuous functions on $(-a, a)$ (respectively $[0, a)$). Thus

$$\mathfrak{M}^1_{(-a,\, a)} \supset \mathfrak{M}^2_{-a,\, a)} \supset \mathfrak{M}^\infty_{(-a,\, a)} \supset C_{(-a,\, a)}\,,$$
$$\mathfrak{M}^1_{[0,\, a)} \supset \mathfrak{M}^2_{[0,\, a)} \supset \mathfrak{M}^\infty_{[0,\, a)} \supset C_{[0,\, a)}\,.$$

We will only be interested in the sets of index 1, 2, or ∞. Hence by p we will always mean one of those numbers.

An important role will be played by additive, homogeneous operators W (respectively V) defined on the functions in $\mathfrak{M}^1_{(-a,a)}$ ($\mathfrak{M}^1_{[0,a)}$) by the formula

$$W[f] = f(x) + \int_{-|x|}^{|x|} K(x, t) f(t)\, dt \qquad \left(V[f] = f(x) + \int_0^x K(x, t) f(t)\, dt\right),$$

where the kernels $K(x, t)$ are assumed to be bounded for $|t| \leqslant |x| \leqslant b < a$ ($t \leqslant x \leqslant b < a$). The operators W (respectively V) will be called Volterra operators given on $\mathfrak{M}^1_{(-a,a)}$ ($\mathfrak{M}^1_{[0,a)}$), and will be written briefly in the form

$$W = E + K \quad (V = E + K),$$

where E is the identity operator and

$$K[f] = \int_{-|x|}^{|x|} K(x, t) f(t)\, dt \qquad \left(K[f] = \int_0^x K(x, t) f(t)\, dt\right).$$

We list the following simple properties of Volterra operators.

I. *The set of all Volterra operators forms a group with respect to multiplication*; that is,

1) the product of Volterra operators is a Volterra operator, and

2) each Volterra operator has an inverse which is also a Volterra operator.

PROOF. Assertion 1) is clear. Assertion 2) follows from the general theory of Volterra equations, where it is shown that

$$W^{-1} = (E+K)^{-1} = E - K + K^2 - \cdots + (-1)^n K^n + \cdots = E + H,$$

and the kernel $H(x, y)$ satisfies a bound which implies that it is bounded for $|y| \leqslant |x| \leqslant b < a$.

II. *The Volterra operators take each set* $\mathfrak{M}^p_{(-a,a)}$ $(\mathfrak{M}^p_{[0,a)})$ *onto itself, and*

$$\| W[f] \|_b^p \leqslant \| f \|_b^p C(b) \quad (\| V[f] \|_b^p \leqslant \| f \|_b^p C(b)),$$

where $C(b)$ *is a constant which does not depend on* $f(x)$.

The property expressed by this inequality is called weak boundedness of the operator.

PROOF. If we set $k(b) = \sup_{|y| \leqslant |x| \leqslant b} |K(x, y)|$, we have

$$\| W[f] \|_b^1 \leqslant \| f \|_b^1 + \left\| \int_{-|x|}^{|x|} K(x, y) f(y)\, dy \right\|_b^1 \leqslant \| f \|_b^1 + \left\| k(b) \int_{-|x|}^{|x|} |f(y)|\, dy \right\|_b^1$$

$$\leqslant \| f \|_b^1 + k(b) \| f \|_b^1 \| 1 \|_b^1 = \| f \|_b^1 (1 + 2bk(b));$$

$$\| W[f] \|_b^2 \leqslant \| f \|_b^2 + \left\| \int_{-|x|}^{|x|} K(x, y) f(y)\, dy \right\|_b^2 \leqslant \| f \|_b^2$$

$$+ \left\| \sqrt{\int_{-|x|}^{|x|} |f(y)|^2 dy \cdot \int_{-|x|}^{|x|} |K(x, y)|^2 dy} \right\|_b^2 \leqslant \| f \|_b^2 + k(b) \| f \|_b^2 \| \sqrt{2|x|} \|_b^2$$

$$= \| f \|_b^2 (1 + \sqrt{2}\, b k(b));$$

$$\| W[f] \|_b^\infty \leqslant \| f \|_b^\infty + \left\| \int_{-|x|}^{|x|} K(x, y) f(y)\, dy \right\|_b^\infty \leqslant \| f \|_b^\infty + \left\| k(b) \int_{-|x|}^{|x|} |f(y)|\, dy \right\|_b^\infty$$

$$\leqslant \| f \|_b^\infty + k(b) \| f \|_b^\infty \| 2|x| \|_b^\infty = \| f \|_b^\infty (1 + 2bk(b)).$$

Thus if $C(b) = 1 + 2bk(b)$, then

$$\| W[f] \|_b^p \leqslant \| f \|_b^p C(b) \qquad (p = 1, 2, \infty),$$

where $C(b)$ does not depend on $f(x)$.

Hence the weak boundedness of Volterra operators is proved. Therefore a Volterra operator takes $\mathfrak{M}^p_{(-a,a)}$ into itself, and since the inverse operator is also a Volterra operator, it also takes $\mathfrak{M}^p_{(-a,a)}$ into itself; that is, the transformation is one-to-one, which is what we wanted to show.

The situation concerning $\mathfrak{M}^p_{[0,a)}$ is completely analogous.

We will say that

1) A sequence $\{f_k(x)\}$, $k = 1, 2, \cdots$, of functions in $\mathfrak{M}^p_{(-a,a)}$ (or in $\mathfrak{M}^p_{[0,a)}$) converges (b_0-converges) in the set $\mathfrak{M}^p_{(-a,a)}$ (or $\mathfrak{M}^p_{[0,a)}$) to $f(x)$ if

$$\lim_{k\to\infty} \| f(x) - f_k(x) \|_b^p = 0$$

for arbitrary $b < a$ (or $b = b_0 < a$).

2) A system $\{\phi_\lambda(x)\}$ of functions in $\mathfrak{M}^p_{(-a,a)}$ (or in $\mathfrak{M}^p_{[0,a)}$), depending on a parameter λ, is complete (b_0-complete) in $\mathfrak{M}^p_{(-a,a)}$ (or $\mathfrak{M}^p_{[0,a)}$) if for each function $f(x)$ in the latter set we can construct a sequence of linear combinations of functions $\phi_\lambda(x)$ which converges (b_0-converges) in $\mathfrak{M}^p_{(-a,a)}$ (or $\mathfrak{M}^p_{[0,a)}$) to $f(x)$.

Weak boundedness of Volterra operators immediately implies that

III. *If a sequence of functions $f_k(x)$ converges (b_0-converges) in $\mathfrak{M}^p_{(-a,a)}$ (or $\mathfrak{M}^p_{[0,a)}$) to $f(x)$, then the sequence of functions $W[f_k]$ (or $V[f_k]$) converges (b_0-converges) in the same set to $W[f]$ (respectively $V[f]$).*

IV. *In the set $\mathfrak{M}^p_{(-a,a)}$ (or $\mathfrak{M}^p_{[0,a)}$) a system of functions $\phi_\lambda(x)$ is complete (b_0-complete) if and only if an arbitrary Volterra operator W (or V) takes this system of functions to a system $W[\phi_\lambda(x)]$ (respectively $V[\phi_\lambda(x)]$) which is complete (b_0-complete) in the same set.*

The proofs of III and IV are reasonably obvious and will be omitted.

We turn now to Lemma 1.1.1 of the preceding section. Since the kernels $\mathfrak{A}(x, y)$ and $\mathfrak{B}(x, y)$ are bounded for $|y| \leqslant |x| \leqslant b < a$, for any number θ the operators W,

$$W[f] = f(x) + \int_{-|x|}^{|x|} K(x, y) f(y)\, dy,$$

where $K(x, y) = \mathfrak{A}(x, y) + \theta\mathfrak{B}(x, y)$, are Volterra operators on the set $\mathfrak{M}^1_{(-a,a)}$. Therefore the inverse operators W^{-1} exist and are of the same form:

$$W^{-1}[f] = f(x) + \int_{-|x|}^{|x|} H(x, y) f(y)\, dy,$$

where $H(x, y)$ is bounded for $|y| \leqslant |x| \leqslant b < a$.

LEMMA 1.2.1. *The operator W,*

$$W[f] = f(x) + \int_{-|x|}^{|x|} K(x, y) f(y)\, dy,$$

where $K(x, y) = \mathfrak{A}(x, y) + \theta\mathfrak{B}(x, y)$, transforms solutions $u(\lambda, x)$ of the equations

1) $$u''(x) + \lambda u(x) = 0$$

into solutions $v(\lambda, x)$ *of the equations*

2) $$v''(x) - q(x)v(x) + \lambda v(x) = 0,$$

changing the initial conditions as follows:

$$v(\lambda, 0) = u(\lambda, 0); \qquad v'_x(\lambda, 0) = u'_x(\lambda, 0) + \theta u(\lambda, 0).$$

The inverse operator W^{-1} *transforms solutions* $v(\lambda, x)$ *of the equations* 2) *into solutions* $u(\lambda, x)$ *of the equations* 1), *changing the initial conditions as follows:*

$$u(\lambda, 0) = v(\lambda, 0); \qquad u'_x(\lambda, 0) = v'_x(\lambda, 0) - \theta v(\lambda, 0).$$

PROOF. Let $u(\lambda, x)$ be a solution of 1). Then

$$u(\lambda, x) = u(\lambda, 0)\cos\sqrt{\lambda}x + u'_x(\lambda, 0)\frac{\sin\sqrt{\lambda}x}{\sqrt{\lambda}}$$

and

$$W[u(\lambda, x)] = u(\lambda, 0)W[\cos\sqrt{\lambda}x] + \frac{u'_x(\lambda, 0)}{\sqrt{\lambda}}W[\sin\sqrt{\lambda}x].$$

Furthermore,

$$W[\cos\sqrt{\lambda}x] = \cos\sqrt{\lambda}x + \int_{-|x|}^{|x|}\mathfrak{A}(x, y)\cos\sqrt{\lambda}y\,dy + \theta\int_{-|x|}^{|x|}\mathfrak{B}(x, y)\cos\sqrt{\lambda}y\,dy,$$

$$W[\sin\sqrt{\lambda}x] = \sin\sqrt{\lambda}x + \int_{-|x|}^{|x|}\mathfrak{A}(x, y)\sin\sqrt{\lambda}y\,dy + \theta\int_{-|x|}^{|x|}\mathfrak{B}(x, y)\sin\sqrt{\lambda}y\,dy,$$

from which according to Lemma 1.1.1 we have

$$W[\cos\sqrt{\lambda}x] = c(\lambda, x) + \theta\frac{s(\lambda, x)}{\sqrt{\lambda}},$$

$$W[\sin\sqrt{\lambda}x] = s(\lambda, x) + \theta\int_{-|x|}^{|x|}\mathfrak{B}(x, y)\sin\sqrt{\lambda}y\,dy,$$

and since the kernel $\mathfrak{B}(x, y)$ is even in y, we have $W[\sin\sqrt{\lambda}x] = s(\lambda, x)$. Therefore

$$W[u(\lambda, x)] = u(\lambda, 0)\, c(\lambda, x) + u(\lambda, 0)\, \theta \frac{s(\lambda, x)}{\sqrt{\lambda}} + u'_x(\lambda, 0) \frac{s(\lambda, x)}{\sqrt{\lambda}}$$

$$= u(\lambda, 0)\, c(\lambda, x) + [u'_x(\lambda, 0) + \theta u(\lambda, 0)] \frac{s(\lambda, x)}{\sqrt{\lambda}}\,.$$

Hence $v(\lambda, x) = W[u(\lambda, x)]$ is indeed the solution of 2) with initial conditions

$$v(\lambda, 0) = u(\lambda, 0); \qquad v'_x(\lambda, 0) = u'_x(\lambda, 0) + \theta + u(\lambda 0)$$

which is what we wanted to show.

A simple corollary of this lemma is the following theorem which will be very important in later arguments.

THEOREM 1.2.1. *For each pair of differential operators* L_1 *and* L_2 *of the form* (1.1.1) *given on the interval* $(-a, a)$, $a \leqslant \infty$, *and for an arbitrary number* θ, *there exists a Volterra operator* W *given on* $\mathfrak{M}^1_{(-a,a)}$ *which takes solutions* $u(\lambda, x)$ *of the equations*

$$1) \qquad L_1[u] + \lambda u = 0$$

into solutions $v(\lambda, x)$ *of the equations*

$$2) \qquad L_2[v] + \lambda v = 0,$$

changing initial conditions as follows:

$$v(\lambda, 0) = u(\lambda, 0); \qquad v'_x(\lambda, 0) = u'_x(\lambda, 0) + \theta u(\lambda, 0).$$

PROOF. According to Lemma 1.2.1, for an arbitrary number θ_1 (θ_2) there exists a Volterra operator W_1 (W_2) which transforms solutions $w(\lambda, x)$ of the equations

$$w''(x) + \lambda w(x) = 0$$

into solutions $u(\lambda, x)$ $(v(\lambda, x))$ of the equations 1) (or 2)), changing initial conditions as follows:

$$u(\lambda, 0) = w(\lambda, 0); \qquad u'_x(\lambda, 0) = w'_x(\lambda, 0) + \theta_1 w(\lambda, 0)$$

$$(v(\lambda, 0) = w(\lambda, 0); \qquad v'_x(\lambda, 0) = w'_x(\lambda, 0) + \theta_2 w(\lambda, 0)).$$

We select θ_1 and θ_2 such that $\theta_2 - \theta_1 = \theta$, and we consider the Volterra operator $W = W_2 W_1^{-1}$. If $u(\lambda, x)$ is a solution of 1), then from the definition of W_1 and W_2 we have

$$W[u(\lambda, x)] = W_2[W_1^{-1}[u(\lambda, x)]] = W_2[w(\lambda, x)] = v(\lambda, x),$$

where

$$w(\lambda, 0) = u(\lambda, 0); \qquad w'_x(\lambda, 0) = u'_x(\lambda, 0) - \theta_1 u(\lambda, 0),$$

$$v(\lambda, 0) = w(\lambda, 0); \qquad v'_x(\lambda, 0) = w'_x(\lambda, 0) + \theta_2 w(\lambda, 0);$$

that is,

$$v(\lambda, 0) = u(\lambda, 0); \quad v'_x(\lambda, 0) = u'_x(\lambda, 0) + (\theta_2 - \theta_1) u(\lambda, 0),$$

or

$$v(\lambda, 0) = u(\lambda, 0); \quad v'_x(\lambda, 0) = u'_x(\lambda, 0) + \theta u(\lambda, 0).$$

Thus the Volterra operator W satisfies the conditions stated in the theorem.

Volterra operators which satisfy the conditions stated in the theorem will be called transformation operators (given on the set $\mathfrak{M}^1_{(-a,a)}$) and will be denoted by $W_{\{L_1 L_2 \theta\}}$ when it is necessary to indicate the differential operators L_1, L_2 and number θ with which the transformation operator is associated.

REMARK. It is easy to show that the operators W constructed in Theorem 1.2.1 depend only on the difference $\theta = \theta_2 - \theta_1$ and not on the numbers θ_1 and θ_2. But this fact will not be utilized.

Up to this point we have assumed that the operator L of the form (1.1.1) was given on $(-a, a)$, $a \leqslant \infty$. Now we will consider an operator L of the form (1.1.1) which is given on the interval $[0, a)$, $a \leqslant \infty$ (that is, the function $q(x)$ is defined for $x \in [0, a)$ and is integrable on each subinterval $(0, b)$ for $b < a$). In this case Lemmas 1.1.1, 1.1.2, and 1.2.1 remain valid if we consider only $x \in [0, a)$. This follows from the fact that in order to define the kernels $\mathfrak{A}(x, y)$ and $\mathfrak{B}(x, y)$ for $x \in [0, a)$ we must know the values of $q(x)$ for $x \in [0, a)$; that is, where it is in fact defined.

Instead of Theorem 1.2.1 we have in this case

THEOREM 1.2.2. *For each pair of differential operators* L_1 *and* L_2 *of the form* (1.1.1) *given on the interval* $[0, a)$, $a \leqslant \infty$, *and for arbitrary numbers* h_1 *and* h_2 *which are both finite or both infinite, there exists a Volterra operator* V *given on* $\mathfrak{M}^1_{[0,a)}$ *which transforms solutions* $u(\lambda, x)$ *of the equations*

1) $$L_1[u] + \lambda u = 0,$$

satisfying the boundary condition

$$u'_x(\lambda, 0) - h_1 u(\lambda, 0) = 0 \qquad (u(\lambda, 0) = 0, \quad \textit{if} \quad h_1 = \infty),$$

into solutions $v(\lambda, x)$ *of the equations*

2) $$L_2[v] + \lambda v = 0$$

satisfying the boundary condition

$$v'_x(\lambda, 0) - h_2 v(\lambda, 0) = 0 \qquad (v(\lambda, 0) = 0, \quad \textit{if} \quad h_2 = \infty),$$

where $v(\lambda, 0) = u(\lambda, 0)$ $(v'_x(\lambda, 0) = u'_x(\lambda, 0)$ *if* $h_1 = h_2 = \infty)$.

PROOF. We first observe that by our previous notation we have

$$u(\lambda, x) = u(\lambda, 0)\, \omega^{(1)}_{h_1}(\lambda, x), \quad v(\lambda, x) = v(\lambda, 0)\, \omega^{(2)}_{h_2}(\lambda, x)$$

($u(\lambda, x) = u'_x(\lambda, 0)\,\omega_\infty^{(1)}(\lambda, x)$, $v(\lambda, x) = v'_x(\lambda, 0)\,\omega_\infty^{(2)}(\lambda, x)$ if $h_1 = h_2 = \infty$, where $\omega_{h_1}^{(1)}(\lambda, x)$ and $\omega_{h_2}^{(2)}(\lambda, x)$ denote solutions of 1) and 2)). Therefore the proof of the theorem reduces to finding a Volterra operator V given on $\mathfrak{M}^1_{[0,a)}$ which takes $\omega_{h_1}^{(1)}(\lambda, x)$ into $\omega_{h_2}^{(2)}(\lambda, x)$.

Suppose $h_1, h_2 \neq \infty$. By Lemma 1.2.1, for $x \in [0, a)$ we have

$$\omega_{h_1}^{(1)}(\lambda, x) = \cos\sqrt{\lambda}x + \int_{-|x|}^{|x|} \{\mathfrak{A}_1(x, y) + h_1\mathfrak{B}_1(x, y)\}\cos\sqrt{\lambda}y\,dy;$$

$$\omega_{h_2}^{(2)}(\lambda, x) = \cos\sqrt{\lambda}x + \int_{-|x|}^{|x|} \{\mathfrak{A}_2(x, y) + h_2\mathfrak{B}_2(x, y)\}\cos\sqrt{\lambda}y\,dy.$$

We introduce Volterra operators V_1 and V_2, given on $\mathfrak{M}^1_{[0,a)}$, by means of the formulas

$$V_1[f] = f(x) + \int_0^x K_1(x, y) f(y)\,dy;$$

$$V_2[f] = f(x) + \int_0^x K_2(x, y) f(y)\,dy,$$

where

$$K_1(x, y) = \mathfrak{A}_1(x, y) + \mathfrak{A}_1(x, -y) + h_1\{\mathfrak{B}_1(x, y) + \mathfrak{B}_1(x, -y)\};$$

$$K_2(x, y) = \mathfrak{A}_2(x, y) + \mathfrak{A}_2(x, -y) + h_2\{\mathfrak{B}_2(x, y) + \mathfrak{B}_2(x, -y)\}.$$

The operators V_1 and V_2 obviously transform the functions $\cos\sqrt{\lambda}x$ into $\omega_{h_1}^{(1)}(\lambda, x)$ and $\omega_{h_2}^{(2)}(\lambda, x)$ respectively. As in the preceding theorem we can show that on the set $\mathfrak{M}^1_{[0,a)}$ the Volterra operator $V = V_2V_1^{-1}$ takes $\omega_{h_1}^{(1)}(\lambda, x)$ into $\omega_{h_2}^{(2)}(\lambda, x)$.

Now suppose $h_1 = h_2 = \infty$. By Lemma 1.2.1 for $x \in [0, a)$ we have

$$s_1(\lambda, x) = \sin\sqrt{\lambda}x + \int_{-|x|}^{|x|} \mathfrak{A}_1(x, y)\sin\sqrt{\lambda}y\,dy;$$

$$s_2(\lambda, x) = \sin\sqrt{\lambda}x + \int_{-|x|}^{|x|} \mathfrak{A}_2(x, y)\sin\sqrt{\lambda}y\,dy$$

or

$$\omega_\infty^{(1)}(\lambda, x) = \frac{s_1(\lambda, x)}{\sqrt{\lambda}} = \frac{\sin\sqrt{\lambda}x}{\sqrt{\lambda}} + \int_{-|x|}^{|x|} \mathfrak{A}_1(x, y)\frac{\sin\sqrt{\lambda}y}{\sqrt{\lambda}}\,dy;$$

$$\omega_\infty^{(2)}(\lambda, x) = \frac{s_2(\lambda, x)}{\sqrt{\lambda}} = \frac{\sin\sqrt{\lambda}x}{\sqrt{\lambda}} + \int_{-|x|}^{|x|} \mathfrak{A}_2(x, y)\frac{\sin\sqrt{\lambda}y}{\sqrt{\lambda}}\,dy.$$

Therefore the operators V_1 and V_2,

$$V_1[f] = f(x) + \int_0^x K_1(x, y) f(y) dy;$$

$$V_2[f] = f(x) + \int_0^x K_2(x, y) f(y) dy,$$

where

$$K_1(x, y) = \mathfrak{A}_1(x, y) - \mathfrak{A}_1(x, -y); \quad K_2(x, y) = \mathfrak{A}_2(x, y) - \mathfrak{A}_2(x, -y),$$

transform the functions $\sin\sqrt{\lambda}x/\sqrt{\lambda}$ into $\omega_\infty^{(1)}(\lambda, x)$ and $\omega_\infty^{(2)}(\lambda, x)$ respectively, and $V = V_2 V_1^{-1}$ is a Volterra operator given on $\mathfrak{M}^1_{[0,a)}$ which takes the function $\omega_\infty^{(1)}(\lambda, x)$ into $\omega_\infty^{(2)}(\lambda, x)$, which is what we wanted to show.

Volterra operators which satisfy the conditions stated in this theorem will also be called transformation operators (given on the set $\mathfrak{M}^1_{[0,a)}$) and will be denoted by $V_{\{L_1 L_2 h_1 h_2\}}$ when it is necessary to indicate the differential operators L_1, L_2 and the numbers h_1, h_2 with which the transformation operator is associated.

We note that while the transformation operators $W_{\{L_1 L_2 \theta\}}$ given on $\mathfrak{M}^1_{(-a,a)}$ transform arbitrary solutions of 1) into solutions of 2) with the appropriate initial conditions (Theorem 1.2.1), the transformation operators $V_{\{L_1 L_2 h_1 h_2\}}$ given on $\mathfrak{M}^1_{[0,a)}$ transform only those solutions which satisfy a fixed boundary condition into solutions which satisfy another fixed boundary condition (Theorem 1.2.2).

Remarks on Theorems 1.2.1 and 1.2.2. The preceding theorems demonstrate the fact that the transformation operators $W = W_{\{L_1 L_2 \theta\}}$ and $V = V_{\{L_1 L_2 h_1 h_2\}}$ are of the form

$$W[f] = f(x) + \int_{-|x|}^{|x|} K(x, y) f(y) dy, \quad V[f] = f(x) + \int_0^x H(x, y) f(y) dy,$$

where the kernels $K(x, y)$ and $H(x, y)$ are bounded for $|y| \leqslant |x| \leqslant b$ (respectively $y \leqslant x \leqslant b$) for arbitrary $b < a$. But in general for $|y| \leqslant |x| < a$ (respectively $y \leqslant x < a$) these kernels are not bounded.

If we assume that the interval $(-a, a)$ (or $[0, a)$) is finite and the functions $q_1(x)$ and $q_2(x)$ in L_1 and L_2 are integrable on the whole interval, then from the bound in Lemma 1.1.1 it is clear that

$$\sup_{|y| \leqslant |x| < a} |K(x, y)| < \infty \qquad (\sup_{y \leqslant x < a} |H(x, y)| < \infty).$$

Since transformation operators are a special type of Volterra operator, they have properties I, II, III, *and* IV *listed at the start of this section.*

These simple properties of transformation operators are at the same time their most important properties. They enable us to apply transformation operators to a wide range of problems concerning differential operators L of the form (1.1.1). It is interesting to note that from the point of view of physics problems we will use only the properties I, II, III, and IV of transformation operators.

The simplicity of the useful properties together with the clarity of the consequent arguments (see, for example, the proof of Theorem 2.3.1) suggests that transformation operators are a natural tool for studying these problems.

CHAPTER 2

APPLICATIONS OF TRANSFORMATION OPERATORS TO SOME QUESTIONS IN SPECTRAL ANALYSIS OF LINEAR DIFFERENTIAL OPERATORS OF THE SECOND ORDER

§1. Theory of Hermann Weyl

As we noted in the Introduction, spectral analysis of linear differential operators of the second order given on an infinite interval or having singular coefficients was formulated by Weyl in 1909–1910 (see [2]).

The present chapter is devoted to the study of some new problems in spectral analysis which arise in modern physics.

We will state without proof the fundamental theorem of Weyl. The theorem can be proved in many ways; one easy proof employs methods developed by Levitan in [4].

Let L given on the interval $[0, a)$, $a \leqslant \infty$, be a second-order linear differential operator of the form

$$L[u] = \frac{d^2}{dx^2} u(x) - q(x) u(x), \tag{2.1.1}$$

where $q(x)$ is a real function which is integrable in each subinterval $(0, b)$ for arbitrary $b < a$.

We let $\phi_\alpha(\lambda, x)$ and $\theta_\alpha(\lambda, x)$ denote the solutions of

$$L[u] + \lambda u = 0 \tag{2.1.2}$$

with initial conditions

$$\varphi_\alpha(\lambda, 0) = \sin\alpha, \qquad \varphi'_\alpha(\lambda, 0) = -\cos\alpha;$$
$$\theta_\alpha(\lambda, 0) = \cos\alpha, \qquad \theta'_\alpha(\lambda, 0) = \sin\alpha,$$

where λ is an arbitrary complex number and α is an arbitrary real number.

In the notation of §1 of Chapter 1 we have

$$\varphi_\alpha(\lambda, x) = \sin\alpha\,\omega_0(\lambda, x) - \cos\alpha\omega_\infty(\lambda, x);$$

$$\theta_\alpha(\lambda, x) = \cos\alpha\,\omega_0(\lambda, x) + \sin\alpha\omega_\infty(\lambda, x).$$

Let $L^2[0, a)$ denote the Hilbert space of square-integrable functions on $[0, a)$.

A. For each nonreal value of λ there exists a solution $\psi(\lambda, x)$ of (2.1.2) belonging to $L^2[0, a)$ of the form

$$\psi(\lambda, x) = \theta_\alpha(\lambda, x) + m(\lambda)\varphi_\alpha(\lambda, x),$$

where the function $m(\lambda)$ is analytic in the upper (lower) half-space and

$$m(\bar{\lambda}) = \overline{m(\lambda)}, \quad \operatorname{Im}[m(\lambda)]\cdot\operatorname{Im}[\lambda] < 0.$$

B. For a given α the function $m(\lambda)$ is not in general unique.

If $m(\lambda)$ is unique, then the boundary condition at zero

$$y(0)\cos\alpha + y'(0)\sin\alpha = 0$$

uniquely determines the boundary value problem for L.

If $m(\lambda)$ is not unique, then in order to determine a boundary value problem we need the boundary condition at zero

$$y(0)\cos\alpha + y'(0)\sin\alpha = 0$$

together with the particular function $m(\lambda)$ which corresponds (in some well defined way) to the choice of a boundary condition at the point a.

C. Suppose that to the boundary value problems for L with boundary conditions at zero

$$y(0)\cos\alpha + y'(0)\sin\alpha = 0,$$
$$y(0)\cos\beta + y'(0)\sin\beta = 0$$

and the same boundary conditions at a if these conditions are necessary there correspond the functions $m_\alpha(\lambda)$ and $m_\beta(\lambda)$. Then we have

$$m_\beta(\lambda) = \frac{1 + hm_\alpha(\lambda)}{h - m_\alpha(\lambda)} \qquad (h = \cot(\beta - \alpha) \neq \infty).$$

D. To each boundary value problem for L there corresponds a nondecreasing resolution function $\rho(\lambda)$ which is associated with the function $m(\lambda)$ by the formula

$$\rho(\lambda_2) - \rho(\lambda_1) = \frac{1}{\pi} \lim_{\delta \to 0} \int_{\lambda_1}^{\lambda_2} \operatorname{Im}[m(u + i\delta)]\,du,$$

where λ_1 and λ_2 are arbitrary points of continuity of $\rho(\lambda)$. The function $\rho(\lambda)$ generates an isometric mapping of $L^2[0, a)$ onto $L^2_\rho(-\infty, \infty)$ by means of the formulas

$$E_f(\lambda) = \int_0^a f(x)\,\varphi_\alpha(\lambda, x)\,dx; \qquad f(x) = \int_{-\infty}^{\infty} E_f(\lambda)\,\varphi_\alpha(\lambda, x)\,d\rho(\lambda),$$

where the integrals are understood in the sense of convergence in the metrics of $L^2_\rho(-\infty, \infty)$ and $L^2[0, a)$ respectively. Parseval's equation is valid in this case:

$$\|f\|^2 = \int_0^a |f(x)|^2\,dx = \int_{-\infty}^{\infty} |E_f(\lambda)|^2\,d\rho(\lambda).$$

(Here $L^2_\rho(-\infty, \infty)$ denotes the Hilbert space of functions $E(\lambda)$ which are square-integrable on $-\infty < \lambda < \infty$ with the weight $d\rho(\lambda)$.)

By the spectrum of a boundary value problem for a given operator L we mean the set of points of increase of the function $\rho(\lambda)$.

In certain special cases it will be useful to establish other formulas for $\rho(\lambda)$. Of particular importance is the case where the operator L of the form (2.1.1) is given on $0 \leqslant x < \infty$ and $q(x)$ satisfies

$$\int_0^\infty (1+x)\,|q(x)|\,dx < \infty. \tag{2.1.3}$$

We will examine this case in detail.

LEMMA 2.1.1. *If a differential operator* L *of the form* (2.1.1) *is given on* $0 \leqslant x < \infty$ *and* $q(x)$ *satisfies* (2.1.3), *then the following assertions are true.*

a) *For complex numbers* s *in the lower half-plane we have the asymptotic formulas*

$$\varphi_\alpha(s^2, x) = \frac{e^{\overline{isx}}}{2i}\,[M(s) + o(1)],$$

where

$$M(s) = i\sin\alpha - \frac{\cos\alpha}{s} + \frac{1}{s}\int_0^\infty e^{-ist}\,q(t)\,\varphi_\alpha(s^2, t)\,dt$$

is analytic in the lower half-plane.

b) *For real numbers* $s \neq 0$ *we have the asymptotic formulas*

$$\varphi_\alpha(s^2, x) = |M(s)|\sin(sx + \delta(s)) + r_\alpha(s^2, x),$$

where

$$r_\alpha(s^2, x) = \int_x^\infty \frac{\sin s(x-t)}{s}\,q(t)\,\varphi_\alpha(s^2, t)\,dt = o(1) \qquad (x \to \infty),$$

and $\delta(s) = \arg M(s)$ *is called the limiting phase of the solution* $\phi_\alpha(s^2, x)$.

PROOF. By Lemma 1.1.1 we have

$$\omega_0(s^2, x) = \cos sx + \int_{-|x|}^{|x|} \mathfrak{A}(x, y) \cos sy\, dy;$$

$$\omega_\infty(s^2, x) = \frac{\sin sx}{s} + \int_{-|x|}^{|x|} \mathfrak{A}(x, y) \frac{\sin sy}{s}\, dy,$$

and by the corollary to the lemma we deduce from (2.1.3) that

$$\sup_{0 \leqslant y \leqslant x < \infty} |\mathfrak{A}(x, y)| < \infty.$$

Therefore

$$|\omega_0(s^2, x)| \leqslant C_0 e^{|\operatorname{Im} s| x}(1 + x) \qquad (0 \leqslant x < \infty), \tag{2.1.4}$$

$$|\omega_\infty(s^2, x)| \leqslant \frac{C_\infty}{s} e^{|\operatorname{Im} s| x}(1 + x) \qquad (0 \leqslant x < \infty). \tag{2.1.4'}$$

Also, by using the kernel $\mathfrak{B}(x, y)$ we obtain

$$|\omega_\infty(s^2, x)| \leqslant C'_\infty e^{|\operatorname{Im} s| x}(1 + x) \qquad (0 \leqslant x < \infty). \tag{2.1.5}$$

Since

$$\varphi_\alpha(s^2, x) = \sin \alpha\omega_0(s^2, x) - \cos \alpha\omega_\infty(s^2, x),$$

we can see from (2.1.4) and (2.1.5) that

$$|\varphi_\alpha(s^2, x)| \leqslant C(1 + x) e^{|\operatorname{Im} s| x} \qquad (0 \leqslant x < \infty), \tag{2.1.6}$$

where the constant C does not depend on s, x, or α.

From (2.1.3) and (2.1.6) we can show that

$$\int_0^\infty e^{-ist} q(t) \varphi_\alpha(s^2, t)\, dt$$

converges absolutely and uniformly with respect to s for s in the lower half-plane and on the real axis. Therefore the function $M(s)$ is analytic in the lower half-plane and continuous on the real axis.

It is also clear that the following limit is attained uniformly in the lower half-plane and on the real axis:

$$\lim_{|s| \to \infty} M(s) = i \sin \alpha. \tag{2.1.7}$$

Furthermore, if $\sin \alpha = 0$, then from (2.1.4′) applied to $\omega_\infty(s^2, x) = \phi_\alpha(s^2, x)$ it can be seen that the following limit is attained uniformly in the lower half-plane and on the real axis:

$$\lim_{|s| \to \infty} sM(s) = -1 \qquad (\sin \alpha = 0). \tag{2.1.7'}$$

Now let s be arbitrary in the lower half-plane. By using a Volterra integral equation for the function $\phi_\alpha(s^2, x)$ we obtain

$$\begin{aligned}\varphi_\alpha(s^2, x) &= \sin\alpha\cos sx - \cos\alpha\,\frac{\sin sx}{s} + \int_0^x \frac{\sin s(x-t)}{s} q(t)\,\varphi_\alpha(s^2, t)\,dt \\ &= \frac{e^{isx}}{2i}\Bigg[i\sin\alpha - \frac{\cos\alpha}{s} + \frac{1}{s}\int_0^x e^{-ist} q(t)\,\varphi_\alpha(s^2, t)\,dt \\ &\qquad - \frac{e^{-2isx}}{s}\int_0^x e^{ist} q(t)\,\varphi_\alpha(s^2, t)\,dt + o(1)\Bigg].\end{aligned}$$

From (2.1.6) it follows that

$$\int_0^x e^{-ist} q(t)\,\varphi_\alpha(s^2, t)\,dt = \int_0^\infty e^{-ist} q(t)\,\varphi_\alpha(s^2, t)\,dt + o(1),$$

$$e^{-2isx}\int_0^x e^{ist} q(t)\,\varphi_\alpha(s^2, t)\,dt = o(1).$$

Therefore

$$\begin{aligned}\varphi_\alpha(s^2, x) &= \frac{e^{isx}}{2i}\left[i\sin\alpha - \frac{\cos\alpha}{s} + \frac{1}{s}\int_0^\infty e^{-ist} q(t)\,\varphi_\alpha(s^2, t)\,dt + o(1)\right] \\ &= \frac{e^{isx}}{2i}[M(s) + o(1)],\end{aligned}$$

and assertion a) is proved.

Similarly for real $s \neq 0$ we obtain

$$\begin{aligned}\varphi_\alpha(s^2, x) &= \sin\alpha\cos sx - \cos\alpha\frac{\sin sx}{s} + \int_0^x \frac{\sin s(x-t)}{s} q(t)\,\varphi_\alpha(s^2, t)\,dt \\ &= \sin\alpha\cos sx - \cos\alpha\,\frac{\sin sx}{s} + \int_0^\infty \frac{\sin s(x-t)}{s}\, q(t)\,\varphi_\alpha(s^2, t)\,dt \\ &\qquad - \int_x^\infty \frac{\sin s(x-t)}{s}\, q(t)\,\varphi_\alpha(s^2, t)\,dt,\end{aligned}$$

or

$$\varphi_\alpha(s^2, x) = \mu(s)\cos sx + \frac{\nu(s)}{s}\sin sx + r_\alpha(s^2, x),$$

where

$$\mu(s) = \sin\alpha - \int_0^\infty \frac{\sin st}{s} q(t)\,\varphi_\alpha(s^2, t)\,dt, \quad \nu(s) = -\cos\alpha + \int_0^\infty \cos st\, q(t)\phi_\alpha(s^2, t)\,dt,$$

$$r_\alpha(s^2, x) = \int_x^\infty \frac{\sin s(x-t)}{s} q(t) \varphi_\alpha(s^2, t)\, dt = o(1) \qquad (x \to \infty).$$

Since

$$\frac{\nu(s)}{s} + i\mu(s) = i \sin\alpha - \frac{\cos\alpha}{s} + \frac{1}{s}\int_0^\infty e^{-ist} q(t) \varphi_\alpha(s^2, t)\, dt = M(s),$$

on using the function

$$\delta(s) = \arg M(s) = \arctan \frac{s\mu(s)}{\nu(s)},$$

we have

$$\varphi_\alpha(s^2, x) = |M(s)| \sin(sx + \delta(s)) + r_\alpha(s^2, x),$$

where $r_\alpha(s^2, x) = o(1)$ as $x \to \infty$. The lemma is proved.

LEMMA 2.1.2. *If a differential operator* L *of the form* (2.1.1) *is given on* $0 \leqslant x < \infty$ *and* $q(x)$ *satisfies* (2.1.3), *then the following assertions are true.*

a) *For complex numbers* s *the equation*

$$L[u] + s^2 u = 0$$

has two linearly independent solutions, of which one grows like $e^{|\operatorname{Im} s| x}$ *and the other decreases like* $e^{-|\operatorname{Im} s| x}$ *as* $x \to \infty$.

b) *The equation*

$$L[u] = 0$$

has two linearly independent solutions $v_1(x)$ *and* $v_2(x)$ *such that*

$$v_1(x) = 1 + o(1), \qquad v_1'(x) = o(1) \qquad (x \to \infty);$$
$$v_2(x) = x + o(x), \qquad v_2'(x) = 1 + o(1) \quad (x \to \infty).$$

PROOF. a) Without loss of generality we can assume that $\operatorname{Im} s < 0$. The asymptotic formulas of the preceding lemma imply that for complex s only one solution of $L[u] + s^2 u = 0$ belongs to $L^2[0, \infty)$. Thus from the solutions $\phi_\alpha(s^2, x)$ and $\theta_\alpha(s^2, x)$ at least one, say $\phi_\alpha(s^2, x)$, does not belong to $L^2[0, \infty)$. Then $M(s) \neq 0$, and for large x

$$|\varphi_\alpha(s^2, x)| \geqslant \frac{1}{3} |M(s)| e^{|\operatorname{Im} s| x}.$$

The function

$$u(x) = \varphi_\alpha(s^2, x) \int_x^\infty \frac{dt}{[\varphi_\alpha(s^2, t)]^2}$$

also is a solution of $L[u] + s^2 u = 0$, and for large x we have

$$|u(x)| \leqslant |M(s)| e^{|\operatorname{Im} s| x} \int_x^\infty \frac{9e^{-2|\operatorname{Im} s| t}}{|M(s)|^2} dt = \frac{9e^{-|\operatorname{Im} s| x}}{2|M(s)||\operatorname{Im} s|},$$

which is what we wanted to show.

b) If a function $u(x)$ satisfies the integral equation

$$u(x) = 1 + \int_x^\infty (x-t) q(t)\, dt, \tag{2.1.8}$$

then it is a solution of

$$L[u] = u''(x) - q(x) u(x) = 0.$$

From (2.1.3) it follows that there exists N, such that

$$\sup_{N \leqslant x < \infty} \int_x^\infty |x-t| \, |q(t)| \, dt < \frac{1}{2}.$$

We will show that (2.1.8) has a solution which is bounded in $[N, \infty)$. In fact, let

$$y_0(x) = 1, \qquad y_n(x) = -\int_x^\infty (x-t) q(t) y_{n-1}(t)\, dt.$$

Then

$$\sup_{N \leqslant x < \infty} |y_n(x)| \leqslant \sup_{N \leqslant t < \infty} |y_{n-1}(t)| \int_N^\infty |N-t| \, |q(t)| \, dt \leqslant \frac{1}{2} \sup_{N \leqslant x < \infty} |y_{n-1}(x)|,$$

and the series

$$y_0(x) + y_1(x) + \cdots + y_n(x) + \cdots$$

converges uniformly for $x \geqslant N$ and its sum $u(x)$ is bounded and satisfies (2.1.8), and thus $L[u] = 0$. Clearly

$$u(x) = 1 + o(1), \qquad u'(x) = o(1) \qquad (x \to \infty).$$

Therefore the equation $L(u) = 0$ has a bounded solution (which we denote by $v_1(x)$) and

$$v_1(x) = 1 + o(1), \qquad v_1'(x) = o(1) \qquad (x \to \infty).$$

But then the function

$$v_2(x) = v_1(x) \int_0^x \frac{dt}{[v_1(t)]^2}$$

also satisfies $L[u] = 0$, and clearly

$$v_2(x) = x + o(x), \qquad v_2'(x) = 1 + o(1) \qquad (x \to \infty).$$

The lemma is proved.

We take note of two simple corollaries of this lemma.

COROLLARY 1. *If L satisfies the conditions of Lemma* 2.1.2 *and*

Im $s \neq 0$, *then the solution of* $L[u] + s^2u = 0$ *which belongs to* $L^2[0, \infty)$ *decreases as* $x \to \infty$ *like* $e^{-|\mathrm{Im}\, s|x}$.

COROLLARY 2. *If* L *satisfies the conditions of Lemma* 2.1.2, *then each solution of* $L[u] = 0$ *has a finite number of zeros.*

Now we will formulate the fundamental results of spectral analysis for operators L which satisfy (2.1.3).

E. Suppose that a differential operator L of the form (2.1.1) is given on $0 \leqslant x < \infty$ and satisfies

$$\int_0^\infty (1 + x)|q(x)|\, dx < \infty. \tag{2.1.3}$$

Then a boundary value problem for L is uniquely determined by the boundary condition at zero

$$y(0)\cos\alpha + y'(0)\sin\alpha = 0.$$

The function $m(\lambda)$ corresponding to this boundary value problem (see part B above) is of the form

$$m(\lambda) = -\frac{N(s)}{M(s)} \qquad (s^2 = \lambda,\ \mathrm{Im}\, s < 0),$$

where

$$M(s) = i\sin\alpha - \frac{\cos\alpha}{s} + \frac{1}{s}\int_0^\infty e^{-ist} q(t)\, \varphi_\alpha(s^2, t)\, dt,$$

$$N(s) = i\cos\alpha + \frac{\sin\alpha}{s} + \frac{1}{s}\int_0^\infty e^{-ist} q(t)\, \theta_\alpha(s^2, t)\, dt.$$

The spectrum of the boundary value problem consists of negative eigenvalues (discrete part) and the whole positive axis $0 \leqslant \lambda < \infty$ (continuous part).

The number of negative eigenvalues equals the number of zeros of the solution $\phi_\alpha(0, x)$; that is, by Corollary 2 of Lemma 2.1.2, some finite number.

In the lower half-plane the function $M(s)$ has a finite number of zeros $s_1, \cdots, s_n$ related to the negative eigenvalues $\lambda_1, \lambda_2, \cdots, \lambda_n$ by the formula

$$\lambda_1 = s_1^2, \quad \lambda_2 = s_2^2, \quad \ldots, \quad \lambda_n = s_n^2.$$

For real $s \neq 0$ we have $M(s) \neq 0$.

The inverse formulas generated by the boundary value problem are of the form

$$f(x) = \sum_{\nu=1}^n E_f(\lambda_\nu) \frac{\varphi_\alpha(\lambda_\nu, x)}{\|\varphi_\alpha(\lambda_\nu, x)\|^2} + \frac{1}{\pi}\int_0^\infty E_f(\lambda)\, \varphi_\alpha(\lambda, x) \frac{d\lambda}{\sqrt{\lambda}\, |M(\sqrt{\lambda})|^2},$$

$$E_f(\lambda) = \int_0^\infty f(x)\,\varphi_\alpha(\lambda, x)\,dx,$$

or

$$f(x) = \sum_{\nu=1}^n E_f(s_\nu^2)\frac{\varphi_\alpha(s_\nu^2, x)}{\|\varphi_\alpha(s_\nu^2, x)\|^2} + \frac{2}{\pi}\int_0^\infty E_f(s^2)\,\varphi_\alpha(s^2, x)\,\frac{ds}{|M(s)|^2},$$

$$E_f(s^2) = \int_0^\infty f(x)\,\varphi_\alpha(s^2, x)\,dx.$$

Hence the resolution function $\rho(\lambda)$ in this case equals

$$\rho(\lambda) = \sum_{\lambda_\nu \leqslant \lambda}\frac{1}{\|\varphi_\alpha(\lambda_\nu, x)\|^2} \qquad \text{for } \lambda < 0,$$

$$\rho(\lambda) = \sum_{\nu=1}^n \frac{1}{\|\varphi_\alpha(\lambda_\nu, x)\|^2} + \frac{1}{\pi}\int_0^\lambda \frac{d\mu}{\sqrt{\mu}\,|M(\sqrt{\mu})|^2} \qquad \text{for } \lambda \geqslant 0.$$

§2. Asymptotic formulas for the resolution function $\rho(\lambda)$

In this section we will consider differential operators L of the form (2.1.1) given on the interval $[0, \infty)$.

According to part D of the preceding section each boundary value problem for L generates inversion formulas of the form

$$E_f(\lambda) = \int_0^\infty f(x)\,\varphi_\alpha(\lambda, x)\,dx, \quad f(x) = \int_{-\infty}^\infty E_f(\lambda)\,\varphi_\alpha(\lambda, x)\,d\rho(\lambda), \tag{2.2.1}$$

where $\rho(\lambda)$ is a nondecreasing function, and the integrals converge in the metrics of $L^2_\rho(-\infty, \infty)$ and $L^2[0, \infty)$ respectively.

In the general case the calculation of $\rho(\lambda)$ is very difficult. In this connection it is natural to consider the asymptotic behavior of $\rho(\lambda)$ for large $|\lambda|$. An upper bound for $\rho(\lambda)$ for large $\lambda > 0$ was given by Levitan in [7].

The present section is devoted to finding precise asymptotic formulas for $\rho(\lambda)$. The fundamental result is given in the following theorem.

THEOREM 2.2.1. *For the resolution function* $\rho(\lambda)$ *appearing in the inversion formulas* (2.2.1) *we have*

$$\rho(\lambda) = \frac{2}{\pi \sin^2\alpha}\sqrt{\lambda} + o(\sqrt{\lambda}), \quad \textit{if} \quad \sin\alpha \neq 0 \quad (\lambda \to +\infty);$$

$$\rho(\lambda) = \frac{2}{3\pi}\sqrt{\lambda^3} + o(\sqrt{\lambda^3}), \quad \textit{if} \quad \sin\alpha = 0 \quad (\lambda \to +\infty).$$

We will give a proof of this theorem for the case $\sin\alpha \neq 0$. The proof for the case $\sin\alpha = 0$ is by analogy.

We observe first of all that for $\sin\alpha \neq 0$ the function $\phi_\alpha(\lambda, x)$ below is generated by the function $\omega_h(\lambda, x)$ introduced in §1 of Chapter 1:

$$\varphi_\alpha(\lambda, x) = \sin\alpha\, \omega_h(\lambda, x), \quad \text{where} \quad h = -\cot\alpha \neq \infty.$$

Therefore the inversion formulas (2.2.1) can be written in the form

$$E_f(\lambda) = \int_0^\infty f(x)\, \omega_h(\lambda, x)\, dx, \qquad f(x) = \int_{-\infty}^\infty E_f(\lambda)\, \omega_h(\lambda, x)\, d\rho_1(\lambda), \tag{2.2.2}$$

$$\|f\|^2 = \int_0^\infty |f(x)|^2\, dx = \int_{-\infty}^\infty |E_f(\lambda)|^2\, d\rho_1(\lambda), \tag{2.2.3}$$

where

$$\rho_1(\lambda) = \sin^2\alpha\rho(\lambda)$$

So instead of the function $\rho(\lambda)$ we can study the function $\rho_1(\lambda)$, which simplifies the calculations.

The idea of the proof consists in comparing the inversion formulas (2.2.2) with the classical inversion formulas of Fourier-Plancherel:

$$C_f(\lambda) = \int_0^\infty f(x) \cos\lambda x\, dx, \qquad f(x) = \frac{2}{\pi} \int_0^\infty C_f(\lambda) \cos\lambda x\, d\lambda, \tag{2.2.4}$$

$$\|f\|^2 = \int_0^\infty |f(x)|^2\, dx = \frac{2}{\pi} \int_0^\infty |C_f(\lambda)|^2\, d\lambda. \tag{2.2.5}$$

We will make use of the transformation operator which takes $\omega_h(\lambda, x)$ into $\cos\sqrt{\lambda}\,x$. According to the notation of the preceding chapter this operator will be denoted by $V_{\{LD^2h0\}}$, where

$$D^2[u] = \frac{d^2}{dx^2} u(x).$$

First we will prove several lemmas.

LEMMA 2.2.1. *On the interval* $-\infty < \lambda \leqslant 0$ *the function* $\rho_1(\lambda)$ *is of bounded variation, and for arbitrary* x

$$\int_{-\infty}^{+0} \cosh x\sqrt{|\lambda|}\, d\rho_1(\lambda) < \infty.$$

PROOF. Let $E(\lambda)$ be an arbitrary function in $L^2_{\rho_1}(-\infty, 0]$; that is,

$$\int_{-\infty}^{+0} |E(\lambda)|^2\, d\rho_1(\lambda) < \infty.$$

According to part D of the preceding section the sequence of functions

$$f_n(x) = \int_{-n}^{+0} |E(\lambda)|\, \omega_h(\lambda, x)\, d\rho_1(\lambda) \qquad (n = 1, 2, \ldots)$$

converges in the metric of $L^2[0, \infty)$ to some function $f(x) \in L^2[0, \infty)$. Hence this sequence is mean square convergent to $f(x)$ in each finite subinterval of $0 \leqslant x < \infty$. From this fact together with weak boundedness of the transformation operator $V = V_{\{LD^2h0\}}$ (see Chapter 1, §2, property III) we conclude that

$$V[f] = \lim_{n \to \infty} V[f_n],$$

where we have mean square convergence in each finite subinterval of $0 \leqslant x < \infty$. From the definition of V we have

$$V[f_n] = V\left[\int_{-n}^{+0} |E(\lambda)|\, \omega_h(\lambda, x)\, d\rho_1(\lambda)\right] = \int_{-n}^{+0} |E(\lambda)|\, V[\omega_h(\lambda, x)]\, d\rho_1(\lambda)$$

$$= \int_{-n}^{+0} |E(\lambda)| \cos x\sqrt{\lambda}\, d\rho_1(\lambda) = \int_{-n}^{+0} |E(\lambda)| \cosh x\sqrt{|\lambda|}\, d\rho_1(\lambda).$$

Thus the sequence of functions

$$\int_{-n}^{+0} |E(\lambda)| \cosh x\sqrt{|\lambda|}\, d\rho_1(\lambda) \qquad (n = 1, 2, \ldots)$$

is mean square convergent to $V[f]$ in each finite subinterval of $0 \leqslant x < \infty$. But since the integrand $|E(\lambda)| \cosh x|\lambda|^{1/2}$ is nonnegative and the function $\rho_1(\lambda)$ is nondecreasing, this convergence is possible if and only if

$$\lim_{n \to \infty} \int_{-n}^{+0} |E(\lambda)| \cosh x\sqrt{|\lambda|}\, d\rho_1(\lambda) < \infty,$$

for any x. Thus for any x the limit

$$\lim_{n \to \infty} \int_{-n}^{+0} E(\lambda) \cosh x\sqrt{|\lambda|}\, d\rho_1(\lambda)$$

exists and is finite for arbitrary $E(\lambda) \in L^2_{\rho_1}(-\infty, 0]$.

We now observe that for fixed x the expressions

$$\int_{-n}^{+0} E(\lambda) \cosh x\sqrt{|\lambda|}\, d\rho_1(\lambda) \qquad (n = 1, 2, \ldots)$$

can be interpreted as linear functionals on $L^2_{\rho_1}(-\infty, 0]$. The preceding arguments show that the sequence of these linear functionals converges to the functional

$$\int_{-\infty}^{+0} E(\lambda)\cosh x\sqrt{|\lambda|}\,d\rho_1(\lambda) = \lim_{n\to\infty}\int_{-n}^{+0} E(\lambda)\cosh x\sqrt{|\lambda|}\,d\rho_1(\lambda)$$

for all elements $E(\lambda)$ of $L^2_{\rho_1}(-\infty, 0]$, and the limit functional has a finite value on each element of $L^2_{\rho_1}(-\infty, 0)$. In functional analysis it is shown that the limit functional is also linear, and therefore the expression

$$\int_{-\infty}^{+0} E(\lambda)\cosh x\sqrt{|\lambda|}\,d\rho_1(\lambda)$$

for arbitrary fixed x determines a linear functional on $L^2_{\rho_1}(-\infty, 0]$. From this fact and from the form of a linear functional in $L^2_{\rho_1}(-\infty, 0]$ it follows that

$$\int_{-\infty}^{+0} (\cosh x\sqrt{|\lambda|})^2\,d\rho_1(\lambda) < \infty,$$

for all x. But then

$$\int_{-\infty}^{+0} \cosh x\sqrt{|\lambda|}\,d\rho_1(\lambda) < \infty,$$

for all x, which is what we wanted to show.

The proof of the lemma is of interest in itself since it characterizes the behavior of $\rho_1(\lambda)$ on the negative semi-axis $-\infty < \lambda \leqslant 0$. In particular, the proof gives us the following corollary.

COROLLARY. *Let* $E(\lambda) \in L^2_{\rho_1}(-\infty, \infty)$ *and* $N < \infty$. *Then the integral*

$$\int_{-\infty}^{N} E(\lambda)\,\omega_h(\lambda, x)\,d\rho_1(\lambda)$$

converges absolutely and uniformly in each finite subinterval of $0 \leqslant x < \infty$.

In fact, since $\rho_1(\lambda)$ is of bounded variation in each finite subinterval, we can assume without loss of generality that $N = 0$. Furthermore, for $\lambda \leqslant 0$ we have

$$\omega_h(\lambda, x) = V_{\{D^2 L 0 h\}}\left[\cos x\sqrt{\lambda}\right] = \cosh x\sqrt{|\lambda|} + \int_0^x H(x, y)\cosh y\sqrt{|\lambda|}\,dy,$$

where the kernel $H(x, y)$ is bounded in each finite portion of the (x, y)-plane. If we set

$$c(b) = \sup_{0\leqslant y\leqslant x\leqslant b} |H(x, y)|,$$

we obtain

$$|\omega_h(\lambda, x)| \leqslant \cosh b\sqrt{|\lambda|}\,(1 + bc(b)) \qquad (0 \leqslant x \leqslant b),$$

which implies that

$$|E(\lambda)\,\omega_h(\lambda, x)| \leqslant (1 + bc(b))\,|E(\lambda)|\cosh b\sqrt{|\lambda|} \quad (0 \leqslant x \leqslant b).$$

It remains to observe that

$$\int_{-\infty}^{+0} |E(\lambda)|\cosh b\sqrt{|\lambda|}\, d\rho_1(\lambda) < \infty$$

(since $E(\lambda) \in L^2_{\rho_1}(-\infty, 0]$ by assumption, and $\cosh b\sqrt{|\lambda|} \in L^2_{\rho_1}(-\infty, 0]$ by the previous lemma) which proves the assertion.

We will say that a function $f(x) \in L^2[0, \infty)$ is finite if it equals zero outside some finite subinterval. For an arbitrary finite function $f(x)$ and an arbitrary complex number λ we set

$$E_f(\lambda) = \int_0^\infty f(x)\,\omega_h(\lambda, x)\,dx, \qquad C_f(\lambda) = \int_0^\infty f(x)\cos\lambda x\,dx,$$

where the integrals are in fact taken over a finite subinterval.

We have

$$\cos x\sqrt{\lambda} = V_{\{LD^*h0\}}[\omega_h(\lambda, x)] = \omega_h(\lambda, x) + \int_0^x H(x, y)\,\omega_h(\lambda, y)\,dy,$$

where $H(x, y)$ is bounded in each finite portion of the (x, y)-plane. Hence for an arbitrary finite $f(x)$ we have

$$\int_0^\infty f(x)\cos x\sqrt{\lambda}\,dx = \int_0^\infty f(x)\,\omega_h(\lambda, x)\,dx + \int_0^\infty f(x)\int_0^x H(x, y)\,\omega_h(\lambda, y)\,dy\,dx,$$

and if we set

$$\hat{f}(y) = \int_y^\infty H(x, y) f(x)\,dx, \qquad g(x) = f(x) + \hat{f}(x)$$

(clearly $\hat{f}(x)$ and $g(x)$ are finite), then we have

$$C_f(\sqrt{\lambda}) = E_g(\lambda). \tag{2.2.6}$$

Furthermore,

$$\left| \|f\|^2 - \|g\|^2 \right| \leqslant \|\hat{f}\|(2\|f\| + \|\hat{f}\|)$$

or by (2.2.3) and (2.2.6)

$$\left| \|f\|^2 - \int_{-\infty}^\infty |C_f(\sqrt{\lambda})|^2\, d\rho_1(\lambda) \right| \leqslant \|\hat{f}\|(2\|f\| + \|\hat{f}\|), \tag{2.2.7}$$

which implies that

$$\left| \|f\|^2 - \int_0^\infty |C_f(\sqrt{\lambda})|^2\, d\rho_1(\lambda) \right| \leqslant \int_{-\infty}^{+0} |C_f(\sqrt{\lambda})|^2\, d\rho_1(\lambda) + \|\hat{f}\|(2\|f\| + \|\hat{f}\|).$$

We introduce the function $\sigma_1(\lambda)$, which is defined by the formula $\sigma_1(\lambda)=\rho_1(\lambda^2)$ for $\lambda>0$, and we consider its odd extension on the negative semi-axis (we assume that the function $\rho_1(\lambda)$ is normalized by the condition $\rho_1(+0)=0$).

In terms of $\sigma_1(\lambda)$ we can now write the previous inequality as

$$\left| \|f\|^2 - \frac{1}{2}\int_{-\infty}^{\infty} |C_f(\lambda)|^2 \, d\sigma_1(\lambda) \right| \leqslant \int_{-\infty}^{+0} |C_f(\sqrt{\lambda})|^2 \, d\rho_1(\lambda) + \|\hat{f}\|(2\|f\| + \|\hat{f}\|). \tag{2.2.8}$$

LEMMA 2.2.2. *For an arbitrary finite function* $f(x)$

$$\lim_{a\to\infty} \frac{1}{4}\int_{-\infty}^{\infty} |C_f(\lambda+a)+C_f(\lambda-a)|^2 \, d\sigma_1(\lambda) = \|f\|^2.$$

PROOF. Let $f(x)$ be a finite function. We consider the family of finite functions $\{f_a(x)\}$, where

$$f_a(x) = f(x)\cos ax \qquad (-\infty < a < \infty).$$

According to (2.2.8) we have

$$\left| \|f_a\|^2 - \frac{1}{2}\int_{-\infty}^{\infty} |C_{f_a}(\lambda)|^2 \, d\sigma_1(\lambda) \right| \leqslant \int_{-\infty}^{+0} |C_{f_a}(\sqrt{\lambda})|^2 \, d\rho_1(\lambda) + \|\hat{f}_a\|(2\|f_a\| + \|\hat{f}_a\|),$$

which implies that

$$\overline{\lim_{a\to\infty}} \left| \|f_a\|^2 - \frac{1}{2}\int_{-\infty}^{\infty} |C_{f_a}(\lambda)|^2 \, d\sigma_1(\lambda) \right| \leqslant \overline{\lim_{a\to\infty}} \int_{-\infty}^{+0} |C_{f_a}(\sqrt{\lambda})|^2 \, d\rho_1(\lambda) + \overline{\lim_{a\to\infty}} \|\hat{f}_a\|(2\|f_a\| + \|\hat{f}_a\|). \tag{2.2.9}$$

And we have

$$C_{f_a}(\sqrt{\lambda}) = \int_0^{\infty} f_a(x)\cos x\sqrt{\lambda}\, dx = \int_0^{\infty} f(x)\cos x\sqrt{\lambda}\cos ax\, dx,$$

which by the Riemann-Lebesgue theorem on convergence of Fourier coefficients to zero gives us

$$\lim_{a\to\infty} C_{f_a}(\lambda) = 0.$$

Also, for $\lambda \leqslant 0$ we have

$$|C_{f_a}(\sqrt{\lambda})| \leqslant \int_0^\infty |f(x)|\cosh x\sqrt{|\lambda|}\,dx \leqslant \|f\|\sqrt{\int_0^N \cosh^2 x\sqrt{|\lambda|}\,dx}$$
$$\leqslant N^{\frac{1}{2}}\|f\|\cosh N\sqrt{|\lambda|},$$

where $[0, N)$ denotes an interval outside of which $f(x)$ is zero.

Thus the sequence $|C_{f_a}(\sqrt{\lambda})|^2$ converges uniformly to zero as $a \to \infty$ in each finite subinterval and is majorized by the functions $N\|f\|^2\cosh^2 N\sqrt{|\lambda|}$, which are integrable, according to Lemma 2.2.1, on $-\infty < \lambda \leqslant 0$ with weight $d\rho_1(\lambda)$.

Therefore in the expression

$$\overline{\lim_{a\to\infty}} \int_{-\infty}^{+0} |C_{f_a}(\sqrt{\lambda})|^2\,d\rho_1(\lambda)$$

we can go to the limit under the integral and obtain

$$\overline{\lim_{a\to\infty}} \int_{-\infty}^{+0} |C_{f_a}(\sqrt{\lambda})|^2\,d\rho_1(\lambda) = 0. \tag{2.2.10}$$

Now we consider the functions

$$\hat{f}_a(y) = \int_y^N H(x, y) f_a(x)\,dx = \int_y^N H(x, y) f(x)\cos ax\,dx$$

(here $[0, N)$ is an interval outside of which $f(x)$ is zero). Since $H(x, y)$ is bounded for $0 \leqslant y \leqslant x \leqslant N$, the functions $\hat{f}_a(y)$ are uniformly bounded and are zero outside $[0, N)$. By the Riemann-Lebesgue theorem we have

$$\lim_{a\to\infty} \hat{f}_a(y) = \lim_{a\to\infty} \int_y^N H(x, y) f(x)\cos ax\,dx = 0;$$

and therefore

$$\lim_{a\to\infty} \|\hat{f}_a\|^2 = \lim_{a\to\infty} \int_y^N |\hat{f}_a(y)|^2\,dy = 0. \tag{2.2.11}$$

Furthermore,

$$\|f_a\|^2 = \int_0^\infty |f(x)|^2\cos^2 ax\,dx = \frac{1}{2}\int_0^\infty |f(x)|^2\,dx + \frac{1}{2}\int_0^\infty |f(x)|^2\cos 2ax\,dx,$$

and by the Riemann-Lebesgue theorem

$$\lim_{a\to\infty} \int_0^\infty |f(x)|^2\cos 2ax\,dx = 0.$$

Hence

$$\lim_{a\to\infty} \|f_a\|^2 = \frac{1}{2}\|f\|^2. \tag{2.2.12}$$

From (2.2.9)–(2.2.12) it follows that

$$\lim_{a\to\infty}\left| \|f_a\|^2 - \frac{1}{2}\int_{-\infty}^{\infty} |C_{f_a}(\lambda)|^2\, d\sigma_1(\lambda)\right| = 0,$$

or

$$\lim_{a\to\infty} \frac{1}{2}\int_{-\infty}^{\infty} |C_{f_a}(\lambda)|^2\, d\sigma_1(\lambda) = \frac{1}{2}\|f\|^2,$$

and since

$$C_{f_a}(\lambda) = \int_0^{\infty} f(x)\cos ax\cos\lambda x\, dx = \frac{1}{2}\int_0^{\infty} f(x)\{\cos(\lambda + a)x + \cos(\lambda - a)x\}\, dx$$
$$= \frac{1}{2}\{C_f(\lambda + a) + C_f(\lambda - a)\},$$

we then have

$$\lim_{a\to\infty} \frac{1}{4}\int_{-\infty}^{\infty} |C_f(\lambda + a) + C_f(\lambda - a)|^2\, d\sigma_1(\lambda) = \|f\|^2.$$

The lemma is proved.

LEMMA 2.2.3. *Uniformly for all* a *and* $\lambda \geqslant 1$

$$\sigma_1(a + \lambda) - \sigma_1(a - \lambda) \leqslant c\lambda,$$

where c *is a constant.*

PROOF. From (2.2.7) we have

$$\frac{1}{2}\int_{-\infty}^{\infty} |C_f(\lambda)|^2\, d\sigma_1(\lambda) + \int_0^{\infty} |C_f(\sqrt{\lambda})|^2\, d\rho_1(\lambda) \leqslant (\|f\| + \|\hat{f}\|)^2,$$

for any finite $f(x)$. If we let

$$f(x) = \frac{1}{h}\left(1 - \frac{x}{h}\right)\cos ax \quad \text{for} \quad 0 \leqslant x \leqslant h, \qquad (h \leqslant 1),$$

$$f(x) = 0 \qquad\qquad \text{for} \quad x > h,$$

we obtain

$$\|f\| = \left(\int_0^h \frac{1}{h^2}\left(1 - \frac{x}{h}\right)^2 \cos^2 ax\, dx\right)^{\frac{1}{2}} \leqslant h^{-\frac{1}{2}},$$

$$|\hat{f}| = \left|\int_y^h H(x, y)\frac{1}{h}\left(1 - \frac{x}{h}\right)\cos ax\, dx\right| \leqslant c_1,$$

where the constant $c_1 = \sup_{0\leqslant y\leqslant x\leqslant 1} |H(x, y)|$ does not depend on a (where $-\infty < a < \infty$) or on h (where $h \leqslant 1$).

Therefore

$$(\|f\|+\|\hat{f}\|)^2 \leqslant (h^{-\frac{1}{2}}+c_1 h^{\frac{1}{2}})^2 \leqslant \frac{(1+c_1)^2}{h}$$

and

$$\frac{1}{2}\int_{-\infty}^{\infty} |C_f(\lambda)|^2 \, d\sigma_1(\lambda) \leqslant (1+c_1)^2 h^{-1} \quad (-\infty < a < \infty,\ h \leqslant 1). \tag{2.2.13}$$

Direct calculation of the Fourier cosine transform $C_f(\lambda)$ of $f(x)$ gives us

$$C_f(\lambda) = \frac{\sin^2 \frac{(\lambda+a)h}{2}}{h^2(\lambda+a)^2} + \frac{\sin^2 \frac{(\lambda-a)h}{2}}{h^2(\lambda-a)^2},$$

which implies that

$$|C_f(\lambda)|^2 \geqslant \left\{\frac{\sin \frac{(\lambda+a)h}{2}}{h(\lambda+a)}\right\}^4 + \left\{\frac{\sin \frac{(\lambda-a)h}{2}}{h(\lambda-a)}\right\}^4 .$$

Then from (2.2.13) we obtain

$$\frac{1}{2}\int_{-\infty}^{\infty} \left[\left\{\frac{\sin \frac{\lambda+ah}{2}}{h(\lambda+a)}\right\}^4 + \left\{\frac{\sin \frac{\lambda-ah}{2}}{h(\lambda-a)}\right\}^4\right] d\sigma_1(\lambda) \leqslant (1+c_1)^2 h^{-1},$$

or

$$\frac{1}{2}\int_{-\infty}^{\infty} \left\{\frac{\sin \frac{\mu h}{2}}{\mu h}\right\}^4 d\,[\sigma_1(\mu-a)+\sigma_1(\mu+a)] \leqslant (1+c_1)^2 h^{-1}.$$

Therefore

$$\frac{1}{2}\int_{-\frac{1}{h}}^{\frac{1}{h}} \left\{\frac{\sin \frac{\mu h}{2}}{\mu h}\right\}^4 d\,[\sigma_1(\mu-a)+\sigma_1(\mu+a)] \leqslant (1+c_1)^2 h^{-1}.$$

We note that uniformly for all $h \leqslant 1$ we have

$$\inf_{|\mu| \leqslant \frac{1}{h}} \left\{\frac{\sin \frac{\mu h}{2}}{\mu h}\right\}^4 \geqslant c_2 > 0.$$

Therefore

$$\frac{c_2}{2}\int_{-\frac{1}{h}}^{\frac{1}{h}} d\,[\sigma_1(\mu-a)+\sigma_1(\mu+a)] \leqslant (1+c_1)^2 h^{-1};$$

that is,

$$\frac{1}{2}\left[\sigma_1\left(\frac{1}{h}-a\right)+\sigma_1\left(\frac{1}{h}+a\right)-\sigma_1\left(-\frac{1}{h}-a\right)-\sigma_1\left(-\frac{1}{h}+a\right)\right]$$

$$\leqslant (1+c_1)^2(c_2 h)^{-1},$$

and since $\sigma_1(\mu)$ is odd we obtain

$$\sigma_1\left(\frac{1}{h}+a\right)-\sigma_1\left(a-\frac{1}{h}\right)\leqslant(1+c_1)^2c_2^{-1}h^{-1},$$

uniformly for all a and all $h \leqslant 1$. If we set $\lambda = h^{-1}$ and $c = (1 + c_1)^2/c_2$, we find that

$$\sigma_1(a+\lambda)-\sigma_1(a-\lambda)\leqslant c\lambda$$

uniformly for all a and all $\lambda \geqslant 1$, which is what we wanted to show.

LEMMA 2.2.4. *If the Fourier cosine transform* $C_f(\lambda)$ *of a finite function* $f(x)$ *satisfies the condition* $C_f(\lambda) = O(\lambda^{-2})$ *as* $|\lambda| \to \infty$, *then*

$$\lim_{a\to\infty}\frac{1}{4}\int_{-\infty}^{\infty}|C_f(\lambda)|^2\,d\,[\sigma_1(a+\lambda)-\sigma_1(a-\lambda)]=\|f\|^2.$$

PROOF. Let $f(x)$ be an arbitrary finite function which satisfies the hypothesis of the lemma. By Lemma 2.2.2 we have

$$\|f\|^2=\lim_{a\to\infty}\frac{1}{4}\int_{-\infty}^{\infty}|C_f(\lambda+a)+C_f(\lambda-a)|^2\,d\sigma_1(\lambda)$$

$$=\lim_{a\to\infty}\frac{1}{4}\left\{\int_{-\infty}^{\infty}[|C_f(\lambda+a)|^2+|C_f(\lambda-a)|^2]\,d\sigma_1(\lambda)\right.$$

$$\left.+\int_{-\infty}^{\infty}[C_f(\lambda+a)\overline{C_f(\lambda-a)}+\overline{C_f(\lambda+a)}C_f(\lambda-a)]\,d\sigma_1(\lambda)\right\}$$

$$=\lim_{a\to\infty}\frac{1}{4}\left\{\int_{-\infty}^{\infty}|C_f(\lambda)|^2\,d\,[\sigma_1(a+\lambda)-\sigma_1(a-\lambda)]\right.$$

$$\left.+2\int_{-\infty}^{\infty}[\operatorname{Re}C_f(\lambda+a)\overline{C_f(\lambda-a)}]\,d\sigma_1(\lambda)\right\}.$$

Therefore it is clear that to prove the lemma it suffices to show that

$$\lim_{a\to\infty}\int_{-\infty}^{\infty}|C_f(\lambda+a)C_f(\lambda-a)|\,d\sigma_1(\lambda)=0,$$

if $C_f(\lambda) = O(\lambda^{-2})$ as $|\lambda| \to \infty$. Since $|C_f(\lambda + a)C_f(\lambda - a)|$ is even and $\sigma_1(\lambda)$ is odd, we have

$$\int_{-\infty}^{\infty}|C_f(\lambda+a)C_f(\lambda-a)|\,d\sigma_1(\lambda)$$

$$=2\int_0^{\infty}|C_f(\lambda+a)C_f(\lambda-a)|\,d\sigma_1(\lambda)\leqslant 2A\int_0^{\infty}\frac{d\sigma_1(\lambda)}{(a+\lambda)^2},$$

where $A = \sup_{-\infty<\lambda<\infty} |C_f(\lambda)| < \infty$. By Lemma 2.2.3 we have $\sigma_1(\lambda) \leqslant c\lambda$ for $\lambda \geqslant 1$, and since $\sigma_1(\lambda)$ is monotone, it follows that for all $\lambda \geqslant 0$

$$\sigma_1(\lambda) \leqslant B + c\lambda,$$

where $B = \sigma_1(1)$. If we integrate by parts and use the preceding bound for $\sigma_1(\lambda)$, we obtain

$$\int_0^\infty \frac{d\sigma_1(\lambda)}{(a+\lambda)^2} = 2\int_0^\infty \frac{\sigma_1(\lambda)}{(a+\lambda)^3}\, d\lambda \leqslant 2\int_0^\infty \frac{B+c\lambda}{(a+\lambda)^3}\, d\lambda,$$

and since

$$\lim_{a\to\infty} \int_0^\infty \frac{B+c\lambda}{(\alpha+\lambda)^3}\, d\lambda = 0,$$

we have

$$\lim_{a\to\infty} \int_0^\infty \frac{d\sigma_1(\lambda)}{(a+\lambda)^2} = 0.$$

But then

$$\lim_{a\to\infty} \int_{-\infty}^\infty |C_f(\lambda+a)C_f(\lambda-a)|\, d\sigma_1(\lambda) = 0,$$

which is what we wanted to show.

LEMMA 2.2.5. *The following equality holds*:

$$\lim_{a\to\infty} [\sigma_1(a+\lambda) - \sigma_1(a-\lambda)] = \frac{4}{\pi}\lambda.$$

PROOF. Assume the contrary. By Lemma 2.2.3 the family of nondecreasing functions $[\sigma_1(a+\lambda) - \sigma_1(a-\lambda)]$ is uniformly bounded in each finite subinterval of $-\infty < \lambda < \infty$. Hence from our assumption it follows that there exists a sequence of numbers $\{a_n\}$ ($a_n \to \infty$ as $n \to \infty$) and a number λ_0 such that

$$\lim_{n\to\infty} [\sigma_1(a_n+\lambda_0) - \sigma_1(a_n-\lambda_0)] \neq \frac{4}{\pi}\lambda_0. \tag{2.2.14}$$

The bound given in Lemma 2.2.3 enables us to apply Helly's theorem, according to which the sequence $\{a_n\}$ contains a subsequence $\{a_{n'}\}$ ($a_{n'} \to \infty$ as $n' \to \infty$) such that the sequence of functions $[\sigma_1(a_{n'}+\lambda) - \sigma_1(a_{n'}-\lambda)]$ converges fundamentally as $n' \to \infty$ to an odd nondecreasing function $\sigma^*(\lambda)$. From Lemma 2.2.4 we obtain

$$\lim_{n'\to\infty}\frac{1}{4}\int_{-\infty}^{\infty}|C_f(\lambda)|^2\,d[\sigma_1(a_{n'}+\lambda)-\sigma_1(a_{n'}-\lambda)]=\|f\|^2,$$

for any finite function $f(x)$ whose Fourier cosine transform $C_f(\lambda)$ satisfies the condition $C_f(\lambda)=O(\lambda^{-2})$ as $|\lambda|\to\infty$. Lemma 2.2.3 allows us to take the limit inside the integral sign, and we obtain

$$\frac{1}{4}\int_{-\infty}^{\infty}|C_f(\lambda)|^2\,d\sigma^*(\lambda)=\|f\|^2.$$

By the classical Parseval identity (2.2.5) we obtain

$$\frac{1}{4}\int_{-\infty}^{\infty}|C_f(\lambda)|^2\,d\sigma^*(\lambda)=\frac{1}{\pi}\int_{-\infty}^{\infty}|C_f(\lambda)|^2\,d\lambda,$$

for any finite function $f(x)$ whose Fourier cosine transform $C_f(\lambda)=O(\lambda^{-2})$ as $|\lambda|\to\infty$. It is easy to show that for arbitrary real μ and $h\neq 0$

$$\left(\frac{\sin h\lambda}{h\lambda}\right)^2\cos\mu\lambda$$

is the Fourier cosine transform of some function with compact support. Therefore

$$\frac{1}{4}\int_{-\infty}^{\infty}\left(\frac{\sin h\lambda}{h\lambda}\right)^4\cos^2\mu\lambda\,d\sigma^*(\lambda)=\frac{1}{\pi}\int_{-\infty}^{\infty}\left(\frac{\sin h\lambda}{h\lambda}\right)^4\cos^2\mu\lambda\,d\lambda,$$

and since μ was arbitrary,

$$\frac{1}{4}\int_{-\infty}^{\infty}\left(\frac{\sin h\lambda}{h\lambda}\right)^4\cos t\lambda\,d\sigma^*(\lambda)=\frac{1}{\pi}\int_{-\infty}^{\infty}\left(\frac{\sin h\lambda}{h\lambda}\right)^4\cos t\lambda\,d\lambda,$$

for any real t and $h\neq 0$.

If we multiply both sides of this equation by $\sin yt/t$ and integrate with respect to t from $-\infty$ to ∞, we obtain (assuming that y is a point of continuity of $\sigma^*(\lambda)$)

$$\frac{1}{4}\int_{-y}^{y}\left(\frac{\sin h\lambda}{h\lambda}\right)^4 d\sigma^*(\lambda)=\frac{1}{\pi}\int_{-y}^{y}\left(\frac{\sin h\lambda}{h\lambda}\right)^4 d\lambda,$$

which as $h\to 0$ implies that

$$\frac{1}{2}\sigma^*(y)=\frac{2}{\pi}y,$$

that is, $\sigma^*(y)=4y/\pi$ for all points of continuity. Since $\sigma^*(y)$ is nondecreasing, the equality is true at all points and $\sigma^*(y)$ is continuous. Therefore the sequence of functions $[\sigma_1(a_{n'}+y)-\sigma_1(a_{n'}-y)]$ converges everywhere to the function $\sigma^*(y)=4y/\pi$ as $n'\to\infty$. In particular,

$$\lim_{n'\to\infty}[\sigma_1(a_{n'}+\lambda_0)-\sigma_1(a_{n'}-\lambda_0)]=\frac{4}{\pi}\lambda_0,$$

which contradicts (2.2.14).

From this contradiction the lemma follows. We will see later that Theorem 2.2.1 is a simple corollary of this lemma. In fact, the lemma is stronger than Theorem 2.2.1. For example, if we note that $\rho(\lambda)=(\sin\alpha)^{-2}\sigma_1(\sqrt{\lambda})$ for $\lambda\geqslant 0$, then from Lemma 2.2.5 we obtain the following corollary.

COROLLARY. *There exists a function $\epsilon(\lambda)$, bounded and converging to zero as $\lambda\to\infty$, such that each interval $(\lambda,\lambda+\epsilon(\lambda)[1+\sqrt{\lambda}])$ $(\lambda\geqslant 0)$ contains points belonging to the spectrum of each boundary value problem for an operator L of the form* (2.1.1).

We note that it is not possible to obtain this corollary just by using the asymptotic formulas in Theorem 2.2.1.

PROOF OF THEOREM 2.2.1. Lemma 2.2.5 implies that for arbitrary $\epsilon>0$ there exists an integer N such that

$$\sigma_1(N+2k+1)-\sigma_1(N+2k-1)=\frac{4}{\pi}+\varepsilon_k\quad(|\varepsilon_k|<\varepsilon)$$

for all $k\geqslant 0$. If we sum these inequalities for k running from 0 to n we find that

$$\sigma_1(N+2n+1)-\sigma_1(N-1)=\frac{4}{\pi}(n+1)+\eta_n(n+1),$$

where $|\eta_n|<\epsilon$. Therefore

$$\overline{\lim_{n\to\infty}}\left|\frac{\sigma_1(N+2n+1)}{N+2n+1}-\frac{2}{\pi}\right|<\varepsilon.$$

Since $\sigma_1(\lambda)$ is nondecreasing and $\epsilon>0$ was arbitrary, it follows that

$$\lim_{\lambda\to\infty}\frac{\sigma_1(\lambda)}{\lambda}=\frac{2}{\pi},$$

or, using the function $\rho_1(\lambda)=\sigma_1(\sqrt{\lambda})$ $(\lambda\geqslant 0)$,

$$\lim_{\lambda\to\infty}\frac{\rho_1(\lambda)}{\sqrt{\lambda}}=\frac{2}{\pi}.$$

Since $\rho(\lambda)=(\sin\alpha)^{-2}\rho_1(\lambda)$, we have

$$\lim_{\lambda\to\infty} \lambda^{-\frac{1}{2}} \rho(\lambda) = \frac{2}{\pi \sin^2 \alpha},$$

which is what we wanted to prove.

§3. Inverse problems in spectral analysis

For spectral theory of differential operators L of the form (2.1.1) the typical problem is one in which we are given L and must establish some spectral property of L. We will call such problems direct spectral analysis problems. The preceding sections of this chapter were devoted to direct problems.

Recently physics has come up with a new kind of problem in which we must use some spectral properties of an operator L of the form (2.1.1) to recover the operator. We will call such problems inverse spectral analysis problems. In order to solve such problems we must first find out which spectral properties uniquely determine an operator L of the form (2.1.1) (that is, determine the function $q(x)$).

This section will be devoted to the fundamental inverse problem of spectral analysis.

We begin with the following general theorem.

THEOREM 2.3.1. *Let* L_1 *and* L_2 *of the form* (2.1.1)

$$L_1[u] = \frac{d^2}{dx^2} u(x) - q_1(x) u(x), \quad L_2[u] = \frac{d^2}{dx^2} u(x) - q_2(x) u(x)$$

be given on $[0, a)$, $a \leqslant \infty$.

If we consider all possible boundary value problems for L_1 *and* L_2, *we obtain two sets* $R(L_1)$ *and* $R(L_2)$ *of resolution functions with which to construct inversion formulas* (*Chapter* 2, §1, *part* D).

If there exists a pair of functions $\rho_1(\lambda) \in R(L_1)$ *and* $\rho_2(\lambda) \in R(L_2)$ *such that* $\rho_1(\lambda) = c\rho_2(\lambda)$ (*where* c *is a constant*), *then* $q_1(x) = q_2(x)$ *almost everywhere in* $[0, a)$.

In other words, the resolution function $\rho(\lambda)$ corresponding to a boundary value problem for L completely determines L.

PROOF. Let $\rho_1(\lambda)$ correspond to the boundary value problem for L_1 with boundary condition

$$y(0)\cos\alpha_1 + y'(0)\sin\alpha_1 = 0,$$

and let $\rho_2(\lambda)$ correspond to the boundary value problem for L_2 with boundary condition

$$y(0)\cos\alpha_2 + y'(0)\sin\alpha_2 = 0.$$

Here it is possible that one or both of these boundary value problems also has a boundary condition at a. We will show first of all that without loss of generality we can assume that

$$\cot \alpha_1 \neq \infty, \qquad \cot \alpha_2 \neq \infty. \tag{2.3.1}$$

Here we let the function $m^{(1)}_{\alpha_1}(\lambda)$ correspond to the first boundary value problem and let $m^{(2)}_{\alpha_2}(\lambda)$ correspond to the second (Chapter 2, §1, part B). We know [4] that

$$-\operatorname{Im}\{m^{(1)}_{\alpha_1}(\lambda)\} = v\int_{-\infty}^{\infty} \frac{d\rho_1(t)}{(u-t)^2+v^2}; \qquad -\operatorname{Im}\{m^{(2)}_{\alpha_2}(\lambda)\} = v\int_{-\infty}^{\infty} \frac{d\rho_2(t)}{(u-t)^2+v^2},$$

where $\lambda = u + iv$. Since $\rho_1(\lambda) = c\rho_2(\lambda)$ by hypothesis, we have

$$\operatorname{Im}[m^{(1)}_{\alpha_1}(\lambda)] = c\operatorname{Im}[m^{(2)}_{\alpha_2}(\lambda)]$$

and

$$m^{(1)}_{\alpha_1}(\lambda) = c\{A + m^{(2)}_{\alpha_2}(\lambda)\},$$

where A is a real constant.

Furthermore, by part C of Chapter 2, §1, there corresponds to the boundary value problem for L_1 with boundary condition

$$y(0)\cos\beta_1 + y'(0)\sin\beta_1 = 0 \tag{2.3.2}$$

the function

$$m^{(1)}_{\beta_1}(\lambda) = \frac{1 + hm^{(1)}_{\alpha_1}(\lambda)}{h - m^{(1)}_{\alpha_1}(\lambda)} = \frac{1 + hc\{A + m^{(2)}_{\alpha_2}(\lambda)\}}{h - c\{A + m^{(2)}_{\alpha_2}(\lambda)\}}, \qquad h = \cot(\beta_1 - \alpha_1),$$

and to the boundary value problem for L_2 with boundary condition

$$y(0)\cos\beta_2 + y'(0)\sin\beta_2 = 0 \tag{2.3.2$'$}$$

the function

$$m^{(2)}_{\beta_2}(\lambda) = \frac{1 + Hm^{(2)}_{\alpha_2}(\lambda)}{H - m^{(2)}_{\alpha_2}(\lambda)}, \qquad H = \cot(\beta_2 - \alpha_2).$$

If we separate out the imaginary parts of these functions we find that

$$\operatorname{Im}[m^{(1)}_{\beta_1}(\lambda)] = \frac{y(\lambda)(1+h^2)}{c\left[\left(\frac{h}{c} - A - x(\lambda)\right)^2 + (y(\lambda))^2\right]};$$

$$\operatorname{Im}[m^{(2)}_{\beta_2}(\lambda)] = \frac{y(\lambda)(1+H^2)}{[(H - x(\lambda))^2 + (y(\lambda))^2]},$$

where

$$\operatorname{Im}[m^{(2)}_{\alpha_2}(\lambda)] = y(\lambda), \quad \operatorname{Re}[m^{(2)}_{\alpha_2}(\lambda)] = x(\lambda).$$

Therefore, if

$$H = \frac{h}{c} - A,$$

then

$$\operatorname{Im}[m_{\beta_1}^{(1)}(\lambda)] = C' \operatorname{Im}[m_{\beta_2}^{(2)}(\lambda)],$$

where C' is a constant. Since the resolution functions generated by the boundary value problems (2.3.2) and (2.3.2′) are uniquely determined by the imaginary parts of the functions $m_{\beta_1}^{(1)}(\lambda)$ and $m_{\beta_2}^{(2)}(\lambda)$ (Chapter 2, §1, part D), the preceding equation shows that to the boundary value problems (2.3.2) and (2.3.2′) there correspond resolution functions which differ only by a constant factor if

$$\beta_1 = \alpha_1 + \operatorname{arc\,cot} h, \quad \beta_2 = \alpha_2 + \operatorname{arc\,cot}\left(\frac{h}{c} - A\right),$$

where h is arbitrary. Clearly we can always choose h such that $\cot \beta_1 \neq \infty$, $\cot \beta_2 \neq \infty$. This proves that the restriction (2.3.1) does not destroy the generality of the proof.

We let $\phi_{\alpha_1}^{(1)}(\lambda, x)$ $(\phi_{\alpha_2}^{(2)}(\lambda, x))$ denote the solution of the equation

$$L_1[u] + \lambda u = 0 \qquad (L_2[u] + \lambda u = 0)$$

with initial conditions

$$\varphi_{\alpha_1}^{(1)}(\lambda, 0) = \sin \alpha_1, \qquad \varphi_{\alpha_1}^{(1)\prime}(\lambda, 0) = -\cos \alpha_1,$$
$$(\varphi_{\alpha_2}^{(2)}(\lambda, 0) = \sin \alpha_2, \qquad \varphi_{\alpha_2}^{(2)\prime}(\lambda, 0) = -\cos \alpha_2).$$

By (2.3.1) we have

$$h_1 = -\cot \alpha_1 \neq \infty, \qquad h_2 = -\cot \alpha_2 \neq \infty,$$

which together with Theorem 1.2.2 implies the existence of a transformation operator $V = V_{\{L_1 L_2 h_1 h_2\}}$ such that

$$V[\varphi_{\alpha_1}^{(1)}(\lambda, x)] = v\, \varphi_{\alpha_2}^{(2)}(\lambda, x), \qquad \text{where } v = \frac{\sin \alpha_1}{\sin \alpha_2}.$$

By virtue of the inversion formulas in part D of Chapter 2, §1, for an arbitrary $f(x) \in L^2[0, a)$ we have

$$E_f(\lambda) = \int_0^a f(x)\, \varphi_{\alpha_1}^{(1)}(\lambda, x)\, dx; \qquad f(x) = \int_{-\infty}^{\infty} E_f(\lambda)\, \varphi_{\alpha_1}^{(1)}(\lambda, x)\, d\rho_1(\lambda),$$

where the integrals converge in the metrics of $L_{\rho_1}^2(-\infty, \infty)$ and $L^2[0, a)$ respectively. From this fact and the weak boundedness of V (see Chapter 1, §2, property III) we conclude that

$$V[f] = \lim_{n \to \infty} V\left[\int_{-n}^{n} E_f(\lambda)\, \varphi_{\alpha_1}^{(1)}(\lambda, x)\, d\rho_1(\lambda)\right] =$$

$$= \lim_{n\to\infty} \int_{-n}^{n} E_f(\lambda) V [\varphi^{(1)}_{\alpha_1}(\lambda, x)] \, d\rho_1(\lambda) = \lim_{n\to\infty} v \int_{-n}^{n} E_f(\lambda) \varphi^{(2)}_{\alpha_2}(\lambda, x) \, d\rho_1(\lambda),$$

or, since $\rho_1(\lambda) = c\rho_2(\lambda)$,

$$V[f] = \lim_{n\to\infty} cv \int_{-n}^{n} E_f(\lambda) \varphi^{(2)}_{\alpha_2}(\lambda, x) \, d\rho_2(\lambda),$$

where we have mean square convergence in each subinterval $(0, b)$ for arbitrary $b < a$. Since $E_f(\lambda) \in L^2_{\rho_1}(-\infty, \infty)$ and $\rho_1(\lambda) = c\rho_2(\lambda)$, we have $E_f(\lambda) \in L^2_{\rho_2}(-\infty, \infty)$, which according to part D of Chapter 2, §1, implies that the integral in the right side of the preceding equation converges in the metric of $L^2[0, a)$ and the function $V[f]$ belongs to this space.

Thus we have Parseval's identity:

$$\| V[f] \|^2 = \int_{-\infty}^{\infty} c^2 v^2 | E_f(\lambda) |^2 \, d\rho_2(\lambda) = cv^2 \int_{-\infty}^{\infty} | E_f(\lambda) |^2 \, d\rho_1(\lambda),$$

and since

$$\int_{-\infty}^{\infty} | E_f(\lambda) |^2 \, d\rho_1(\lambda) = \| f \|^2,$$

we have

$$\| V[f] \|^2 = cv^2 \|f\|^2,$$

for arbitrary $f(x) \in L^2[0, a)$.

This equation shows that $V_1 = (cv^2)^{-1/2} V_{\{L_1 L_2 h_1 h_2\}}$ is a unitary operator in $L^2[0, a)$. For a unitary operator V_1 it is known that $V_1^* = V_1^{-1}$, where V_1^* is the adjoint and V_1^{-1} is the inverse.

We now observe that V is a Volterra operator:

$$V[f] = f(x) + \int_0^x K(x, y) f(y) \, dy.$$

Therefore

$$V_1^*[f] = (cv^2)^{-\frac{1}{2}} \left[f(x) + \int_x^a \overline{K(y, x)} f(y) \, dy \right],$$

$$V_1^{-1}[f] = (cv^2)^{\frac{1}{2}} \left[f(x) + \int_0^x H(x, y) f(y) \, dy \right]$$

and

$$(cv^2)^{-\frac{1}{2}} \left[f(x) + \int_x^a \overline{K(y, x)} f(y) \, dy \right] = (cv^2)^{\frac{1}{2}} \left[f(x) + \int_0^x H(x, y) f(y) dy \right];$$

that is,

$$\int_x^a \overline{K(y, x)}\, f(y)\, dy - cv^2 \int_0^x H(x, y)\, f(y)\, dy = (cv^2 - 1) f(x),$$

for arbitrary $f(x) \in L^2[0, a)$.

The last equation implies that $K(x, y) \equiv H(x, y) \equiv 0$, $cv^2 = 1$ and thus $V[f] = f(x)$. If we set $f(x) = \phi^{(1)}_{\alpha_1}(\lambda, x)$ we obtain

$$\varphi^{(1)}_{\alpha_1}(\lambda, x) = V[\varphi^{(1)}_{\alpha_1}(\lambda, x)] = v\varphi^{(2)}_{\alpha_2}(\lambda, x);$$

therefore $\sin \alpha_1 = \sin \alpha_2$, $\phi^{(1)}_{\alpha_1}(\lambda, x) \equiv \phi^{(2)}_{\alpha_2}(\lambda, x)$ and $q_1(x) = q_2(x)$ almost everywhere in $[0, a)$.

The theorem is proved.

In some special situations we can use the preceding theorem to show that L is uniquely determined by some of its spectral properties which do not explicitly involve the function $\rho(\lambda)$. We will discuss two such cases which are of interest in physics.

In spectral analysis it is shown that if one boundary value problem for an operator L of the form (2.1.1) has a discrete spectrum, then every boundary value problem for L has a discrete spectrum. Thus it makes sense to speak of an operator which has a discrete spectrum, meaning that the spectrum of each of its boundary value problems is discrete.

THEOREM 2.3.2. *Let* L_1 *and* L_2 *of the form* (2.1.1) *be given on* $[0, a)$, $a \leqslant \infty$:

$$L_1[u] = \frac{d^2}{dx^2} u(x) - q_1(x) u(x), \qquad L_2[u] = \frac{d^2}{dx^2} u(x) - q_2(x) u(x),$$

and suppose that they have discrete spectra.

We consider the following boundary value problems:

(α_1) *and* (β_1) *for* L_1 *with boundary conditions at zero*

$$y(0)\cos\alpha_1 + y'(0)\sin\alpha_1 = 0, \tag{α_1}$$

$$y(0)\cos\beta_1 + y'(0)\sin\beta_1 = 0 \tag{β_1}$$

and the same condition repeated at a *if necessary.*

(α_2) *and* (β_2) *for* L_2 *with boundary conditions at zero*

$$y(0)\cos\alpha_2 + y'(0)\sin\alpha_2 = 0, \tag{α_2}$$

$$y(0)\cos\beta_2 + y'(0)\sin\beta_2 = 0 \tag{β_2}$$

and the same condition repeated at a *if necessary.*

If $\cot(\beta_1 - \alpha_1) \neq \infty$, $\cot(\beta_2 - \alpha_2) \neq \infty$ *and the spectrum of problem* (α_1) *coincides with the spectrum of problem* (α_2) *and the spectrum of* (β_1)

coincides with the spectrum of (β_2) *then* $q_1(x) = q_2(x)$ *almost everywhere in* $[0, a)$.

In other words, an operator L with discrete spectrum is uniquely determined by the spectra of two boundary value problems with distinct boundary conditions at zero and the same conditions repeated at a if necessary.

PROOF. For the problems (α_1), (β_1) and (α_2), (β_2) let the corresponding functions be $m_{\alpha_1}^{(1)}(\lambda)$, $m_{\beta_1}^{(1)}(\lambda)$ and $m_{\alpha_2}^{(2)}(\lambda)$, $m_{\beta_2}^{(2)}(\lambda)$ (see Chapter 2, §1, part B). Since L_1 and L_2 have discrete spectra, the above functions are meromorphic with simple real poles at the eigenvalues of the respective boundary value problems. Since the spectra of the problems (α_1) and (α_2) coincide, the poles of $m_{\alpha_1}^{(1)}(\lambda)$ and $m_{\alpha_2}^{(2)}(\lambda)$ coincide. Furthermore, from part B of Chapter 2, §1, it follows that

$$m_{\beta_1}^{(1)}(\lambda) = \frac{1 + h_1 m_{\alpha_1}^{(1)}(\lambda)}{h_1 - m_{\alpha_1}^{(1)}(\lambda)}, \qquad m_{\beta_2}^{(2)}(\lambda) = \frac{1 + h_2 m_{\alpha_2}^{(2)}(\lambda)}{h_2 - m_{\alpha_2}^{(2)}(\lambda)},$$

where $h_1 = \cot(\beta_1 - \alpha_1) \neq \infty$, $h_2 = \cot(\beta_2 - \alpha_2) \neq \infty$. Therefore the poles of $m_{\beta_1}^{(1)}(\lambda)$ $(m_{\beta_2}^{(2)}(\lambda))$ coincide with the zeros of $h_1 - m_{\alpha_1}^{(1)}(\lambda)$ $(h_2 - m_{\alpha_2}^{(2)}(\lambda))$. Since the spectra of the problems (β_1) and (β_2) coincide, the zeros of $h_1 - m_{\alpha_1}^{(1)}(\lambda)$ and $h_2 - m_{\alpha_2}^{(2)}(\lambda)$ coincide.

Thus the functions $m_{\alpha_1}^{(1)}(\lambda)$ and $m_{\alpha_2}^{(2)}(\lambda)$ have the same simple real poles, and the functions $h_1 - m_{\alpha_1}^{(1)}(\lambda)$ and $h_2 - m_{\alpha_2}^{(2)}(\lambda)$ have the same simple real zeros.

Now it is clear that

$$g(z) = \frac{h_1 - m_{\alpha_1}^{(1)}(z)}{h_2 - m_{\alpha_2}^{(2)}(z)}$$

is an entire analytic function with no zeros. Therefore

$$\psi(z) = \ln g(z) = \ln \left| \frac{h_1 - m_{\alpha_1}^{(1)}(z)}{h_2 - m_{\alpha_2}^{(2)}(z)} \right| + i \arg \frac{h_1 - m_{\alpha_1}^{(1)}(z)}{h_2 - m_{\alpha_2}^{(2)}(z)}$$

is also an entire analytic function.

According to part A of Chapter 2, §1, we have $\operatorname{Im}[m(z)] \operatorname{Im}[z] < 0$, which implies for $\operatorname{Im}[z] > 0$ that

$$0 < \arg[h_i - m_{\alpha_i}^{(i)}(z)] < \pi \qquad (i = 1, 2),$$

and for $\operatorname{Im}[z] < 0$ that

$$-\pi < \arg[h_i - m_{\alpha_i}^{(i)}(z)] < 0 \qquad (i = 1, 2).$$

Consequently the function

$$\operatorname{Im}[\psi(z)] = \arg \frac{h_1 - m^{(1)}_{\alpha_1}(z)}{h_2 - m^{(2)}_{\alpha_2}(z)}$$

is bounded in the whole plane:

$$|\operatorname{Im}[\psi(z)]| < k < \infty$$

and

$$|\psi(z) - i(k+1)| > 1$$

for all z. Then

$$\frac{1}{\psi(z) + i(k+1)}$$

is an entire analytic function which is bounded in the whole plane, and by Liouville's theorem this function is constant. Hence $\psi(z) \equiv \text{const}$ and $g(z) \equiv \text{const}$; that is,

$$\frac{h_1 - m^{(1)}_{\alpha_1}(z)}{h_2 - m^{(2)}_{\alpha_2}(z)} = c,$$

where c is a constant. Since the functions $m^{(1)}_{\alpha_1}(z)$ and $m^{(2)}_{\alpha_2}(z)$ are real on the real axis, the constant c is real and

$$\operatorname{Im}[m^{(1)}_{\alpha_1}(z)] = c \operatorname{Im}[m^{(2)}_{\alpha_2}(z)].$$

From part D of Chapter 2, §1, it now follows that $\rho_1(\lambda) = c\rho_2(\lambda)$, where $\rho_1(\lambda)$ and $\rho_2(\lambda)$ are the resolution functions generated by the boundary value problems (α_1) and (α_2). Thus by Theorem 2.3.1 we have $q_1(x) = q_2(x)$ almost everywhere in $[0, a)$, which is what we wanted to show.

An operator L given on a finite interval $[0, a]$ for which the function $q(x)$ is integrable over the whole interval is a simple example of an operator of the form (2.1.1) with a discrete spectrum.

The problem of recovering an operator from the spectra of its boundary value problems is known in the literature as the inverse Sturm-Liouville problem. It was first posed by Ambarzumian [13], and he solved it in a special case. A reasonably complete solution of the inverse Sturm-Liouville problem was given by Borg in [14]. But the methods used by these authors cannot produce a proof of Theorem 2.3.2 for the general case.

We consider the class of operators L of the form (2.1.1):

$$L[u] = \frac{d^2}{dx^2} u(x) - q(x) u(x),$$

given on $0 \leqslant x < \infty$ and satisfying

$$\int_0^\infty (1+x)|q(x)|\,dx < \infty. \tag{2.3.3}$$

It is known (see Chapter 2, §1, part E) in this case that boundary value problems for L are determined by giving a boundary condition at zero, and the problems have only a finite number of negative eigenvalues. For all real $s \neq 0$ we have an asymptotic formula for $\phi_\alpha(s^2, x)$:

$$\varphi_\alpha(s^2, x) = |M(s)| \sin(sx + \delta(s)) + o(1) \qquad (x \to \infty),$$

where the function $\delta(s)$ is called the limiting phase.

The limiting phase $\delta(s)$ plays a fundamental role in scattering theory, and the question of whether the limiting phase uniquely determines L is of interest in physics.

THEOREM 2.3.3. *Suppose that operators* L_1 *and* L_2 *of the form* (2.1.1)

$$L_1[u] = \frac{d^2}{dx^2} u(x) - q_1(x) u(x), \quad L_2[u] = \frac{d^2}{dx^2} u(x) - q_2(x) u(x)$$

are given on $0 \leqslant x < \infty$ *and satisfy*

$$\int_0^\infty (1 + x) |q_1(x)| \, dx < \infty, \qquad \int_0^\infty (1 + x) |q_2(x)| \, dx < \infty.$$

For some α *let the following conditions be satisfied*:

1) *The boundary value problems for* L_1 *and* L_2 *with boundary condition*

$$y(0) \cos\alpha + y'(0) \sin\alpha = 0$$

have the same negative eigenvalues $\lambda_1, \cdots, \lambda_n$.

2) *The norms of the eigenfunctions* $\phi_\alpha^{(1)}(\lambda_\nu, x)$ *of the boundary value problem for* L_1 *equal the norms of the corresponding eigenfunctions* $\phi_\alpha^{(2)}(\lambda_\nu, x)$ *for* L_2:

$$\| \varphi_\alpha^{(1)}(\lambda_\nu, x) \| = \| \varphi_\alpha^{(2)}(\lambda_\nu, x) \| \qquad (\nu = 1, 2, \ldots, n).$$

3) *For real* $s \neq 0$ *the limiting phases* $\delta^{(1)}(s)$ *for the solutions* $\phi_\alpha^{(1)}(\lambda_\nu, x)$ *equal the limiting phases* $\delta^{(2)}(s)$ *for the solutions* $\phi_\alpha^{(2)}(\lambda_\nu, x)$:

$$\delta^{(1)}(s) = \delta^{(2)}(s).$$

Then $q_1(x) = q_2(x)$ *almost everywhere in* $[0, \infty)$.

PROOF. Let $\rho_1(\lambda)$ and $\rho_2(\lambda)$ be the resolution functions corresponding to the boundary value problems with boundary condition $y(0)\cos\alpha + y'(0)\sin\alpha = 0$ for L_1 and L_2 respectively. According to part E of Chapter 2, §1, for $\lambda \leqslant 0$ we have

$$\rho_1(\lambda) = \sum_{\lambda_\nu \leqslant \lambda} \frac{1}{\| \varphi_\alpha^{(1)}(\lambda_\nu, x) \|^2}, \qquad \rho_2(\lambda) = \sum_{\lambda_\nu \leqslant \lambda} \frac{1}{\| \varphi_\alpha^{(2)}(\lambda_\nu, x) \|^2},$$

which, by virtue of conditions 1) and 2) of the present theorem, implies that

$\rho_1(\lambda) = \rho_2(\lambda)$ for $\lambda \leqslant 0$. If we can show that

$$\rho_1(\lambda) = \rho_2(\lambda) \tag{2.3.4}$$

also for positive values of λ, then by Theorem 2.3.1 we will know that $q_1(x) = q_2(x)$ almost everywhere in $[0, \infty)$.

We note that for $\lambda \geqslant 0$

$$\rho_1(\lambda) = \sum_{\nu=1}^{n} \frac{1}{\|\varphi_\alpha^{(1)}(\lambda_\nu, x)\|^2} + \frac{1}{\pi}\int_0^\lambda \frac{d\mu}{\sqrt{\mu}\,|M_1(\sqrt{\mu})|^2},$$

$$\rho_2(\lambda) = \sum_{\nu=1}^{n} \frac{1}{\|\varphi_\alpha^{(2)}(\lambda_\nu, x)\|^2} + \frac{1}{\pi}\int_0^\lambda \frac{d\mu}{\sqrt{\mu}\,|M_2(\sqrt{\mu})|^2}$$

(see Chapter 2, §1, part E), where the functions

$$M_1(\sqrt{\lambda}) = M_1(s) = i\sin\alpha - \frac{\cos\alpha}{s} + \frac{1}{s}\int_0^\infty q_1(t)\,e^{-ist}\,\varphi_\alpha^{(1)}(s^2, t)\,dt, \tag{2.3.5}$$

$$M_2(\sqrt{\lambda}) = M_2(s) = i\sin\alpha - \frac{\cos\alpha}{s} + \frac{1}{s}\int_0^\infty q_2(t)\,e^{-ist}\,\varphi_\alpha^{(2)}(s^2, t)\,dt \tag{2.3.5'}$$

are analytic for $s = \sqrt{\lambda}$ in the lower half-plane (Lemma 2.1.1). Then for the proof of (2.3.4) for positive λ it suffices to show that

$$M_1(s) = M_2(s) \quad (\mathrm{Im}\, s \leqslant 0).$$

We will show that this equation follows from conditions 1) and 3) in the hypothesis.

First we show that the functions $sM(s)$ corresponding to boundary value problems for an operator L satisfying (2.3.3) have bounded amplitude in a neighborhood of zero. In fact, the bound (2.1.6) for $\phi_\alpha(s^2, x)$ obtained in the proof of Lemma 2.1.1 implies the existence of the limit

$$\lim_{s\to 0} sM(s) = -\cos\alpha + \int_0^\infty q(t)\,\varphi_\alpha(0, t)\,dt.$$

If this limit is not zero, then the boundedness of the amplitude of $sM(s)$ in a neighborhood of zero is clear. Suppose then that the limit is zero; that is,

$$-\cos\alpha + \int_0^\infty q(t)\,\varphi_\alpha(0, t)\,dt = 0. \tag{2.3.6}$$

We know that $\phi_\alpha(0, x)$ is a solution of $L[u] = 0$; therefore

$$\varphi_\alpha'(0, x)\Big|_0^N - \int_0^N q(x)\,\varphi_\alpha(0, x)\,dx = 0,$$

and considering the initial conditions for $\phi_\alpha(0, x)$ we obtain

$$\varphi'_\alpha(0, N) + \cos\alpha - \int_0^N q(x)\varphi_\alpha(0, x)\,dx = 0.$$

Therefore from (2.3.6) it follows that

$$\lim_{N\to\infty} \varphi'_\alpha(0, N) = 0.$$

By virtue of Lemma 2.1.2 we now see that $\phi_\alpha(0, x)$ is a bounded solution of $L[u] = 0$, and since that lemma guarantees that there is only one bounded solution, the solution $\theta_\alpha(0, x)$ is not bounded and

$$\lim_{s\to 0} sN(s) = \sin\alpha + \int_0^\infty q(t)\,\theta_\alpha(0, t)\,dt \neq 0,$$

where

$$N(s) = i\cos\alpha + \frac{\sin\alpha}{s} + \frac{1}{s}\int_0^\infty e^{-ist} q(t)\,\theta_\alpha(s^2, t)\,dt.$$

Therefore the amplitude of $sN(s)$ must be bounded in a neighborhood of zero.

Furthermore, by part E of Chapter 2, §1, for the function $m(\lambda)$ corresponding to the boundary value problem with boundary condition $y(0)\cos\alpha + y'(0)\sin\alpha = 0$ we have

$$m(\lambda) = \frac{N(s)}{M(s)} = \frac{sN(s)}{sM(s)} \qquad (s^2 = \lambda,\ \operatorname{Im} s < 0),$$

or

$$sM(s) = \frac{sN(s)}{m(s^2)}.$$

We now observe that the function $m(\lambda)$ is regular and has no zeros in some small disk which has its center at zero and which is cut along the positive real axis starting at zero. This follows from the fact that the zeros and poles of $m(\lambda)$ are the eigenvalues of boundary value problems for L, and by Corollary 2 of Lemma 2.1.2 there are only finitely many eigenvalues.

Also $\operatorname{Im}[m(\lambda)]\operatorname{Im}[\lambda] < 0$ (see Chapter 2, §1, part A). Therefore the change in the amplitude of $m(\lambda)$ in the disk with the cut does not exceed 2π. Hence in a neighborhood of zero in the lower half-plane the amplitude of $m(s^2)$ is bounded.

Thus the numerator and denominator of $sN(s)/m(s^2)$ have bounded amplitude in a neighborhood of zero, which implies that the amplitude of $sM(s)$ is also bounded.

Now we can complete the proof of the theorem. We consider the function

$$g(s) = \frac{sM_1(s)}{sM_2(s)},$$

where $M_1(s)$ and $M_2(s)$ are given by formulas (2.3.5) and (2.3.5′) for $\operatorname{Im}[s] < 0$. The functions $M_1(s)$ and $M_2(s)$ are regular in the lower half-plane and share the same simple zeros. This follows from condition 1) in the hypothesis, since by part E of Chapter 2, §1, the zeros $s_1, \cdots, s_n$ of these functions are related to the eigenvalues $\lambda_1, \cdots, \lambda_n$ by the formula

$$s_1^2 = \lambda_1, \quad s_2^2 = \lambda_2, \ldots, \quad s_n^2 = \lambda_n.$$

In the proof of Lemma 2.1.1 we established the continuity of the functions $sM_1(s)$ and $sM_2(s)$ on the real axis and the existence of the limit

$$\lim_{|s| \to \infty} \frac{sM_1(s)}{sM_2(s)} = 1$$

(see (2.1.7) and (2.1.7′)). And by part E of Chapter 2, §1, for real $s \neq 0$ we have $M_1(s) \neq 0$ and $M_2(s) \neq 0$.

It follows that $g(s)$ is regular in the lower half-plane, continuous up to the real axis (except perhaps at 0), has no zeros for $s \neq 0$, and

$$\lim_{|s| \to \infty} g(s) = 1.$$

Now we use condition 3) of the hypothesis. Since $\delta^{(1)}(s) = \arg M_1(s)$ and $\delta^{(2)}(s) = \arg M_2(s)$, we see that $g(s)$ is real on the real axis by virtue of condition 3).

We introduce the function

$$h(s) = g\left(\frac{1}{s}\right) \qquad (h(0) = \lim_{|s| \to \infty} g(s) = 1).$$

This function is regular in the lower half-plane, is continuous and real on the real axis, and has no zeros. Therefore the function can be extended by symmetry onto the whole plane, and we obtain an entire function with no zeros. Then $\ln h(s)$ is an entire function. The imaginary part of $\ln h(s)$ equals the amplitude of $h(s)$; that is,

$$\operatorname{Im}[\ln h(s)] = \arg \frac{\frac{1}{s} M_1\left(\frac{1}{s}\right)}{\frac{1}{s} M_2\left(\frac{1}{s}\right)}.$$

Since the amplitude of $sM_1(s)$ and $sM_2(s)$ have been shown to be bounded in a neighborhood of zero, it follows that $\operatorname{Im}[\ln h(s)]$ is bounded outside of some disk and hence everywhere.

Now as in the proof of Theorem 2.3.2 we conclude that $\ln h(s)$ is constant. That is, $h(s) \equiv \text{const}$, and since $h(0) = 1$, we have $h(s) \equiv 1$. Therefore $sM_1(s)/sM_2(s) \equiv 1$; that is, $M_1(s) = M_2(s)$. The proof of the theorem is complete.

We note that if one of the conditions 1), 2) or 3) in the hypothesis is not fulfilled, the operators L_1 and L_2 cannot coincide. This follows from the work of Bargmann [16], where we find an example of two distinct operators which satisfy 1) and 3).

Theorem 2.3.3 has an immediate corollary which presents a sufficient condition under which a limiting phase $\delta(s)$ uniquely determines L.

COROLLARY. *Suppose that an operator L of the form (2.1.1) satisfies (2.3.3). If the boundary value problem with boundary condition $y(0)\cos\alpha + y'(0)\sin\alpha = 0$ has no negative eigenvalues, then the limiting phase $\delta(s)$ of the solutions $\phi_\alpha(s^2, x)$ uniquely determines the operator.*

This result was first obtained by Levinson [15]. He considered the case where the boundary condition is $y(0) = 0$, which is interesting from the point of view of physics. This allowed him to replace (2.3.3) by the weaker condition

$$\int_0^\infty x\,|q(x)|\,dx < \infty.$$

Our method of proof for Theorem 2.3.3 can also be generalized in this direction. But we will not do so here.

We conclude this chapter with the following observations.

1. The number of special cases in which L is uniquely determined by spectral properties could be expanded. But the question of which cases are of the most interest should be settled in the light of considerations from physics.

2. All results in this chapter can be generalized to the case where L is given on $(-a, a)$, $a \leqslant \infty$. Then instead of transformation operators V given on $\mathfrak{M}^1_{[0,a)}$ we would use transformation operators W given on $\mathfrak{M}^1_{(-a,a)}$.

SUPPLEMENT TO CHAPTER 2

SOLUTION OF THE INVERSE STURM-LIOUVILLE PROBLEM

Let $q(x)$ be a real function which is integrable on the finite interval $0 \leqslant x \leqslant a$, and let

$$L[u] = \frac{d^2}{dx^2}u(x) - q(x)\,u(x) \qquad (0 \leqslant x \leqslant a)$$

be the corresponding differential operator.

We denote by $\Lambda\{q(x), a, h, H\}$ the set of eigenvalues (spectrum) of the boundary value problem for L with boundary conditions

$$hu(0) - u'(0) = 0, \qquad Hu(a) + u'(a) = 0,$$

where h and H are arbitrary real numbers.(3)

By Theorem 2.3.2 the spectra

$$\Lambda\{q(x), a, h, H_1\}, \quad \Lambda\{q(x), a, h, H_2\}, \quad H_1 \neq H_2$$

of two boundary value problems with the same boundary condition at zero and distinct boundary conditions at a uniquely determine the function $q(x)$ and the numbers a, h, H_1 and H_2. But Theorem 2.3.2 does not give a method of actually recovering $q(x)$ and a, h, H_1 and H_2 from the spectra.

This Supplement is devoted to finding an effective solution of the inverse Sturm-Liouville problem; that is, it is devoted to the construction of an algorithm for recovering $q(x)$ and a, h, H_1 and H_2 from the spectra of the two corresponding boundary value problems.

To this end, let the sets

$$\{\lambda_0 < \lambda_1 < \cdots < \lambda_n < \lambda_{n+1} < \cdots\} = \Lambda\{q(x), a, h, H_1\},$$
$$\{\mu_0 < \mu_1 < \cdots < \mu_n < \mu_{n+1} < \cdots\} = \Lambda\{q(x), a, h, H_2\}$$

be given and let $q(x)$ and a, h, H_1 and H_2 be unknown.

The sequences $\{\lambda_n\}_0^\infty$ and $\{\mu_n\}_0^\infty$ must satisfy the well-known asymptotic formulas [4]

$$\sqrt{\lambda_n} = c\left[n + A_0 + \frac{1}{\pi n} A_1 + o\left(\frac{1}{n}\right)\right], \tag{1}$$

$$\sqrt{\mu_n} = c\left[n + B_0 + \frac{1}{\pi n} B_1 + o\left(\frac{1}{n}\right)\right], \tag{1'}$$

where $c = \pi/a$, and the other constants are expressed in terms of h, H_1, H_2, and $g = ½ \int_0^a q(x)\,dx$ as follows:

h	H_1	H_2	A_0	B_0	A_1	B_1	$A_1 - B_1$
∞	∞	$\neq \infty$	1	$\frac{1}{2}$	g	$H_2 + g$	$-H_2$
∞	$\neq \infty$	∞	$\frac{1}{2}$	1	$H_1 + g$	g	H_1
∞	$\neq \infty$	$\neq \infty$	$\frac{1}{2}$	$\frac{1}{2}$	$H_1 + g$	$H_2 + g$	$H_1 - H_2$
$\neq \infty$	∞	$\neq \infty$	$\frac{1}{2}$	0	$h + g$	$h + H_2 + g$	$-H_2$
$\neq \infty$	$\neq \infty$	∞	0	$\frac{1}{2}$	$h + H_1 + g$	$h + g$	H_1
$\neq \infty$	$\neq \infty$	$\neq \infty$	0	0	$h + H_1 + g$	$h + H_2 + g$	$H_1 - H_2$

(3) For $h = \infty$ $(H = \infty)$ the boundary condition becomes $u(0) = 0$ $(u(a) = 0)$.

If we know the sets $\{\lambda_n\}$ and $\{\mu_n\}$, we can of course find the constants c, A_0, B_0, A_1 and B_1, so we assume that we know these constants and also the number

$$a = \frac{\pi}{c} = \pi \lim_{n\to\infty} \frac{n}{\sqrt{\lambda_n}} = \pi \lim_{n\to\infty} \frac{n}{\sqrt{\mu_n}}.$$

From the table it is clear that the constants A_0 and B_0 indicate which of the numbers h, H_1 and H_2 are infinite. We will examine the case $A_0 = B_0 = 0$ in detail; that is, the case where none of the numbers h, H_1 or H_2 is infinite. The other cases are analogous.

We let $\omega_h(\lambda, x)$ denote the solution of the equation

$$u''(x) - q(x)u(x) + \lambda u(x) = 0$$

with initial conditions $u(0) = 1$, $u'(0) = h$. Then the set $\{\lambda_n\}$ $(\{\mu_n\})$ is the set of roots of the equation

$$H_1\omega_h(\lambda, a) + \omega_h'(\lambda, a) = 0 \qquad (H_2\omega_h(\lambda, a) + \omega_h'(\lambda, a) = 0),$$

where we let $\omega_h'(\lambda, a)$ denote the derivative of $\omega_h(\lambda, x)$ with respect to x at the point $x = a$.

LEMMA 1. *If we know the spectra* $\{\lambda_n\}_0^\infty$ *and* $\{\mu_n\}_0^\infty$, *we can recover the functions*

$$\mathcal{L}(z) = H_1\omega_h(z^2, a) + \omega_h'(z^2, a), \tag{2}$$

$$M(z) = H_2\omega_h(z^2, a) + \omega_h'(z^2, a) \tag{2'}$$

by means of the formulas

$$\mathcal{L}(z) = \frac{\pi}{c}(\lambda_0 - z^2) \prod_{n=1}^{\infty} \frac{\lambda_n - z^2}{c^2 n^2},$$

$$M(z) = \frac{\pi}{c}(\mu_0 - z^2) \prod_{n=1}^{\infty} \frac{\mu_n - z^2}{c^2 n^2}.$$

PROOF. It suffices to prove the lemma for one of the functions, say $\mathcal{L}(z)$. Since $\omega_h(z^2, a)$ and $\omega_h'(z^2, a)$ are entire functions of exponential type, $\mathcal{L}(z)$ is also an entire function of exponential type with zeros at the points $\pm\lambda_n^{1/2}$. Therefore

$$\mathcal{L}(z) = p \prod_{n=0}^{\infty} \left(1 - \frac{z^2}{\lambda_n}\right) = H_1\omega_h(z^2, a) + \omega_h'(z^2, a),$$

where p is some constant. (The factor $e^{p_1 z}$ is missing because $\mathcal{L}(z)$ is even.)

By using known asymptotic formulas for $\omega_h(z^2, a)$ and $\omega_h'(z^2, a)$ with $z = it$ $(-\infty < t < \infty)$, we obtain

$$H_1\omega_h(z^2, a) + \omega_h'(z^2, a) \sim -it \sin iat \qquad (t \to \infty),$$

and thus

$$p = \lim_{t\to\infty} \frac{-it \sin iat}{\prod_{n=0}^{\infty}\left(1 + \frac{t^2}{\lambda_n}\right)} = \lim_{t\to\infty} \frac{\lambda_0 a t^2}{\lambda_0 + t^2} \prod_{n=1}^{\infty} \frac{\left(1 + \frac{a^2t^2}{n^2\pi^2}\right)}{\left(1 + \frac{t^2}{\lambda_n}\right)}.$$

Since

$$\left(1 + \frac{a^2t^2}{n^2\pi^2}\right)\left(1 + \frac{t^2}{\lambda_n}\right)^{-1} = \frac{a^2\lambda_n}{n^2\pi^2} \frac{(n^2\pi^2 + a^2t^2)}{a^2\lambda_n + a^2t^2} = \frac{a^2\lambda_n}{n^2\pi^2}\left(1 + \frac{n^2\pi^2 - \lambda_n a^2}{a^2\lambda_n + a^2t^2}\right)$$

and $a = \pi/c$, we have

$$p = \frac{\lambda_0\pi}{c} \lim_{t\to\infty} \prod_{n=1}^{\infty} \frac{\lambda_n}{c^2n^2}\left(1 + \frac{c^2n^2 - \lambda_n}{\lambda_n + t^2}\right).$$

By virtue of formula (1) and the assumption that $A_0 = 0$ we have

$$\lim_{n\to\infty} (c^2n^2 - \lambda_n) = \frac{2}{\pi} A_1, \tag{3}$$

which guarantees absolute convergence of each of the products

$$\prod_{n=1}^{\infty} \frac{\lambda_n}{c^2n^2}, \quad \prod_{n=1}^{\infty}\left(1 + \frac{c^2n^2 - \lambda_n}{\lambda_n + t^2}\right)$$

separately. Thus

$$p = \frac{\lambda_0\pi}{c} \prod_{n=1}^{\infty} \frac{\lambda_n}{c^2n^2} \lim_{t\to\infty} \prod_{n=1}^{\infty}\left(1 + \frac{c^2n^2 - \lambda_n}{\lambda_n + t^2}\right).$$

From (1) and (3) we obtain

$$\lim_{t\to\infty} \prod_{n=1}^{\infty}\left(1 + \frac{c^2n^2 - \lambda_n}{\lambda_n + t^2}\right) = 1,$$

which implies that

$$p = \frac{\pi\lambda_0}{c} \prod_{n=1}^{\infty} \frac{\lambda_n}{c^2n^2}.$$

For the function $\mathcal{L}(z)$ we have

$$\mathcal{L}(z) = \frac{\pi\lambda_0}{c}\left(1 - \frac{z^2}{\lambda_0}\right) \prod_{n=1}^{\infty} \frac{\lambda_n}{c^2n^2}\left(1 - \frac{z^2}{\lambda_n}\right),$$

or

$$\mathcal{L}(z) = \frac{\pi}{c}(\lambda_0 - z^2)\prod_{n=1}^{\infty}\frac{\lambda_n - z^2}{c^2 n^2},$$

which is what we wanted to show.

LEMMA 2. *The following identity is true*:

$$\int_0^a \omega_h(\lambda^2, x)\,\omega_h(\mu^2, x)\,dx = \frac{\mathcal{L}(\mu)M(\lambda) - \mathcal{L}(\lambda)M(\mu)}{(A_1 - B_1)(\mu^2 - \lambda^2)}.$$

PROOF. We know that

$$(\mu^2 - \lambda^2)\int_0^a \omega_h(\lambda^2, x)\,\omega_h(\mu^2, x)\,dx = \omega_h'(\lambda^2, a)\,\omega_h(\mu^2, a) - \omega_h'(\mu^2, a)\,\omega_h(\lambda^2, a).$$

On the other hand, from (2) and (2′), which define $\mathcal{L}(z)$ and $M(z)$, we have

$$\mathcal{L}(\mu)M(\lambda) - \mathcal{L}(\lambda)M(\mu) = (H_1 - H_2)[\omega_h'(\lambda^2, a)\,\omega_h(\mu^2, a) - \omega_h'(\mu^2, a)\,\omega_h(\lambda^2, a)].$$

Therefore

$$\int_0^a \omega_h(\lambda^2, x)\,\omega_h(\mu^2, x)\,dx = \frac{\mathcal{L}(\mu)M(\lambda) - \mathcal{L}(\lambda)M(\mu)}{(H_1 - H_2)(\mu^2 - \lambda^2)},$$

which implies the desired identity if we note that from the table we have $H_1 - H_2 = A_1 - B_1$ (for the case $A_0 = B_0 = 0$).

For the solution of the inverse Sturm-Liouville problem it clearly suffices to recover at least one function $\omega_h(\lambda^2, x)$, which in fact can be found if we recover the kernel $K(x, t)$ of the transformation operator $V = V_{\{D^2 L_{0h}\}}$. The lemmas proved above allow us to find a (nonlinear) integral equation whose unique solution is the kernel $K(x, t)$.

For according to the definition of V we have

$$\omega_h(\lambda^2, x) = V[\cos\lambda x] = \cos\lambda x + \int_0^a K(x, t)\cos\lambda t\,dt,$$

where $K(x, t)$ is bounded and equals zero for $x < t$. Therefore

$$\int_0^a \omega_h(\lambda^2, x)\,\omega_h(\mu^2, x)\,dx = \int_0^a \cos\lambda x\cos\mu x\,dx + \int_0^a\int_0^a K(x,t)\cos\mu t\cos\lambda x\,dt\,dx$$
$$+ \int_0^a\int_0^a K(x, t)\cos\lambda t\cos\mu x\,dt\,dx$$
$$+ \int_0^a\left\{\int_0^a K(x, t)\cos\lambda t\,dt\cdot\int_0^a K(x, t)\cos\mu t\,dt\right\}dx,$$

or

$$\int_0^a \omega_h(\lambda^2, x)\,\omega_h(\mu^2, x)dx - \int_0^a \cos\lambda x\cos\mu x\,dx$$
$$= \int_0^a\int_0^a \Big[K(x, y) + K(y, x) + \int_0^a K(z, x)\,K(z, y)\,dz\Big]\cos\lambda x\cos\mu y\,dx\,dy.$$

We rewrite this equation in the form

$$\int_0^a\int_0^a F(x, y)\cos\lambda x\cos\mu y\,dx\,dy = G(\lambda, \mu), \tag{4}$$

where

$$G(\lambda, \mu) = \int_0^a [\omega_h(\lambda^2, x)\,\omega_h(\mu^2, x) - \cos\lambda x\cos\mu x]\,dx;$$
$$F(x, y) = K(x, y) + K(y, x) + \int_0^a K(z, x)\,K(z, y)\,dz. \tag{5}$$

By Lemma 2 and the identity

$$\int_0^a \cos\lambda x\cos\mu x\,dx = \frac{\mu\sin\mu a\cos\lambda a - \lambda\sin\lambda a\cos\mu a}{\mu^2 - \lambda^2}$$

we have

$$G(\lambda, \mu) = \frac{\mathscr{L}(\mu)\,M(\lambda) - \mathscr{L}(\lambda)\,M(\mu)}{(A_1 - B_1)(\mu^2 - \lambda^2)} - \frac{\mu\sin\mu a\cos\lambda a - \lambda\sin\lambda a\cos\mu a}{\mu^2 - \lambda^2}. \tag{6}$$

This formula and Lemma 1 enable us to recover $G(\lambda, \mu)$ if we know the spectra $\{\lambda_n\}_0^\infty$ and $\{\mu_n\}_0^\infty$. If we know $G(\lambda, \mu)$, we can then find an expansion of $F(x, y)$ in a double Fourier series (or integral).

In fact, using formula (4) for calculating the Fourier coefficients of $F(x, y)$, we obtain

$$F(x, y) = \frac{1}{a^2}\sum_{m=-\infty}^{\infty}\sum_{n=-\infty}^{\infty} G\left(\frac{n\pi}{a}, \frac{m\pi}{a}\right)\cos\frac{n\pi}{a}x\cos\frac{m\pi}{a}y. \tag{7}$$

Therefore we can consider $F(x, y)$ to be known. But then (5) can be interpreted as an equation in which the unknown is $K(x, y)$.

For $x > y$ this equation is of the form

$$F(x, y) = K(x, y) + \int_x^a K(z, x)\,K(z, y)\,dz. \tag{8}$$

This follows from the fact that $K(x, t)$ is zero for $x < t$.

From (6), (7) and (8) we deduce the following lemma.

LEMMA 3. *For $x > y$ the kernel $K(x, y)$ of the transformation* *operator* $V_{\{D^2 L_0 h\}}$ *satisfies the equation*

$$F(x, y) = K(x, y) + \int_x^a K(z, x) K(z, y) dz,$$

where

$$F(x, y) = \frac{1}{a^2} \sum_{m=-\infty}^{\infty} \sum_{n=-\infty}^{\infty} G\left(\frac{n\pi}{a}, \frac{m\pi}{a}\right) \cos \frac{n\pi}{a} x \cos \frac{m\pi}{a} y$$

and

$$G(\lambda, \mu) = \frac{\mathscr{L}(\mu) M(\lambda) - \mathscr{L}(\lambda) M(\mu)}{(A_1 - B_1)(\mu^2 - \lambda^2)} - \frac{\mu \sin \mu a \cos \lambda a - \lambda \sin \lambda a \cos \mu a}{\mu^2 - \lambda^2}.$$

Thus we have found an equation which has as one solution the unknown kernel $K(x, y)$. By means of this equation we will be able to recover the kernel $K(x, y)$ and thereby solve the inverse Sturm-Liouville problem.

To this end we introduce the following operator U which is defined on all functions $\Phi(x, y)$ which are bounded in the triangle $0 \leqslant y \leqslant x \leqslant a$:

$$U[\Phi] = F(x, y) - \int_x^a \Phi(z, x) \Phi(z, y) dz. \tag{9}$$

Then (8) assumes the form

$$K(x, y) = U[K(x, y)].$$

Since the unknown $K(x, y)$ automatically exists and is bounded, the function $F(x, y)$ defined by (7) is also bounded:

$$\sup_{\substack{0 \leqslant x \leqslant a \\ 0 \leqslant y \leqslant a}} |F(x, y)| = \| F \| < \infty.$$

Let c be an arbitrary positive number such that

$$c \geqslant \frac{4}{3} \| F \|$$

and

$$\delta = \min \left\{ a, \frac{1}{4c} \right\}.$$

We let $M_{\{c,\delta\}}$ denote the complete normed space of functions $\Phi(x, y)$ which are defined in the trapezoid $a - \delta \leqslant x \leqslant a,\ 0 \leqslant y \leqslant x \leqslant a$ and which satisfy

$$\| \Phi \|_\delta = \sup_{\substack{a-\delta \leqslant x \leqslant a \\ 0 \leqslant y \leqslant x \leqslant a}} |\Phi(x, y)| \leqslant c,$$

where $\| \Phi \|_\delta$ is the norm of the function $\Phi(x, y) \in M_{\{c,\delta\}}$.

We will consider the operator U on th. functions of this space.(4)

(4) By virtue of (9) this is reasonable.

According to (9) we have

$$|U[\Phi]| \leqslant \|F\| + \|\Phi\|_\delta^2 (a - x),$$

and since $\|\Phi\|_\delta \leqslant c$, $\|F\| \leqslant 3c/4$ and $a - x \leqslant \delta \leqslant 1/4c$, we have

$$|U[\Phi]| \leqslant \frac{3}{4} c + \frac{1}{4} c = c,$$

that is,

$$\|U[\Phi]\|_\delta \leqslant c,$$

which implies that U maps $M_{\{c,\delta\}}$ into itself.

Furthermore,

$$U[\Phi] - U[\Phi_1] = \int_x^a [\Phi(z, x)\Phi(z, y) - \Phi_1(z, x)\Phi_1(z, y)]\,dz$$

$$= \int_x^a [\Phi(z, x) - \Phi_1(z, x)]\,\Phi(z, y)\,dz + \int_x^a [\Phi(z, y) - \Phi_1(z, y)]\,\Phi_1(z, x)\,dz,$$

which implies that

$$|U[\Phi] - U[\Phi_1]| \leqslant \|\Phi - \Phi_1\|_\delta \|\Phi\|_\delta (a - x) + \|\Phi - \Phi_1\|_\delta \|\Phi_1\|_\delta (a - x),$$

and since $\|\Phi\|_\delta \leqslant c$, $\|\Phi_1\|_\delta \leqslant c$ and $a - x \leqslant \delta \leqslant 1/4c$, we have

$$|U[\Phi] - U[\Phi_1]| \leqslant \frac{1}{2} \|\Phi - \Phi_1\|_\delta .$$

Therefore

$$\|U[\Phi] - U[\Phi_1]\|_\delta \leqslant \frac{1}{2} \|\Phi - \Phi_1\|_\delta ,$$

which implies that the operator U acts on the space $M_{\{c,\delta\}}$ in such a way that the distance between any two functions is reduced by at least half.

By virtue of the theorem of Tihonov and Cacciopoli (see [23], Russian pp. 146–147) the operator has precisely one fixed point, and this point can be found by iteration.

Thus if

$$K_0(x, y) = F(x, y),$$

$$K_n(x, y) = F(x, y) - \int_x^a K_{n-1}(z, x) K_{n-1}(z, y)\,dz,$$

then as $n \to \infty$ the sequence of functions $\{K_n(x, y)\}$ converges uniformly in the region $a - \delta \leqslant x \leqslant a$, $0 \leqslant y \leqslant x \leqslant a$ to a function $K(x, y)$ which is the unique solution of the equation

$$K(x, y) = F(x, y) - \int_x^a K(z, x) K(z, y)\,dz \quad (a - \delta \leqslant x \leqslant a;\ 0 \leqslant y \leqslant x \leqslant a),$$

satisfying

$$\sup_{\substack{a-\delta\leqslant x\leqslant a\\ 0\leqslant y\leqslant x\leqslant a}} |K(x, y)| \leqslant C.$$

Since the constant C can be chosen arbitrarily large, the limit function $K(x, y)$ coincides with the unknown kernel for $a - \delta \leqslant x \leqslant a$, $0 \leqslant y \leqslant x \leqslant a$.

Hence the kernel $K(x, y)$ of the transformation operator $V_{\{D^2 L_0 h\}}$ has been recovered in $a - \delta \leqslant x \leqslant a$. We now consider the values $x \leqslant a - \delta$. According to (8) we have

$$K(x, y) = F(x, y) - \int_x^{a-\delta} K(z, x)K(z, y)\,dz - \int_{a-\delta}^{a} K(z, x)K(z, y)\,dz,$$

or

$$K(x, y) = F_1(x,y) - \int_x^{a-\delta} K(z, x)K(z, y)\,dz, \tag{10}$$

where the function

$$F_1(x, y) = F(x, y) - \int_{a-\delta}^{a} K(z, x)K(z, y)\,dz$$

is known.

If we solve (10) by the method of successive approximations we recover $K(x, y)$ in some new region $a - \delta - \delta_1 \leqslant x \leqslant a - \delta$, $0 \leqslant y \leqslant x \leqslant a - \delta$, and so forth.

After a finite number of steps we have completely recovered $K(x, y)$ and thereby solved the inverse Sturm-Liouville problem.([5])

In fact, if

$$\omega_h(\lambda^2, x) = \cos\lambda x + \int_0^x K(x, t)\cos\lambda t\,dt,$$

then

$$h = \frac{\partial}{\partial x}\omega_h(0, x)\Big|_{x=0}, \quad q(x) = \frac{\frac{\partial^2}{\partial x^2}\omega_h(0, x)}{\omega_h(0, x)}$$

and

$$H_1 = -\left.\frac{\frac{\partial}{\partial x}\omega_h(\lambda_0, x)}{\omega_h(\lambda_0, x)}\right|_{x=a}, \qquad H_2 = -\left.\frac{\frac{\partial}{\partial x}\omega_h(\mu_0, x)}{\omega_h(\mu_0, x)}\right|_{x=a}.$$

([5]) Let $K(x, y)$ be the unknown kernel which we know exists and is bounded. We set

$$k = \sup_{0\leqslant y\leqslant x\leqslant a} |K(x, y)|.$$

It is easy to show that in order to recover $K(x,y)$ in the whole triangle $0 \leqslant y \leqslant x \leqslant a$ we need not perform more than $k + k^2 a$ steps, and so the number of steps is finite.

Up to this point we have assumed that $A_0 = B_0 = 0$.

In the general case we perform the following operations in order to solve the inverse Sturm-Liouville problem:

1) Calculate the limits

$$c = \lim_{n\to\infty} \frac{\sqrt{\lambda_n}}{n} = \lim_{n\to\infty} \frac{\sqrt{\mu_n}}{n},$$

$$A_0 = \frac{1}{c} \lim_{n\to\infty} [\sqrt{\lambda_n} - cn], \qquad B_0 = \frac{1}{c} \lim_{n\to\infty} [\sqrt{\mu_n} - cn],$$

$$A_1 = \frac{\pi}{c} \lim_{n\to\infty} n\, [\sqrt{\lambda_n} - c(n + A_0)], \qquad B_1 = \frac{\pi}{c} \lim_{n\to\infty} n\, [\sqrt{\mu_n} - c(n + B_0)]$$

and define the number

$$a = \frac{\pi}{c}.$$

2) Construct the functions $\mathcal{L}(z)$, $M(z)$ and $P(\lambda, \mu)$ by means of the formulas

$$\mathcal{L}(z) = \prod_{n=0}^{\infty} \frac{\lambda_n - z^2}{c^2 (n + A_0)^2} \quad \left(\mathcal{L}(z) = \frac{\pi}{c}(\lambda_0 - z^2) \prod_{n=1}^{\infty} \frac{\lambda_n - z^2}{c^2 n^2}, \quad \text{if} \quad A_0 = 0\right);$$

$$M(z) = \prod_{n=0}^{\infty} \frac{\mu_n - z^2}{c^2 (n + B_0)^2} \quad \left(M(z) = \frac{\pi}{c}(\mu_0 - z^2) \prod_{n=1}^{\infty} \frac{\mu_n - z^2}{c^2 n^2}, \quad \text{if} \quad B_0 = 0\right);$$

$$P(\lambda, \mu) = \frac{\mathcal{L}(\mu) M(\lambda) - \mathcal{L}(\lambda) M(\mu)}{(A_1 - B_1)(\mu^2 - \lambda^2)}, \qquad \text{if} \quad A_0 = B_0;$$

$$P(\lambda, \mu) = \frac{\mathcal{L}(\mu) M(\lambda) - \mathcal{L}(\lambda) M(\mu)}{(\mu^2 - \lambda^2)}, \qquad \text{if} \quad A_0 \neq B_0.$$

3) Formulate the equation

$$F(x, y) = K(x, y) + \int_x^a K(z, x) K(z, y)\, dz,$$

where $F(x, y)$ is defined as follows:

$$F(x, y) = \frac{1}{a^2} \sum_{m=-\infty}^{\infty} \sum_{n=-\infty}^{\infty} G\left(\frac{m\pi}{a}, \frac{n\pi}{a}\right) \cos \frac{m\pi}{a} x \cos \frac{n\pi}{a} y$$

and

$$G(\lambda, \mu) = P(\lambda, \mu) - \frac{\mu \sin \mu a \cos \lambda a - \lambda \sin \lambda a \cos \mu a}{\mu^2 - \lambda^2},$$

if $A_0 B_0 = 0$;

$$F(x, y) = \frac{1}{a^2} \sum_{m=-\infty}^{\infty} \sum_{n=-\infty}^{\infty} G\left(\frac{m\pi}{a}, \frac{n\pi}{a}\right) \sin \frac{m\pi}{a} x \sin \frac{n\pi}{a} y$$

and

$$G(\lambda, \mu) = \lambda\mu P(\lambda, \mu) - \frac{\lambda \cos \lambda a \sin \mu a - \mu \cos \mu a \sin \lambda a}{\mu^2 - \lambda^2},$$

if $A_0 B_0 \neq 0$.

4) By some method or other solve the equation in the preceding paragraph for $K(x, y)$.

To obtain the solution we can always apply the method of successive approximations, and we find $K(x, y)$ first in some region $a - \delta \leqslant x \leqslant a$, $0 \leqslant y \leqslant x \leqslant a$. Then by writing a second equation, as we did above, we can find $K(x, y)$ (again by the method of successive approximations) in an adjacent region, and so forth. In this manner we can find $K(x, y)$ throughout the triangle $0 \leqslant y \leqslant x \leqslant a$ after a finite number of steps.

5) The function $K(x, y)$ is the kernel of the transformation operator $V_{\{D^2 L_{0h}\}}$ if $A_0 B_0 = 0$, and of the transformation operator $V_{\{D^2 L_{\infty\infty}\}}$ if $A_0 B_0 \neq 0$. The remainder of the proof is clear. We remark only that we still might not be able to find the functions $\omega_h(\lambda, x)$ $(\omega_\infty(\lambda, x))$, but we make use of the identities

$$K(x, x) = h + \frac{1}{2}\int_0^x q(t)\,dt, \qquad \text{if} \quad A_0 B_0 = 0;$$

$$K(x, x) = \frac{1}{2}\int_0^x q(t)\,dt, \qquad \text{if} \quad A_0 B_0 \neq 0,$$

in order to calculate $q(x)$.

We will not prove the identities.

It is interesting to note that we never use the fact that $q(x)$, h, H_1 and H_2 are real. Therefore the above solution of the inverse Sturm-Liouville problem is valid when $q(x)$, h, H_1 and H_2 are complex; that is, the function $q(x)$ and the numbers h, H_1 and H_2 are uniquely determined by the spectra of two boundary value problems also in the complex case.

Kreĭn presented another method for solving the inverse Sturm-Liouville problem. The method is based on a generalization of the classical moment problem (see [18]).

In conclusion we consider the question of whether $q(x)$ is determined by the limiting phase $\delta(s)$ of the solutions $\omega_\infty(s^2, x)$.

Let the differential operator

$$L[u] = \frac{d^2}{dx^2} u(x) - q(x) u(x)$$

be given on $0 \leqslant x < \infty$, and let $q(x)$ be real and equal to zero outside some finite subinterval $(0, a)$. For simplicity we assume that $q(x)$ is nonnegative.

Suppose further that $\omega_\infty(s^2, x)$ is the solution of $L[u] + s^2 u = 0$ with initial conditions $u(0) = 0$, $u'(0) = 1$. We have

$$\omega_\infty(s^2, x) = \frac{\sin sx}{s} + \int_0^x \frac{\sin s(x-t)}{s} q(t)\,\omega_\infty(s^2, t)\,dt,$$

and so for $x \geqslant a$

$$s\omega_\infty(s^2, x) = \sin sx\left[1 + \int_0^a \cos st\, q(t)\,\omega_\infty(s^2, t)\,dt\right] - \cos sx \int_0^a \sin st\, q(t)\,\omega_\infty(s^2, t)\,dt.$$

Therefore for real s and $x \geqslant a$

$$\begin{aligned} s\omega_\infty(s^2, x) &= |A(s)| \sin[sx + \delta(s)], \\ s\omega'_\infty(s^2, x) &= s|A(s)| \cos[sx + \delta(s)], \end{aligned} \tag{11}$$

where

$$A(s) = 1 + \int_0^a e^{-ist} q(t)\,\omega_\infty(s^2, t)\,dt, \quad \delta(s) = \arg A(s).$$

The function $\delta(s)$ is called the limiting phase.

From Lemma 2.1.1 it follows that in the lower half-plane and on the real axis $A(s)$ is holomorphic, has no zeros,(5) and $A(s) = 1 + O(s^{-1})$ as $|s| \to \infty$. Consequently the function

$$\ln A(z) = \ln|A(z)| + i\delta(z)$$

is also holomorphic in the lower half-plane and approaches zero as $|z| \to \infty$.

If we know the imaginary part $\delta(t)$ of $\ln A(z)$ on $-\infty < t < \infty$, we can recover the whole function by means of the well-known formula

$$\ln A(z) = \frac{1}{\pi}\int_{-\infty}^{\infty} \frac{\delta(t)}{z-t}\,dt,$$

from which for real $z = s$ we obtain

$$\ln|A(s)| = \frac{1}{\pi}\int_{-\infty}^{\infty} \frac{\delta(t)}{s-t}\,dt \qquad (-\infty < s < \infty). \tag{12}$$

We will now show how to recover $q(x)$ from the limiting phase $\delta(s)$.

We consider L only in the interval $0 < x < a$. The kernel $K(x, y)$ of the transformation operator $V_{\{D^2 L_{\infty\infty}\}}$ for $0 \leqslant y \leqslant x \leqslant a$ must satisfy the integral equation

$$F(x, y) = K(x, y) + \int_x^a K(z, x)\,K(z, y)\,dz, \tag{13}$$

where

(5) The absence of zeros follows from the nonnegativity of $q(x)$.

$$F(x,y)=\frac{1}{a^2}\sum_{m=-\infty}^{\infty}\sum_{n=-\infty}^{\infty}G\left(\frac{m\pi}{a},\frac{n\pi}{a}\right)\sin\frac{m\pi}{a}x\sin\frac{n\pi}{a}y,$$

$$G(\lambda,\mu)=\lambda\mu\int_0^a\omega_\infty(\lambda^2,x)\,\omega_\infty(\mu^2,x)\,dx-\int_0^a\sin\lambda x\sin\mu x\,dx.$$

This equation can be established just like the corresponding equation in Lemma 3.

Since

$$(\mu^2-\lambda^2)\int_0^a\omega_\infty(\lambda^2,x)\,\omega_\infty(\mu^2,x)\,dx=\omega'_\infty(\lambda^2,a)\,\omega_\infty(\mu^2,a)-\omega_\infty(\lambda^2,a)\,\omega'_\infty(\mu^2,a),$$

formulas (11) and (12) enable us to recover $G(\lambda,\mu)$, and hence $F(x,y)$, by means of the limiting phase $\delta(s)$.

Thus if we know $\delta(s)$, we can establish (13), in which $F(x,y)$ is known and $K(x,y)$ is unknown. The rest of the solution is clear (see 4) and 5)).

Chapter 3

APPLICATION OF TRANSFORMATION OPERATORS TO THE THEORY OF GENERALIZED SHIFTS

§1. Additional properties of transformation operators

Let L be a differential operator given on $(-a,a)$ (or $[0,a)$), $a\leqslant\infty$, of the form

$$L[u]=\frac{d^2}{dx^2}u(x)-q(x)\,u(x),\tag{3.1.1}$$

where $q(x)$ is a real function which is integrable on each subinterval $(-b,b)$ (or $[0,b)$) for arbitrary $b<a$.

Let $f(x)$ be an arbitrary function in $\mathfrak{M}^1_{(-a,a)}$ ($\mathfrak{M}^1_{[0,a)}$).(6)

We denote by $L^{-1}[f]$ the solution of the equation

$$L[v]=f(x)$$

with initial conditions $v(0)=v'(0)=0$. By the method of variation of constants we obtain

$$v(x)=L^{-1}[f]=\int_0^x[X^1(x)\,X^0(t)-X^1(t)\,X^0(x)]\,f(t)\,dt,\tag{3.1.2}$$

---

(6) For the definition of the sets $\mathfrak{M}^1_{(-a,a)}$, $\mathfrak{M}^1_{[0,a)}$ and the concept of convergence (completeness) in these sets see §2 in Chapter 1.

where $X^0(t)$ and $X'(t)$ denote, as in §1 of Chapter 1, the solutions of $L[u] = 0$ with initial conditions

$$X^0(0) = 1, \quad \frac{d}{dt} X^0(t)|_{t=0} = 0; \qquad X^1(0) = 0, \quad \frac{d}{dt} X^1(t)|_{t=0} = 1.$$

Formula (3.1.2) defines an additive homogeneous operator L^{-1} on the set $\mathfrak{M}^1_{(-a,a)}$ (or $\mathfrak{M}^1_{[0,a)}$ if L is given on $[0, a)$). This operator takes functions in $\mathfrak{M}^1_{(-a,a)}$ ($\mathfrak{M}^1_{[0,a)}$) into twice differentiable functions, and if the sequence of functions $f_k(x) \in \mathfrak{M}^1_{(-a,a)}$ ($f_k(x) \in \mathfrak{M}^1_{[0,a)}$) converges in this set to $f(x)$ as $k \to \infty$, then the sequence of functions $L^{-1}[f_k]$ converges as $k \to \infty$ to $L^{-1}[f]$ uniformly in each subinterval $(-b, b)$ (or $[0, b)$) for arbitrary $b < a$.

We consider two operators L_1 and L_2 of the form (3.1.1). The transformation operator $W_{\{L_1 L_2 \theta\}}$ (or $V_{\{L_1 L_2 h_1 h_2\}}$ if L_1 and L_2 are given on $[0, a)$) enables us to express L_1 and L_1^{-1} in terms of L_2 and L_2^{-1} according to the formulas

$$L_1 = W_{\{L_2 L_1 \theta\}} L_2 W^{-1}_{\{L_2 L_1 \theta\}}; \quad L_1^{-1} = W_{\{L_2 L_1 \theta\}} L_2^{-1} W^{-1}_{\{L_2 L_1 \theta\}}$$

$$\left(L_1 = V_{\{L_2 L_1 h_2 h_1\}} L_2 V^{-1}_{\{L_2 L_1 h_2 h_1\}}, \quad L_1^{-1} = V_{\{L_2 L_1 h_2 h_1\}} L_2^{-1} V^{-1}_{\{L_2 L_1 h_2 h_1\}}\right),$$

which are analogous to the linear algebra formulas for expressing the transformation of matrix operators under a change of basis.

We will prove this for the case where L_1 and L_2 are given on $(-a, a)$, $a \leqslant \infty$. We denote (as in §1 of Chapter 1) by $e_1(s^2, x)$ ($e_2(s^2, x)$) the solution of the equation $L_1[u] + s^2 u = 0$ ($L_2[u] + s^2 u = 0$) with initial conditions $u(0) = 1$ and $u'_x(0) = is$.

By means of an easy calculation we can see that L_1^{-1} on $e_1(s^2, x)$ equals

$$L_1^{-1}[e_1(s^2, x)] = \frac{e_1(s^2, x) - X_1^0(x) - isX_1^1(x)}{s^2};$$

that is,

$$L_1^{-1}[e_1(s^2, x)] = W\left[\frac{e_2(s^2, x) + \theta\omega_{2\infty}(s^2, x) - X_2^0(x) - (is - \theta) X_2^1(x)}{s^2}\right]$$

$$= W L_2^{-1} [e_2(s^2, x) - \theta\omega_{2\infty}(s^2, x)] = W L_2^{-1} W^{-1} [e_1(s^2, x)],$$

where $W = W_{\{L_1 L_2 \theta\}}$. Since the operators are additive and homogeneous we have

$$L_1^{-1}[\varphi] = W L_2^{-1} W^{-1} [\varphi]$$

on the set of all functions $\phi(x)$ which are linear combinations of the functions $e_1(s^2, x)$ (where s is a parameter).

Now let $f(x)$ be an arbitrary function in $\mathfrak{M}^1_{(-a,a)}$. The family of functions $\{e^{isx}\}$, $-\infty < s < \infty$, is complete in $\mathfrak{M}^1_{(-a,a)}$. Using the notation

$$D^2[u] = \frac{d^2}{dx^2} u(x),$$

we have

$$e_1(s^2, x) = W_{\{D^2 L_1 0\}}[e^{isx}],$$

which by virtue of property IV of transformation operators (Chapter 1, §2) implies that the family $\{e_1(s^2, x)\}$ $(-\infty < s < \infty)$ is also complete in $\mathfrak{M}^1_{(-a,a)}$. Thus there exists a sequence of linear combinations $\phi_n(x)$ of the functions $e_1(s^2, x)$ which converges in $\mathfrak{M}^1_{(-a,a)}$ to $f(x)$ as $n \to \infty$. But then the sequence of functions $L_1^{-1}[\phi_n]$ as $n \to \infty$ will converge to $L_1^{-1}[f]$ uniformly in each subinterval $(-b, b)$ for arbitrary $b < a$. Therefore

$$L_1^{-1}[f] = \lim_{n\to\infty} L_1^{-1}[\varphi_n] = \lim_{n\to\infty} W L_2^{-1} W^{-1}[\varphi_n], \tag{3.1.3}$$

where $W = W_{\{L_2 L_1 \theta\}}$. By property III of transformation operators (Chapter 1, §2) the sequence $W^{-1}[\phi_n]$ as $n \to \infty$ converges to $W^{-1}[f]$ in the set $\mathfrak{M}^1_{(-a,a)}$. Then as $n \to \infty$ the sequence $L_2^{-1}W^{-1}[\phi_n]$ converges to $L_2^{-1}W^{-1}[f]$ uniformly in each subinterval $(-b, b)$ for arbitrary $b < a$. So by property III again we know that as $n \to \infty$ the sequence $WL_2^{-1}W^{-1}[\phi_n]$ converges to $WL_2^{-1}W^{-1}[f]$ uniformly in each interval $(-b, b)$, $b < a$. Therefore in (3.1.3) we can take the limit inside the operator $WL_2^{-1}W^{-1}$, and we get

$$L_1^{-1}[f] = WL_2^{-1}W^{-1}[f].$$

Now let $f(x)$ be a twice differentiable function whose second derivative is in $\mathfrak{M}^1_{(-a,a)}$. Then

$$f(x) = L_1^{-1}[L_1[f]] + f(0) X_1^0(x) + f'(0) X_1^1(x),$$

or

$$f(x) - f(0) X_1^0(x) - f'(0) X_1^1(x) = L_1^{-1}[L_1[f]] = WL_2^{-1}W^{-1}[L_1[f]],$$

where $W = W_{\{L_2 L_1 \theta\}}$. Therefore

$$\begin{aligned} L_2^{-1}W^{-1}[L_1[f]] &= W^{-1}[f(x) - f(0) X_1^0(x) - f'(0) X_1^1(x)] \\ &= W^{-1}[f] - f(0) X_2^0(x) - (f'(0) - \theta f(0)) X_2^1(x) \end{aligned}$$

and

$$W^{-1}[f] = f(0) X_2^0(x) + (f'(0) - \theta f(0)) X_2^1(x) + L_2^{-1}W^{-1}[L_1[f]].$$

On the right side of this equation stands a twice differentiable function whose second derivative is in $\mathfrak{M}^1_{(-a,a)}$. Consequently we can apply L_2, and we get

$$L_2W^{-1}[f] = W^{-1}[L_1[f]].$$

We apply W to both sides of this equation to obtain

$$WL_2W^{-1}[f] = L_1[f].$$

We have proved the following theorem.

THEOREM 3.1.1. *Let operators* L_1 *and* L_2 *of the form* (3.1.1) *be given on the interval* $(-a, a)$, $a \leqslant \infty$.

1) *For an arbitrary* $f(x) \in \mathfrak{M}^1_{(-a,a)}$ *we have*

$$L_1^{-1}[f] = WL_2^{-1}W^{-1}[f].$$

2) *For an arbitrary twice differentiable* $f(x)$ *whose second derivative is in* $\mathfrak{M}^1_{(-a,a)}$ *we have*

$$L_1[f] = WL_2W^{-1}[f]$$

almost everywhere in $(-a, a)$, *where* $W = W_{\{L_2L_1\theta\}}$ *is a transformation operator defined on the set* $\mathfrak{M}^1_{(-a,a)}$ (*the number* $\theta \neq \infty$ *is arbitrary*).

Along the way we have proved that transformation operators W defined on $\mathfrak{M}^1_{(-a,a)}$ transform twice differentiable functions $f(x)$ $(f''(x) \in \mathfrak{M}^1_{(-a,a)})$ into twice-differentiable functions $W[f]$ $((d^2/dx^2)W[f] \in \mathfrak{M}^1_{(-a,a)})$.

If operators L_1 and L_2 of the form (3.1.1) are given on $[0, a)$, $a \leqslant \infty$, then instead of Theorem 3.1.1 we have

THEOREM 3.1.1′. *Let operators* L_1 *and* L_2 *of the form* (3.1.1) *be given on the interval* $[0, a)$, $a \leqslant \infty$.

1) *For an arbitrary* $f(x) \in \mathfrak{M}^1_{[0,a)}$ *we have*

$$L_1^{-1}[f] = VL_2^{-1}V[f];$$

2) *For an arbitrary twice differentiable* $f(x)$ $(f''(x) \in \mathfrak{M}^1_{[0,a)})$ *satisfying the condition* $f'(0) - h_1f(0) = 0$ $(f(0) = 0$ *if* $h_1 = \infty)$ *we have*

$$L_1[f] = VL_2V^{-1}[f]$$

almost everywhere in $[0, a)$, *where* $V = V_{\{L_2L_1h_2h_1\}}$ $(h_2 = \infty$ *if* $h_1 = \infty)$ *is a transformation operator defined on the set* $\mathfrak{M}^1_{[0,a)}$.

The proof of this theorem is completely analogous to the proof of Theorem 3.1.1, and is omitted. We observe only that from the proof of Theorem 3.1.1′ it follows that the transformation operator $V = V_{\{L_1L_2h_1h_2\}}$ transforms twice differentiable functions $f(x)$ $(f''(x) \in \mathfrak{M}^1_{[0,a)})$ satisfying the condition $f'(0) - h_1f(0) = 0$ $(f(0) = 0$ if $h_1 = h_2 = \infty)$ into twice differentiable functions $V[f(x)]$ $((d^2/dx^2)V[f(x)] \in \mathfrak{M}^1_{[0,a)})$ satisfying

$$\left(\frac{d}{dx} V\left[f(x)\right] - h_2 V\left[f(x)\right]\right)\Big|_{x=0} = 0.$$

We will examine several details of the theorems just proved. Let L be a differentiable operator of the form (3.1.1) given on the interval $(-a, a)$, $a \leqslant \infty$.

If we set

$$D[u] = \frac{d}{dx} u(x), \qquad D^2[u] = \frac{d^2}{dx^2} u(x),$$

then by Theorem 3.1.1 we can write L in the form

$$L[f] = WD^2W^{-1}[f] = WDW^{-1}WDW^{-1}[f],$$

where $W = W_{\{D^2 L\theta\}}$. This formula leads to the concept of a square root of L, which we define by the formula

$$L^{\frac{1}{2}}[f] = WDW^{-1}[f] \qquad (W = W_{\{D^2 L\theta\}}) \tag{3.1.4}$$

on all twice differentiable functions $f(x)$ whose second derivative belongs to $\mathfrak{M}^1_{(-a,a)}$.

We note that (3.1.4) does not define $L^{1/2}$ uniquely; by varying the parameter θ we can obtain different operators. Thus if we speak of the square root of an operator L we must say which value of θ we choose in (3.1.4).

If the operator L can be applied to a function $f(x)$ n times, then the operators $L^{k/2}$ can be defined on this function for all $k \leqslant 2n$ according to the formula

$$L^{\frac{k}{2}}[f] = WD^kW^{-1}[f] \quad (W = W_{\{D^2 L\theta\}}),$$

which implies that $W^{-1}[f]$ is $2n$ times differentiable.

The concept of the square root of an operator L of the form (3.1.1) given on $[0, a)$, $a \leqslant \infty$, is obtained by analogy. But it will not be needed and so we will not discuss it.

To conclude this section we prove another important theorem.

THEOREM 3.1.2. *The transformation operators* $W_{\{L_1 L_2 \theta\}}$ $(V_{\{L_1 L_2 h_1 h_2\}})$ *take the set* $C_{(-a,a)}$ $(C_{[0,a)})$ *onto itself; that is, they take continuous functions to continuous functions.*

PROOF. We consider, for example, the case where L_1 and L_2 of the form (3.1.1) are given on $[0, a)$, $a \leqslant \infty$, so that $V_{\{L_1 L_2 h_1 h_2\}}$ is defined on $\mathfrak{M}^1_{[0,a)}$. Let $f(x) \in C_{[0,a)}$ and $V = V_{\{L_1 L_2 h_1 h_2\}}$. Then

$$V[f] = f(x) + \int_0^x K(x, t) f(t)\, dt,$$

where $K(x, t)$ is bounded for $0 \leqslant t \leqslant x \leqslant b < a$ for arbitrary $b < a$.

The theorem will be proved if we show that for arbitrary $\epsilon > 0$ and $b < a$ there exists a continuous function $\phi_{\epsilon, b}(x)$ such that

$$\sup_{0 \leqslant x \leqslant b} |V[f(x)] - \varphi_{\varepsilon, b}(x)| < \varepsilon.$$

We have

$$V[f] = V[f_1] + V[f - f_1]$$

$$= V[f_1] + f(x) - f_1(x) + \int_0^x K(x, t)[f(t) - f_1(t)]\, dt,$$

which implies that

$$\sup_{0 \leqslant x \leqslant b} |V[f] - V[f_1] - f(x) + f_1(x)| \leqslant k(b) \int_0^b |f(t) - f_1(t)|\, dt,$$

where $k(b) = \sup_{0 \leqslant t \leqslant x \leqslant b} |K(x, t)|$.

If we let $f_1(x)$ be a twice differentiable function which satisfies the condition

$$f_1(0) = f_1'(0) = 0, \qquad \int_0^b |f(t) - f_1(t)|\, dt < \frac{\varepsilon}{k(b) + 1}$$

(such a function always exists), we obtain

$$\sup_{0 \leqslant x \leqslant b} |V[f] - (V[f_1] + f(x) - f_1(x))| < \varepsilon.$$

By the remark following Theorem 3.1.1′ the function $V[f_1]$ is twice differentiable and continuous. Therefore the function

$$\varphi_{\varepsilon, b}(x) = V[f_1] + f(x) - f_1(x)$$

is also continuous and

$$\sup_{0 \leqslant x \leqslant b} |V[f] - \varphi_{\varepsilon, b}(x)| < \varepsilon,$$

which is what we wanted to show.

The proof for the case where L_1 and L_2 are given on $(-a, a)$, $a \leqslant \infty$, is completely analogous.

REMARK. The preceding theorem is a corollary of the fact that the kernels $\mathfrak{A}(x, y)$ and $\mathfrak{B}(x, y)$ from Lemma 1.1.1 are continuous. We will not prove that these kernels are continuous, since we will not need that fact.

§2. Generalized shift operators

Up to this point we have assumed that our differential operators and transformation operators act on functions of a single variable. In this section we will consider functions of several variables. We first introduce some notation.

Let H be an operator defined on functions of a single variable, and let $u(x_1, \cdots, x_n)$ be a function of n independent variables. By the notation $\underset{x_i}{H}[u]$ we mean the result of applying H to the function $u(x_1, \cdots, x_n)$ as if it were a function of the single variable x_i with all of the other variables fixed.

For example,

$$\underset{x_i}{D}[u] = \frac{\partial}{\partial x_i} u(x_1, x_2, \ldots, x_n); \qquad \underset{x_i}{D^2}(u) = \frac{\partial^2}{\partial x_i^2} u(x_1, x_2, \ldots, x_n);$$

$$\underset{x_i}{L}[u] = \frac{\partial^2}{\partial x_i^2} u(x_1, x_2, \ldots, x_n) - q(x_i)\, u(x_1, x_2, \ldots, x_n);$$

$$\underset{x_i}{W}[u] = u(x_1, x_2, \ldots, x_n) + \int_{-|x_i|}^{|x_i|} K(x_i, y)\, u(x_1, x_2, \ldots, x_{i-1}, y, x_{i+1}, \ldots, x_n)\, dy$$

and so forth.

We remark that all differentiable operators L of the form (3.1.1) in this section are assumed to be given on the whole real line.

LEMMA 3.2.1. *Let L_i, $i = 1, \cdots, n$, be differential operators of the form*

$$\underset{t}{L_i}[v] = \frac{d^2}{dt^2} v(t) - q_i(t) v(t).$$

Consider the following linear partial differential equations:

$$\sum_{i=1}^{n} \alpha_i \left(\frac{\partial^2}{\partial x_i^2} v - q_i(x_i)\, v \right) \equiv \sum_{i=1}^{n} \alpha_i \underset{x_i}{L_i}[v] = f(x_1, x_2, \ldots, x_n);$$

$$\sum_{i=1}^{n} \alpha_i \frac{\partial^2}{\partial x_i^2} u \equiv \sum_{i=1}^{n} \alpha_i \underset{x_i}{D^2}[u] = \varphi(x_1, x_2, \ldots, x_n),$$

where $\alpha_1, \alpha_2, \cdots, \alpha_n$ are arbitrary constants,

$$\varphi(x_1, x_2, \ldots, x_n) = \underset{x_1}{W_1^{-1}} \underset{x_2}{W_2^{-1}} \cdots \underset{x_n}{W_n^{-1}} [f(x_1, x_2, \ldots, x_n)]$$

and

$$\underset{x_i}{W_i} = \underset{x_i}{W}_{\{D^2 L_i\, \theta_i\}}$$

(the numbers $\theta_1, \cdots, \theta_n$ are arbitrary).

When the function $v(x_1, x_2, \cdots, x_n)$ runs through the set of all solutions *of equation* 1), *the function*

$$u(x_1, x_2, \dots, x_n) = \underset{x_1}{W_1^{-1}} \underset{x_2}{W_2^{-1}} \cdots \underset{x_n}{W_n^{-1}} [v(x_1, x_2, \dots, x_n)]$$

runs through the set of all solutions of equation 2), *and vice versa.*

PROOF. According to Theorem 3.1.1 we can write equation 1) in the form

$$\sum_{i=1}^{n} \alpha_i \underset{x_i}{W_i} \underset{x_i}{D^2} \underset{x_i}{W_1^{-1}} [v] = f(x_1, x_2, \dots, x_n).$$

If we apply to both sides of this equation the operators $\underset{x_1}{W_1^{-1}}, \dots, \underset{x_n}{W_n^{-1}}$ and take into account the fact that $\underset{x_i}{D^2}$ and $\underset{x_j}{W_j^{-1}}$ commute when $i \neq j$, we find that

$$\sum_{i=1}^{n} \alpha_i \underset{x_i}{D^2} [\underset{x_1}{W_1^{-1}} \underset{x_2}{W_2^{-1}} \cdots \underset{x_n}{W_n^{-1}} [v]] = \underset{x_1}{W_1^{-1}} \underset{x_2}{W_2^{-1}} \cdots \underset{x_n}{W_n^{-1}} [f(x_1, x_2, \dots, x_n)],$$

or

$$\sum_{i=1}^{n} \alpha_i \underset{x_i}{D^2} [u] = \varphi(x_1, x_2, \dots, x_n),$$

which implies our assertion.

We will only be interested in very simple partial differential equations of the form

$$\underset{x}{L_1} [v] = \underset{y}{L_2} [v],$$

which lead to the generalized shift operators of Delsarte [8].

An ordinary shift operator U_x^y is defined on the set of all functions $f(t)$ of a real variable by the formula

$$U_x^y [f] = f(x - y).$$

If $f(t)$ is twice continuously differentiable, then $u(x, y) = U_x^y [f]$ is a solution of the partial differential equation

$$\frac{\partial^2}{\partial x^2} u(x, y) = \frac{\partial^2}{\partial y^2} u(x, y) \tag{3.2.1}$$

with initial conditions

$$u(x, y)|_{y=0} = f(x), \quad \frac{\partial}{\partial y} u(x, y)|_{y=0} = -f'(x). \tag{3.2.1'}$$

Thus the shift operator U_x^y can be thought of as a natural generalization of the solution of (3.2.1) with initial conditions (3.2.1′) to nondifferentiable functions.

This point of view leads to the concept of the generalized shift operator

R_x^y associated with the partial differential equation

$$U_x^y[f] = f(x-y). \tag{3.2.2}$$

and defined at first on twice continuously differentiable functions $f(x)$ as the solution $v(x, y)$ of equation (3.2.2) with initial conditions

$$v(x, y)|_{y=0} = f(x), \quad \underset{y}{L_2^{\frac{1}{2}}}[v(x, y)]|_{y=0} = -\underset{x}{L_1^{\frac{1}{2}}}[f(x)]. \tag{3.2.2$'$}$$

Here L_1 and L_2 are differential operators of the form (3.1.1):

$$L_1^{\frac{1}{2}} = W_{\{D^2L_1\theta_1\}} DW^{-1}_{\{D^2L_1\theta_1\}}, \quad L_2^{\frac{1}{2}} = W_{\{D^2L_2\theta_2\}} DW^{-1}_{\{D^2L_2\theta_2\}}$$

and the numbers θ_1 and θ_2 are arbitrary but fixed.

According to Lemma 3.2.1 the function

$$u(x, y) = \underset{x}{W_1^{-1}}\underset{y}{W_2^{-1}}[v(x, y)] = \underset{x}{W_1^{-1}}\underset{y}{W_2^{-1}}R_x^y[f];$$

(where $W_1 = W_{\{D^2L_1\theta_1\}}$ and $W_2 = W_{\{D^2L_2\theta_2\}}$) is a solution of equation (3.2.1). Furthermore,

$$u(x, y)|_{y=0} = \lim_{y\to 0} \underset{x}{W_1^{-1}}\underset{y}{W_2^{-1}}[v(x, y)] = \underset{x}{W_1^{-1}}[v(x, 0)] = \underset{x}{W_1^{-1}}[f(x)]$$

and

$$\begin{aligned}\frac{\partial}{\partial y}u(x, y)|_{y=0} &= \lim_{y\to 0} \underset{x}{W_1^{-1}}\frac{\partial}{\partial y}\underset{y}{W_2^{-1}}[v(x, y)]\\ &= \lim_{y\to 0} \underset{x}{W_1^{-1}}\underset{y}{W_2^{-1}}\underset{y}{W_2}\underset{y}{D}\underset{y}{W_2^{-1}}[v(x, y)] = \lim_{y\to 0}\underset{x}{W_1^{-1}}\underset{y}{L_2^{\frac{1}{2}}}[v(x, y)],\end{aligned}$$

and by (3.2.2$'$) we have

$$\begin{aligned}\frac{\partial}{\partial y}u(x, y)|_{y=0} &= -\underset{x}{W_1^{-1}}\underset{x}{L_1^{\frac{1}{2}}}[f(x)] = -\underset{x}{W_1^{-1}}\underset{x}{L_1^{\frac{1}{2}}}\underset{x}{W_1}[\underset{x}{W_1^{-1}}[f(x)]]\\ &= -\underset{x}{D}\underset{x}{W_1^{-1}}[f(x)] = -(\underset{x}{W_1^{-1}}[f])'.\end{aligned}$$

Therefore

$$u(x, y) = U_x^y[W_1^{-1}[f]];$$

that is,

$$U_x^y[W_1^{-1}[f]] = \underset{x}{W_1^{-1}}\underset{y}{W_2^{-1}}R_x^y[f],$$

or

$$R_x^y[f] = \underset{x}{W_1}\underset{y}{W_2}U_x^y[W_1^{-1}[f]].$$

The right side of this equation makes sense for all $f(t) \in \mathfrak{M}^1_{(-\infty,\infty)}$, and therefore we can look at the formula as an extension of the operator R_x^y to all functions $f(t)$ which are integrable on each finite interval of the real line.

These considerations lead to the following definition.

DEFINITION 3.2.1. By the *generalized shift operator of the first kind* associated with the differential operators L_1 and L_2 of the form (3.1.1) we mean the operator R_x^y which takes each function $f(t) \in \mathfrak{M}^1_{(-\infty,\infty)}$ to the function $R_x^y[f]$ of two variables defined by the formula

$$R_x^y[f] = \underset{x}{W_1}\underset{y}{W_2}U_x^y[W_1^{-1}[f]],$$

where

$$W_1 = W_{\{D^2L_1\theta_1\}}, \quad W_2 = W_{\{D^2L_2\theta_2\}}, \quad U_x^y[\varphi(t)] = \varphi(x-y).$$

If $f(t)$ is twice differentiable, then $R_x^y[f]$ can be defined as the solution of the partial differential equation

$$\underset{x}{L_1}[v] = \underset{y}{L_2}[v]$$

with initial conditions

$$v(x, y)\big|_{y=0} = f(x), \quad \underset{y}{L_2^{\frac{1}{2}}}[v(x, y)]\big|_{y=0} = -\underset{x}{L_1^{\frac{1}{2}}}[f(x)],$$

where

$$L_1^{\frac{1}{2}} = W_1 D W_1^{-1}, \quad L_2^{\frac{1}{2}} = W_2 D W_2^{-1}.$$

We note that in order to determine the generalized shift operator R_x^y we have to know not only the operators L_1 and L_2 but also which of the square roots we are taking; that is, we have to fix the numbers θ_1 and θ_2.

We arrived at the concept of a generalized shift operator of the first kind by considering equation (3.2.2), which is a natural generalization of equation (3.2.1). In order to preserve the analogy with the ordinary shift operator U_x^y we had to introduce into the initial conditions (3.2.2′) the square root of L, a concept which is justified only by the desire to preserve the analogy. This makes the assignment of initial conditions for (3.2.2) artificial, and therefore generalized shift operators of the first kind appear artificial.

We can avoid the artificiality of the generalization if we start with the symmetric shift operator S_x^y defined by the formula

$$S_x^y[f(t)] = \frac{1}{2}[f(x+y) + f(x-y)]$$

on the set of all $f(t) \in \mathfrak{M}^1_{(-\infty,\infty)}$. If $f(t)$ is twice continuously differentiable, then $u(x, y) = S_x^y[f]$ is the solution of the partial differential equation

$$\frac{\partial^2}{\partial x^2}u(x, y) = \frac{\partial^2}{\partial y^2}u(x, y)$$

with initial conditions

$$u(x, y)\big|_{y=0} = f(x), \quad \frac{\partial}{\partial y} u(x, y)\big|_{y=0} = 0.$$

We define the generalized shift operator of the second kind T_x^y on all functions $f(t) \in \mathfrak{M}^1_{(-\infty,\infty)}$ by the formula

$$T_x^y[f(t)] = \underset{x}{W_1}\underset{y}{W_2} S_x^y [W_1^{-1}[f]],$$

where

$$W_1 = W_{\{D^2 L_1 \theta_1\}}, \quad W_2 = W_{\{D^2 L_2 \theta_2\}}.$$

Let $f(t)$ be twice continuously differentiable. Then by Lemma 3.2.1 the function

$$v(x, y) = T_x^y[f] = \underset{x}{W_1}\underset{y}{W_2} S_x^y [W_1^{-1}[f]]$$

is a solution of the partial differential equation

$$\underset{x}{L_1}[v] = \underset{y}{L_2}[v].$$

Here

$$v(x, y)\big|_{y=0} = \lim_{y \to 0} \underset{x}{W_1}\underset{y}{W_2} S_x^y [W_1^{-1}[f]] = \underset{x}{W_1}\underset{x}{W_1^{-1}}[f] = f(x).$$

Furthermore,

$$\frac{v(x, y) - v(x, 0)}{y} = \frac{v(x, y) - f(x)}{y} = \underset{x}{W_1}\left[\frac{\underset{y}{W_2} S_x^y [W_1^{-1}[f]] - \underset{x}{W_1^{-1}}[f]}{y}\right],$$

or, setting $\phi(x) = \underset{x}{W_1^{-1}}[f(x)]$,

$$\frac{v(x, y) - v(x, 0)}{y} = \underset{x}{W_1}\left[\frac{\underset{y}{W_2}[\varphi(x + y) + \varphi(x - y)] - 2\varphi(x)}{2y}\right]. \tag{3.2.3}$$

The expression to the right of $\underset{x}{W_1}$ can be written in the form

$$\frac{\underset{y}{W_2}[\varphi(x + y) + \varphi(x - y) - 2\varphi(x)] + 2\varphi(x)\underset{y}{W_2}[1] - 2\varphi(x)}{2y}$$

$$= \frac{1}{2y}\underset{y}{W_2}[\varphi(x + y) + \varphi(x - y) - 2\varphi(x)] + \frac{1}{y}\varphi(x)(\underset{y}{W_2}[1] - 1).$$

From the definition of the transformation operator $W_2 = W_{\{D^2 L_2 \theta_2\}}$ we have

$$\underset{y}{W_2}[1] = \omega_{\theta_2}^{(2)}(0, y),$$

where $\omega_{\theta_2}^{(2)}(0, y)$ is the solution of $\underset{y}{L_2}[u] = 0$ with initial conditions $u(0) = 1$ and $u'(0) = \theta_2$. Therefore

$$\lim_{y \to 0} \varphi(x) \frac{\underset{y}{W_2}[1] - 1}{y} = \varphi(x) \lim_{y \to 0} \frac{\omega_{\theta_2}^{(2)}(0, y) - \omega_{\theta_2}^{(2)}(0, 0)}{y} = \theta_2 \varphi(x). \tag{3.2.4}$$

Also

$$\lim_{y\to 0}\frac{1}{2y}\underset{y}{W_2}[\varphi(x+y)+\varphi(x-y)-2\varphi(x)]$$

$$=\lim_{y\to 0}\left\{\frac{\varphi(x+y)+\varphi(x-y)-2\varphi(x)}{2y}\right.$$

$$\left.+\frac{1}{2y}\int_{-|y|}^{|y|}K(y,t)[\varphi(x+t)+\varphi(x-t)-2\varphi(x)]\,dt\right\},$$

where $K(y, t)$ is the kernel corresponding to $\underset{y}{W_2}$. Since $K(y, t)$ is bounded in each finite portion of the plane and $\phi(x)$ is differentiable, we have

$$\lim_{y\to 0}\frac{1}{2y}\int_{-|y|}^{|y|}K(y,t)\,[\varphi(x+t)+\varphi(x-t)-2\varphi(x)]\,dt=0,$$

$$\lim_{y\to 0}\frac{\varphi(x+y)+\varphi(x-y)-2\varphi(x)}{2y}=\frac{1}{2}(\varphi'(x)-\varphi'(x))=0.$$

Therefore

$$\lim_{y\to 0}\frac{1}{2y}\underset{y}{W_2}[\varphi(x+y)+\varphi(x-y)-2\varphi(x)]=0. \tag{3.2.5}$$

Equations (3.2.4) and (3.2.5) show that the expression to the right of $\underset{x}{W_1}$ in (3.2.3) converges to $\theta_2\phi(x)$ as $y\to 0$. Therefore

$$\frac{\partial}{\partial y}v(x,y)\,|_{y=0}=\theta_2\underset{x}{W_1}[\varphi(x)]=\theta_2 f(x).$$

We see that on a twice differentiable function $f(t)$ the generalized shift $T_x^y[f]$ is the solution of (3.2.2) with initial conditions $v(x, 0) = f(x)$, $v_y'(x, 0) = \theta_2 f(x)$.

DEFINITION 3.2.2. By the *generalized shift operator of the second kind* associated with the differential operators L_1 and L_2 of the form (3.1.1) we mean the operator T_x^y which takes each function $f(t)\in\mathfrak{M}^1_{(-\infty,\infty)}$ to the function $T_x^y[f]$ of two variables defined by the formula

$$T_x^y[f]=\underset{x}{W_1}\underset{y}{W_2}S_x^y[W_1^{-1}[f]],$$

where

$$W_1=W_{\{D^2L_1\theta_1\}},\quad W_2=W_{\{D^2L_2\theta_2\}},$$

$$S_x^y[\varphi(t)]=\frac{1}{2}[\varphi(x+y)+\varphi(x-y)].$$

If $f(t)$ is twice differentiable, the operator $T_x^y[f]$ can be defined as the solution of the partial differential equation

$$\underset{x}{L_1}[v]=\underset{y}{L_2}[v]$$

with initial conditions

$$v(x, y)\big|_{y=0} = f(x), \qquad \frac{\partial}{\partial y} v(x, y)\big|_{y=0} = \theta_2 f(x).$$

Generalized shift operators allow us to generalize classical harmonic analysis in several directions (see [6], [9] and [11]). The most interesting generalizations are those of almost periodic functions and positive definite functions. In order to illustrate the features of these generalizations and to show the roles played by transformation operators, we will conclude this section by looking at generalized positive definite functions associated with generalized shift operators of the first kind.

DEFINITION 3.2.3. Let R_x^y be a generalized shift operator of the first kind of the form

$$R_x^y[f] = \underset{x}{W}\underset{y}{W} \dot{U}_x^y [W^{-1}[f]],$$

where $W = W_{\{D^2 L\theta\}}$. A continuous function $f(t)$ is said to be *positive definite* relative to R_x^y if for arbitrary points $x_1, \cdots, x_n$ and arbitrary complex numbers $\xi_1, \cdots, \xi_n$ we have

$$\sum_{\mu=1}^{n} \sum_{\nu=1}^{n} R_{x_\mu}^{x_\nu}[f]\, \xi_\mu \bar{\xi}_\nu \geqslant 0.$$

For $L \equiv D^2$ and $\theta = 0$ we obtain ordinary positive definite functions, which by a well-known theorem of Bochner (see [12], Russian pp. 213–218) are characterized by the fact that they can be written in the form

$$f(x) = \int_{-\infty}^{\infty} e^{isx}\, d\sigma(s),$$

where $\sigma(s)$ is a nondecreasing bounded function on $-\infty < s < \infty$.

We can establish a similar characterization of functions which are positive definite relative to R_x^y.

THEOREM 3.2.1. *A function $f(x)$ is positive definite in the sense of Definition 3.2.3 if and only if it is of the form*

$$f(x) = \int_{-\infty}^{\infty} \varphi(s, x)\, d\sigma(s),$$

where $\sigma(s)$ is a nondecreasing bounded function on $-\infty < s < \infty$, and $\phi(s, x)$ is the solution of

$$L[u] + s^2 u = 0$$

with initial conditions

$$\phi(s, 0) = 1; \quad \phi_x'(s, 0) = is + \theta.$$

PROOF. Let

$$f(x)=\int_{-\infty}^{\infty}\varphi(s,x)\,d\sigma(s).$$

Since this integral converges uniformly when x varies over a finite interval, $f(x)$ is continuous. Furthermore,

$$R_x^y[f]=R_x^y\left[\int_{-\infty}^{\infty}\varphi(s,x)\,d\sigma(s)\right]=\int_{-\infty}^{\infty}R_x^y[\varphi(s,t)]\,d\sigma(s)$$

(the validity of moving R_x^y inside the integral can be verified by using the weak boundedness of the transformation operators in terms of which R_x^y is defined).

From the definition of $W=W_{\{D^2L\theta\}}$ it follows that

$$W[e^{isx}]=\varphi(s,x),\quad W^{-1}[\varphi(s,x)]=e^{isx}. \tag{3.2.6}$$

Therefore

$$\begin{aligned}R_x^y[\varphi(s,t)]&=\underset{x}{W}\underset{y}{W}U_x^y[W^{-1}[\varphi(s,t)]]=\underset{x}{W}\underset{y}{W}[e^{is(x-y)}]\\&=\underset{x}{W}[e^{isx}]\cdot\underset{y}{W}[e^{-isy}]=\varphi(s,x)\,\varphi(-s,y).\end{aligned}$$

Since $q(t)$ for the operator L is real, $\overline{\phi(s,x)}$ is the solution of the equation $L[u]+s^2u=0$ for which $\overline{\phi(s,0)}=1$ and $\overline{\phi_x'(s,0)}=-is+\theta$; that is, $\phi(-s,x)=\overline{\phi(s,x)}$, and therefore

$$R_x^y[\varphi(s,t)]=\varphi(s,x)\,\overline{\varphi(s,y)}.$$

Consequently

$$R_x^y[f(t)]=\int_{-\infty}^{\infty}\varphi(s,x)\,\overline{\varphi(s,y)}\,d\sigma(s)$$

and

$$\begin{aligned}\sum_{\nu=1}^{n}\sum_{\mu=1}^{n}R_{x_\mu}^{x_\nu}[f]\,\xi_\mu\bar{\xi}_\nu&=\int_{-\infty}^{\infty}\sum_{\nu=1}^{n}\sum_{\mu=1}^{n}\varphi(s,x_\mu)\,\overline{\varphi(s,x_\nu)}\,\xi_\mu\bar{\xi}_\nu\,d\sigma(s)\\&=\int_{-\infty}^{\infty}\left|\sum_{\mu=1}^{n}\varphi(s,x_\mu)\,\xi_\mu\right|^2d\sigma(s)\geqslant 0,\end{aligned}$$

which proves sufficiency.

Now let $f(x)$ be positive definite in the sense of Definition 3.2.3. Then the function

$$g(x)=W^{-1}[f]$$

is an ordinary positive definite function. In fact, continuity of $g(x)$ follows from the continuity of $f(x)$ by virtue of Theorem 3.1.3. Furthermore,

$$g(x-y)=U_x^y[g]=U_x^y[W^{-1}[f]]=\underset{x}{W^{-1}}\underset{y}{W^{-1}}[\underset{x}{W}\underset{y}{W}U_x^y[W^{-1}[f]]]$$
$$=\underset{x}{W^{-1}}\underset{y}{W^{-1}}R_x^y[f(t)],$$

which implies that

$$\sum_{\mu=1}^{n}\sum_{\nu=1}^{n} g(x_\mu-x_\nu)\xi_\mu\bar{\xi}_\nu=\sum_{x=x_\mu}\sum_{y=x_\nu}\underset{x}{W^{-1}}\underset{y}{W^{-1}}R_x^y[f(t)]\xi_\mu\bar{\xi}_\nu.$$

We have

$$\underset{x}{W^{-1}}[F(x)]=F(x)+\int_{-\infty}^{\infty}K(x,\ t)F(t)\,dt,$$

where the kernel $K(x, t)$ is real, bounded in each finite portion of the (x, t)-plane, and satisfies $K(x, t) = 0$ for $|t| > |x|$. Therefore, setting

$$k(t)=\sum_{\nu=1}^{n}K(x_\nu,\ t)\xi_\nu,$$

we obtain

$$\sum_{\mu=1}^{n}\sum_{\nu=1}^{n} g(x_\mu-x_\nu)\xi_\mu\bar{\xi}_\nu=\sum_{x=x_\mu}\sum_{y=x_\nu}\underset{x}{W^{-1}}\underset{y}{W^{-1}}R_x^y[f(t)]\xi_\mu\bar{\xi}_\nu$$
$$=\sum_{\mu=1}^{n}\sum_{\nu=1}^{n}R_{x_\mu}^{x_\nu}[f]\xi_\mu\bar{\xi}_\nu+\int_{-\infty}^{\infty}\int_{-\infty}^{\infty}k(x)\overline{k(y)}R_x^y[f(t)]\,dx\,dy.$$

(Here the integrals are actually taken over finite intervals.)

The function $f(x)$ is positive definite relative to R_x^y. Therefore the first term on the right side of the above equation is nonnegative. The second term is the continuous analogue of the form

$$\sum_{\nu=1}^{n}\sum_{\mu=1}^{n}R_{x_\mu}^{x_\nu}[f]\xi_\nu\bar{\xi}_\mu$$

and can be represented as the limit of such forms; therefore it is also nonnegative.

Thus $g(x)$ is continuous and

$$\sum_{\mu=1}^{n}\sum_{\nu=1}^{n} g(x_\mu-x_\nu)\xi_\mu\bar{\xi}_\nu\geqslant 0.$$

Hence $g(x)$ is an ordinary positive definite function, and by Bochner's theorem

$$g(x)=\int_{-\infty}^{\infty}e^{isx}\,d\sigma(s),$$

where $\sigma(s)$ is nondecreasing and bounded on the whole real line. Since W is weakly bounded, we now find that

$$f(x) = \underset{x}{W}[g(x)] = \int_{-\infty}^{\infty} \underset{x}{W}[e^{isx}]\, d\sigma(s);$$

that is, by (3.2.6)

$$f(x) = \int_{-\infty}^{\infty} \varphi(s, x)\, d\sigma(s),$$

which is what we wanted to show.

The proof of this theorem does not present any serious difficulties because the question of the form of a function which is positive definite in the sense of Definition 3.2.3 can be reduced by means of transformation operators to the already solved (by Bochner's theorem) question of the form of a function which is positive definite relative to the ordinary shift operator U_x^y.

The situation is different for functions which are positive definite relative to a generalized shift operator T_x^y of the second kind. A generalization of this type was first investigated by Povzner [11].

In this case we also have a theorem analogous to Bochner's theorem. For bounded functions the theorem was proved by Povzner, and in the general case by Kreĭn [17]. Kreĭn's proof was based on very general theory and did not involve transformation operators.

A proof of the Povzner-Kreĭn theorem based on the theory of transformation operators was first given by Levitan [5].

§3. The Taylor-Delsarte formula

Let $\Phi(t)$ be a function with $2n$ continuous derivatives. Then

$$S_x^y[\Phi(t)] = \frac{1}{2}[\Phi(x+y) + \Phi(x-y)]$$

is even and $2n$ times differentiable with respect to y. By Taylor's formula we have

$$S_x^y[\Phi(t)] = \sum_{k=0}^{n-1} \frac{y^{2k}}{2k!} \underset{x}{D^{2k}}[\Phi(x)] + A_{2n}(x, y),$$

where the Lagrange form of the remainder equals

$$A_{2n}(x, y) = \frac{y^{2n}}{2n!} S_x^z[\underset{t}{D^{2n}}[\Phi(t)]] \qquad (0 < z < y).$$

We can interpret Taylor's formula as an expansion of the symmetric shift operator S_x^y in powers of the operator $D^2 = d^2/dx^2$.

In this section we will examine a generalization of Taylor's formula which was presented by Delsarte [8] and which provides an expansion of a generalized shift operator of the second kind in powers of a differential operator L of the form (3.1.1). We naturally assume that $q(x)$ in the operator $\underset{t}{L}$ is continuous.

Let L_1 and L_2 be differential operators of the form (3.1.1), and let T_x^y be the associated generalized shift operator:

$$T_x^y[F(t)] = \underset{x}{W_1}\underset{y}{W_2} S_x^y [W_1^{-1}[F]],$$

where $W_1 = W_{\{D^2 L_1 \theta_1\}}$ and $W_2 = W_{\{D^2 L_2 \theta_2\}}$. Let $F(t)$ be a function to which L_1 can be applied n times. Since the function $q_1(t)$ in L_1 is continuous, we have

$$\underset{t}{L_1^k}[F(t)] = \underset{t}{W_1}\underset{t}{D^{2k}}\underset{t}{W_1^{-1}}[F(t)] \tag{3.3.1}$$

(for almost all t this follows from Theorem 3.1.1) for all t if $k \leqslant n$. Consequently the function

$$\Phi(t) = \underset{t}{W_1^{-1}}[F(t)] \tag{3.3.2}$$

has $2n$ continuous derivatives. But then the function

$$\underset{x}{W_1} S_x^y[\Phi(t)] = \frac{1}{2}\underset{x}{W_1}[\Phi(x+y) + \Phi(x-y)]$$

$$= \frac{1}{2}[\Phi(x+y) + \Phi(x-y)] + \frac{1}{2}\int_{-|x|}^{|x|} K_1(x, t)[\Phi(t+y) + \Phi(t-y)]\,dt$$

for any fixed x has $2n$ continuous derivatives in y and

$$\underset{y}{D^{2k}}\underset{x}{W_1} S_x^y[\Phi(t)] = \frac{1}{2}[\Phi^{(2k)}(x+y) + \Phi^{(2k)}(x-y)]$$

$$+\frac{1}{2}\int_{-|x|}^{|x|} K_1(x, t)[\Phi^{(2k)}(t+y) + \Phi^{(2k)}(t-y)]\,dt. \tag{3.3.3}$$

If we let x be arbitrary but fixed and apply the ordinary Taylor formula, we obtain

$$\underset{x}{W_1} S_x^y[\Phi(t)] = \sum_{k=0}^{n-1}\frac{y^{2k}}{2k!}\underset{y}{D^{2k}}[\underset{x}{W_1} S_x^y[\Phi(t)]]|_{y=0} + A_{2n}(x, y),$$

where

$$A_{2n}(x, y) = \frac{y^{2n}}{2n!}\underset{z}{D^{2n}}\underset{x}{W_1} S_x^z[\Phi(t)]|_{z=\theta(y)}$$

$$(0 < \theta(y) < y, \quad \text{if} \quad y > 0, \text{ and } y < \theta(y) < 0, \quad \text{if} \quad y < 0).$$

From (3.3.3) with $y = 0$ we have

$$\underset{y}{D^{2k}}\underset{x}{W_1}S_x^y[\Phi(t)]|_{y=0}=\Phi^{(2k)}(x)+\int_{-|x|}^{|x|}K_1(x,t)\Phi^{(2k)}(t)\,dt=\underset{x}{W_1}[\underset{x}{D^{2k}}[\Phi(x)]];$$

that is, by (3.3.1) and (3.3.2)

$$\underset{y}{D^{2k}}\underset{x}{W_1}S_x^y[\Phi(t)]|_{y=0}=\underset{x}{W_1}\underset{x}{D^{2k}}\underset{x}{W_1^{-1}}[F(x)]=\underset{x}{L_1^k}[F(x)].$$

Hence the Taylor formula for $\underset{x}{W_1}S_x^y[\Phi(t)]$ can be written in the form

$$\underset{x}{W_1}S_x^y[\Phi(t)]=\sum_{k=0}^{n-1}\frac{y^{2k}}{2k!}\underset{x}{L_1^k}[F(x)]+A_{2n}(x,y). \tag{3.3.4}$$

Similarly

$$\underset{z}{D^{2n}}\underset{x}{W_1}S_x^z[\Phi(t)]=\underset{x}{W_1}S_x^z[\Phi^{(2n)}(t)]=\underset{z}{W_2^{-1}}\underset{x}{W_1}\underset{z}{W_2}S_x^z[\underset{t}{W_1}\underset{t}{L_1^n}[F(t)]],$$

and from the definition of T_x^y we have

$$\underset{z}{D^{2n}}\underset{x}{W_1}S_x^z[\Phi(t)]=\underset{z}{W_2^{-1}}T_x^z[\underset{t}{L_1^n}[F(t)]].$$

This equation enables us to write the remainder term $A_{2n}(x,y)$ in the form

$$A_{2n}(x,y)=\frac{y^{2n}}{2n!}\underset{z}{W_2^{-1}}T_x^z[\underset{t}{L_1^n}[F(t)]]|_{z=\theta(y)}.$$

Since

$$T_x^y[F(t)]=\underset{y}{W_2}\underset{x}{W_1}S_x^y[\Phi(t)],$$

when we apply $\underset{y}{W_2}$ to both sides of (3.3.4) we obtain

$$T_x^y[F(t)]=\sum_{k=0}^{n-1}\underset{y}{W_2}\left[\frac{y^{2k}}{2k!}\right]\underset{x}{L_1^k}[F(x)]+B_{2n}(x,y), \tag{3.3.5}$$

where

$$B_{2n}(x,y)=\underset{y}{W_2}A_{2n}(x,y)=\underset{y}{W_2}\left\{\frac{y^{2n}}{2n!}\underset{z}{W_2^{-1}}T_x^z[\underset{t}{L_1^n}[F(t)]]\right\},$$

and $z=\theta(y)$ and $0<|\theta(y)|<|y|$.

The function y^{2k} is the solution of $\underset{y}{D^2}[u]=2\kappa(2\kappa-1)y^{2(k-1)}$ with the initial conditions which in the notation of §1 of this chapter can be written

$$y^{2k}=2k(2k-1)(\underset{y}{D^2})^{-1}[y^{2(k-1)}].$$

If we use this equation a sufficient number of times, we obtain

$$y^{2k}=2k!\,(\underset{y}{D^2})^{-1}[1],$$

which implies that

$$\underset{y}{W_2}\left[\frac{y^{2k}}{2k!}\right] = \underset{y}{W_2}(\underset{y}{D^2})^{-k}\underset{y}{W_2^{-1}}[\underset{y}{W_2}[1]];$$

that is (by the first assertion in Theorem 3.1.1)

$$\underset{y}{W_2}\left[\frac{y^{2k}}{2k!}\right] = \underset{y}{L_2^{-k}}\underset{y}{W_2}[1].$$

From the definition of W_2 we have

$$\underset{y}{W_2}[1] = \omega_{\theta_2}^{(2)}(0, y),$$

where $\omega_{\theta_2}^{(2)}(0, y)$ is the solution of $\underset{y}{L_2}[u] = 0$ with initial conditions $u(0)=1$ and $u'(0) = \theta_2$. Therefore

$$\underset{y}{W_2}\left[\frac{y^{2k}}{2k!}\right] = \underset{y}{L_2^{-k}}[\omega_{\theta_2}^{(2)}(0, y)],$$

and formula (3.3.5) takes the form

$$T_x^y[F(t)] = \sum_{k=0}^{n-1}\underset{y}{L_2^{-k}}[\omega_{\theta_2}^{(2)}(0, y)]\,\underset{x}{L_1^k}[F(x)] + B_{2n}(x, y), \tag{3.3.6}$$

where

$$B_{2n}(x, y) = \underset{y}{W_2}\left[\frac{y^{2n}}{2n!}\underset{z}{W_2^{-1}}T_x^z[\underset{t}{L_1^n}[F(t)]]\right], \tag{3.3.6'}$$

and $z = \theta(y)$ and $0 < |\theta(y)| < |y|$.

This is the Taylor-Delsarte formula for an arbitrary function $F(t)$ to which the operator $\underset{t}{L_1}$ can be applied n times.

The following lemma is useful for studying the remainder term $B_{2n}(x, y)$.

LEMMA 3.3.1. *A generalized shift operator T_x^y of the second kind can be written in the form*

$$T_x^y[f(t)] = \frac{1}{2}[f(x+y) + f(x-y)] + \int_{x-y}^{x+y} T(x, y, t)\,f(t)\,dt,$$

where the kernel $T(x, y, t)$ is bounded in each finite portion of (x, y, t)-space.

PROOF. Let

$$T_x^y[f(t)] = \underset{x}{W_1}\underset{y}{W_2}S_x^y[\underset{t}{W_1^{-1}}[f(t)]], \tag{3.3.7}$$

where $W_1 = W_{\{D^2 L_1\theta_1\}}$, $W_2 = W_{\{D^2 L_2\theta_2\}}$ and

$$\underset{t}{L_1}[u] = \frac{d^2}{dt^2}u(t) - q_1(t)u(t), \qquad \underset{t}{L_2}[u] = \frac{d^2}{dt^2}u(t) - q_2(t)u(t).$$

We introduce the differential operator L_a which depends on the parameter a as follows:

$$\underset{t}{L_a}[u] = \frac{d^2}{dt^2} u(t) - q_1(a+t)\, u(t).$$

It is clear that the transformation operator $W_a = W_{\{L_a D^2 0\}}$ also depends on a and is of the form

$$\underset{t}{W_a}[g(t)] = g(t) + \int_{-|t|}^{|t|} K(a, t, u)\, g(u)\, du, \tag{3.3.8}$$

where the kernel $K(a, t, u)$ is bounded in each finite portion of (a, t, u)-space.

The proof of the lemma is based on the identity

$$T_x^y[f(t)]_{x=a} = \underset{y}{W_2} S_0^y [\underset{t}{W_a}[f(a+t)]], \tag{3.3.9}$$

which we will establish first of all.

If $f(t)$ satisfies the equation $\underset{t}{L_1}[u] + \lambda u = 0$ for some value of λ, then from (3.3.7) we obtain

$$T_x^y[f(t)]_{x=a} = f(a)\, \omega_{\theta_2}^{(2)}(\lambda, y),$$

where $\omega_{\theta_2}^{(2)}(\lambda, y)$ denotes the solution of $\underset{y}{L_2}[u] + \lambda u = 0$ with initial conditions $u(0) = 1$ and $u'(0) = \theta_2$.

The definition of L_a implies that the function $f(a+t)$ satisfies $\underset{t}{L_a}[u] + \lambda u = 0$. Therefore

$$\underset{y}{W_2} S_0^y [W_a [f(a+t)]] = f(a)\, \omega_{\theta_2}^{(2)}(\lambda, y)$$

and

$$T_x^y[f(t)]_{x=a} = \underset{y}{W_2} S_0^y [\underset{t}{W_a}[f(a+t)]],$$

which implies that (3.3.9) is valid for functions $f(t)$ which are solutions of $\underset{t}{L_1}[u] + \lambda u = 0$ for some value of λ. Since linear combinations of such solutions can be used to approximate an arbitrary function in $\mathfrak{M}^1_{(-\infty,\infty)}$, and since the operators in (3.3.9) are weakly bounded, it follows that (3.3.9) is true for all functions in $\mathfrak{M}^1_{(-\infty,\infty)}$.

We now prove the lemma. By (3.3.8) we have

$$\begin{aligned} S_0^y [\underset{t}{W_a}[f(a+t)]] &= \frac{1}{2} f(a+y) + \frac{1}{2} \int_{-|y|}^{|y|} K(a, y, t) f(a+t)\, dt \\ &+ \frac{1}{2} f(a-y) + \frac{1}{2} \int_{-|y|}^{|y|} K(a, -y, t) f(a+t)\, dt \\ &= \frac{1}{2} \{\underset{y}{W_a}[f(a+y)] + \underset{y}{\hat{W}_a}[f(a-y)]\}, \end{aligned}$$

where the operator $\underset{y}{\hat{W}}_a$ is defined by the formula

$$\underset{y}{\hat{W}}_a [\varphi(y)] = \varphi(y) + \int_{-|y|}^{|y|} K(a, -y, -t)\varphi(t)\,dt.$$

Thus

$$\underset{y}{W}_2 S_0^y [\underset{t}{W}_a [f(a+t)]] = \frac{1}{2} \{\underset{y}{W}_2 \underset{y}{W}_a [f(a+y)] + \underset{y}{W}_2 \underset{y}{\hat{W}}_a [f(a-y)]\}.$$

and since the product of Volterra operators is a Volterra operator,

$$\underset{y}{W}_2 S_0^y [\underset{t}{W}_a [f(a+t)]] = \frac{1}{2} \{f(a+y) + f(a-y)\}$$
$$+ \frac{1}{2} \int_{-|y|}^{|y|} A_1(a, y, t) f(a+t)\,dt + \frac{1}{2} \int_{-|y|}^{|y|} A_2(a, y, t) f(a-t)\,dt,$$

where the kernels $A_1(a, y, t)$ and $A_2(a, y, t)$ are bounded in each finite portion of (a, y, t)-space.

Finally, if we set

$$T(a, y, t) = \frac{1}{2} \frac{y}{|y|} [A_1(a, y, t-a) - A_2(a, y, a-t)],$$

we find that

$$\underset{y}{W}_2 S_0^y [\underset{t}{W}_a [f(a+t)]] = \frac{1}{2} [f(a+y) + f(a-y)] + \int_{a-y}^{a+y} T(a, y, t) f(t)\,dt,$$

which by (3.3.9) implies that

$$T_x^y [f(t)] = \frac{1}{2} [f(x+y) + f(x-y)] + \int_{x-y}^{x+y} T(x, y, t) f(t)\,dt,$$

which is what we wanted to show.

REMARK. Let $f(x)$ be a twice continuously differentiable function. According to Definition 3.2.2 the function $T_x^y [f(t)]$ is the solution of the partial differential equation

$$\frac{\partial^2}{\partial x^2} u(x, y) - q_1(x) u(x, y) = \frac{\partial^2}{\partial y^2} u(x, y) - q_2(y) u(x, y)$$

with initial conditions

$$u(x, 0) = f(x); \quad u'_y(x, 0) = \theta_2 f(x).$$

On the basis of the lemma just proved we can find a representation of this solution.

On the other hand, following Povzner [10], we can use Riemann's method for integrating hyperbolic partial differential equations and can find a representation of the solution $u(x, y)$, and therefore of the operator T_x^y, in a form which corresponds to Lemma 3.3.1.

THEOREM 3.3.1. *For the remainder term in the Taylor-Delsarte formula*

$$B_{2n}(x, y) = T_x^y[F(t)] - \sum_{k=0}^{n-1} \underset{y}{L_2^{-k}}[\omega_{0_2}^{(2)}(0, y)] \underset{x}{L_1^k}[F(x)]$$

we have the bound

$$|B_{2n}(x, y)| \leqslant c(x, y) \frac{y^{2n}}{2n!} \sup_{x-|y|<t<x+|y|} |\underset{t}{L_1^n}[F(t)]|,$$

where the coefficient $c(x, y)$ *is bounded in each finite portion of the* (x, y)-*plane and does not depend on* $F(t)$ *or* n.

PROOF. By (3.3.6′) we have

$$B_{2n}(x, y) = \underset{y}{W_2}\left\{\frac{y^{2n}}{2n!} \underset{z}{W_2^{-1}} T_x^z[\underset{t}{L_1^n}[F(t)]]\right\} \quad (z = \theta(y),\ 0 < |\theta(y)| < |y|),$$

where $W_2 = W_{\{D^2 L_2 \theta_2\}}$ is of the form

$$\underset{y}{W_2}[f(y)] = f(y) + \int_{-|y|}^{|y|} K_2(y, t) f(t)\, dt.$$

Since

$$|\underset{y}{W_2}[f(y)]| \leqslant \left(1 + \int_{-|y|}^{|y|} |K_2(y, t)|\, dt\right) \sup_{-|y|<t<|y|} |f(t)| = c_2(y) \sup_{-|y|<t<|y|} |f(t)|,$$

it follows that

$$|B_{2n}(x, y)| \leqslant c_2(y) \sup_{-|y|<t<|y|} \left|\frac{t^{2n}}{2n!} \underset{z}{W_2^{-1}} T_x^z[\underset{t}{L_1^n}[F(t)]]\right|,$$

where $z = \theta(t)$ and $0 < |\theta(t)| < |t|$. Therefore

$$|B_{2n}(x, y)| \leqslant c_2(y) \frac{y^{2n}}{2n!} \sup_{-|y|<z<|y|} |\underset{z}{W_2^{-1}} T_x^z[\underset{t}{L_1^n}[F(t)]]|.$$

Furthermore,

$$\underset{z}{W_2^{-1}}[\varphi(z)] = \varphi(z) + \int_{-|z|}^{|z|} H_2(z, t) \varphi(t)\, dt,$$

which implies that

$$\sup_{-|y|<z<|y|} |\underset{z}{W_2^{-1}}[\varphi(z)]| \leqslant c_1(y) \sup_{-|y|<t<|y|} \varphi|(t)|,$$

where

$$c_1(y) = \sup_{-|y|<z<|y|}\left(1 + \int_{-|z|}^{|z|} |H_2(z, t)|\, dt\right).$$

Consequently

$$|B_{2n}(x, y)| \leqslant c_1(y)\, c_2(y)\, \frac{y^{2n}}{2n!} \sup_{-|y|<z<|y|} |T_x^z [\underset{t}{L_1^n} [F(t)]]|. \qquad (3.3.10)$$

Now we use Lemma 3.3.1. According to this lemma

$$T_x^z [f(t)] = \frac{1}{2} [f(x+z) + f(x-z)] + \int_{x-z}^{x+z} T(x, z, t) f(t)\, dt$$

and therefore

$$\sup_{-|y|<z<|y|} |T_x^z [f(t)]| \leqslant c_3(x, y) \sup_{x-|y|<t<x+|y|} |f(t)|,$$

where

$$c_3(x, y) = \sup_{x-|y|<z<x+|y|}\left(1 + \int_{x-|z|}^{x+|z|} |T(x, z, t)|\, dt\right).$$

Thus for (3.3.10) we obtain

$$|B_{2n}(x, y)| \leqslant c(x, y) \sup_{x-|y|<t<x+|y|} |\underset{t}{L_1^n} [F(t)]|\, \frac{y^{2n}}{2n!},$$

where the coefficients $c(x, y) = c_1(x) c_2(y) c_3(x, y)$ is bounded in each finite portion of the (x, y)-plane and does not depend on $F(t)$ or n.

The theorem is proved.

The bound we obtained for the remainder $B_{2n}(x, y)$ is far from precise. But it does give an estimate of $|B_{2n}(x, y)|$ for large n and is sometimes useful. We can use it in proving the following theorem.

THEOREM 3.3.2. *Suppose that* $q(t)$ *in the differential operator*

$$\underset{t}{L}[u] = \frac{d^2}{dt^2} u(t) - q(t) u(t),$$

is even: $q(t) = q(-t)$. *Then we can expand the even function* $F(x)$ *in a Taylor-Delsarte series*

$$F(x) = \sum_{k=0}^{\infty} \underset{x}{L}^{-k} [\omega_0(0, x)]\, \underset{t}{L}^k [F(t)]\,|_{t=0},$$

which is uniformly convergent in each finite interval of the real line $-\infty < x < \infty$, *if and only if we can apply to* $F(t)$ *the operator* $\underset{t}{L}$ *an unbounded number of times and*

$$\overline{\lim_{n\to\infty}} \left\{ \frac{\sup_{|t|<a} |L_t^n [F(t)]|}{2n!} \right\}^{\frac{1}{2n}} = 0,$$

for arbitrary $a > 0$.

PROOF. SUFFICIENCY. We introduce the generalized shift operator of the second kind

$$T_x^y [f(t)] = \underset{x}{W} \underset{y}{W} S_x^y [W^{-1}[f]],$$

where $W = W_{\{D^2 L_0\}}$. Since $q(t)$ is even, the solution $\omega_0(\lambda, t)$ of $L[u] + \lambda u = 0$ with initial conditions $u(0) = 1$ and $u'(0) = 0$ is also even.

Since

$$W[\cos\sqrt{\lambda}\, t] = \omega_0(\lambda, t), \qquad W^{-1}[\omega_0(\lambda, t)] = \cos\sqrt{\lambda}\, t$$

and an arbitrary even function $f(t) \in \mathfrak{M}^1_{(-\infty,\infty)}$ can be approximated in the set by linear combinations of the functions $\{\cos\sqrt{\lambda}\, t\}$ and $\{\omega_0(\lambda, t)\}$ (λ is a parameter), the operators W and W^{-1} take even functions to even functions. Therefore if $f(t)$ is even we have

$$T_0^y [f(t)] = \underset{y}{W} S_0^y [W^{-1}[f(t)]] = \underset{y}{W} \underset{y}{W^{-1}} [f(y)] = f(y).$$

Now suppose $\underset{t}{L}$ can be applied to $F(t)$ an unbounded number of times. By the Taylor-Delsarte formula we have

$$F(y) = T_0^y [F(t)] = \sum_{k=0}^{n-1} \underset{y}{L^{-k}} [\omega_0(0, y)] \underset{t}{L^k} [F(t)]|_{t=0} + B_{2n}(0, y),$$

where (Theorem 3.3.1)

$$|B_{2n}(0, y)| \leqslant c(0, y) \frac{y^{2n}}{2n!} \sup_{|t|<|y|} |\underset{t}{L^n} [F(t)]|.$$

Therefore

$$\sup_{|y|<a} |B_{2n}(0, y)| \leqslant c(a) \frac{a^{2n}}{2n!} \sup_{|t|<a} |\underset{t}{L^n} [F(t)]|,$$

where $c(a) = \sup_{|y|<a} c(0, y) < \infty$ for arbitrary $a > 0$.

From the above inequality we have

$$\overline{\lim_{n\to\infty}} \sup_{|y|<a} |B_{2n}(0, y)| \leqslant c(a) \overline{\lim_{n\to\infty}} \left\{ a \left[\frac{\sup_{|t|<a} |L_t^n [F(t)]|}{2n!} \right]^{\frac{1}{2n}} \right\}^{2n},$$

and if $F(t)$ satisfies the conditions in the theorem, we also have

$$\overline{\lim_{n\to\infty}} \sup_{|y|<a} |B_{2n}(0, y)| = 0,$$

for arbitrary $a > 0$. Consequently the Taylor-Delsarte series for $F(y)$ converges uniformly in each finite interval, which is what we wanted to show.

NECESSITY. Let

$$F(x) = \sum_{k=0}^{\infty} c_k \underset{x}{L}^k [\omega_0(0, x)],$$

where the series converges uniformly in each finite interval. Then by the weak boundedness of W^{-1}

$$\Phi(x) = \underset{x}{W}^{-1}[F(x)] = \sum_{k=0}^{\infty} c_k \underset{x}{W}^{-1} \underset{x}{L}^k [\omega_0(0, x)] = \sum_{k=0}^{\infty} c_k \frac{x^{2k}}{2k!},$$

where the power series converges everywhere. Then $\Phi(x)$ is an entire analytic function, and

$$\overline{\lim_{n\to\infty}} \left\{ \frac{\sup_{|x|<a} |\underset{x}{D}^{2n}[\Phi(x)]|}{2n!} \right\}^{\frac{1}{2n}} = 0,$$

for arbitrary $a > 0$. Since $\Phi(x)$ is infinitely differentiable, by Theorem 3.3.1 we can apply $\underset{x}{L}$ to $\underset{x}{W}[\Phi(x)] = F(x)$ an unbounded number of times, and

$$\underset{x}{L}^n[F(x)] = \underset{x}{W} \underset{x}{D}^{2n}[\Phi(x)],$$

where the transformation operator W is of the form

$$\underset{x}{W}[f(x)] = f(x) + \int_{-|x|}^{|x|} K(x, t) f(t)\, dt.$$

Therefore

$$\sup_{|x|<a} |\underset{x}{L}^n[F(x)]| \leqslant k(a) \sup_{|x|<a} |\underset{x}{D}^{2n}[\Phi(x)]|,$$

where

$$k(a) = \sup_{|x|<a} \left(1 + \int_{-|x|}^{|x|} |K(x, t)|\, dt\right) < \infty$$

for arbitrary $a > 0$. Consequently

$$\overline{\lim_{n\to\infty}} \left\{ \frac{\sup_{|x|<a} |\underset{x}{L}^n[F(x)]|}{2n!} \right\}^{\frac{1}{2n}} \leqslant \overline{\lim_{n\to\infty}} \left\{ \frac{k(a) \sup_{|x|<a} |\underset{x}{D}^{2n}[\Phi(x)]|}{2n!} \right\}^{\frac{1}{2n}} = 0,$$

and necessity is proved.

The theorem is proved.

From the proof it is clear that $\Phi(x) = \underset{x}{W}^{-1}[F(x)]$ is an entire analytic function if and only if $F(x)$ satisfies the conditions of Theorem 3.3.2.

An entire function

$$\Phi(x) = \sum_{k=0}^{\infty} c_k \frac{x^{2k}}{2k!}$$

is said to be of exponential type (see [19], Russian pp. 144–154) if

$$\overline{\lim_{k\to\infty}} |c_k|^{\frac{1}{2k}} = \sigma < \infty,$$

where the number σ is called the order of $\Phi(x)$.

From this definition and the foregoing results we have the following

COROLLARY. *Let L be a differentiable operator which satisfies the conditions of Theorem* 3.3.2, *and let $F(x)$ be an even function. Then the function*

$$\Phi(x) = W_x\{LD^{\circ}0\} [F(x)]$$

is an entire function of exponential type of order σ if and only if the following conditions are satisfied:

1) *The operator $\underset{x}{L}$ can be applied to $F(x)$ an unbounded number of times.*

2) *For arbitrary $a \geqslant 0$*

$$\overline{\lim_{n\to\infty}} \left\{ \frac{\sup_{|x|<a} |L_x^n [F(x)]|}{2n!} \right\}^{\frac{1}{2n}} = 0.$$

3)
$$\overline{\lim_{n\to\infty}} |L_t^n [F(t)]|_{t=0}|^{\frac{1}{2n}} = \sigma.$$

Functions which satisfy conditions 1), 2) and 3) share many properties with ordinary entire functions of exponential type (see [19]). For example, we can prove for these functions an analogue of the Paley-Wiener Theorem [19] and in some cases we can generalize the classical Bernstein inequality [19]. We will generalize the Bernstein inequality in §4 of Chapter 4.

Received 8 MAY 51

BIBLIOGRAPHY

1. V. A. Steklov, *A cooling problem for a nonhomogeneous rigid rod,* Soobšč. Har'kov. Mat. Obšč. (2) 5 (1896), 136–181. (Russian). Cf. *Problème de refroidissement d'une barre hétérogène,* Ann. Fac. Sci. Univ. Toulouse (2) 3 (1901), 281–313.

2. H. Weyl, *Über gewöhnliche Differentialgleichungen mit Singularitäten und die zugehörige Entwicklungen willkürlicher Funktionen,* Math. Ann. 68 (1910), 220–269.

3. E. C. Titchmarsh, *Eigenfunction expansions associated with second-order differential equations,* Clarendon Press, Oxford, 1946. MR 8, 458

4. B. M. Levitan, *Expansion in eigenfunctions of differential equations of the second order,* GITTL, Moscow, 1950. (Russian) MR 12, 183.

5. ———, *The application of generalized displacement operators to second-order linear differential equations,* Uspehi Mat. Nauk 4 (1949), no. 1 (29), 3–112; English transl., Amer. Math. Soc. Transl. (1) 10 (1962), 408–541. MR 11, 116.

6. ———, *A generalization of almost periodic functions,* Mat. Sb. 24 (66) (1949), 321–346; English transl., Amer. Math. Soc. Transl. (1) 10 (1962), 378–407. MR 11, 174.

7. B. M. Levitan, *On a decomposition theorem for characteristic functions of differential equations of the second order,* Dokl. Akad. Nauk SSSR 71 (1950), 605–608. (Russian) MR 11, 720.

8. J. Delsarte, *Sur une extension de la formule de Taylor,* J. Math. Pures Appl. (9) **17** (1938), 213–231.

9. ———, *Une extension nouvelle de la théorie des fonctions presque-périodiques de Bohr,* Acta Math. 69 (1939), 259–317.

10. A. Ja. Povzner, *On differential equations of Sturm-Liouville type on a half-axis,* Mat. Sb. **23** (65) (1948), 3–52; English transl., Amer. Math. Soc. Transl. (1) **4** (1962), 24–101. MR **10**, 299.

11. ———, *On equations of the Sturm-Liouville type on a semi-axis,* C. R. (Dokl.) Acad. Sci. URSS **53** (1946), 295–298. MR **8**, 469.

12. N. I. Ahiezer and I. M. Glazman, *Theory of linear operators in Hilbert space,* GITTL, Moscow, 1950; English transl., Ungar, New York, 1961. MR **13**, 358.

13. V. A. Ambarzumian [Ambartsumian], *Über eine Frage der Eigenwerttheorie,* Z. Phys. **53** (1929), 690–695.

14. G. Borg, *Eine Umkehrung der Sturm-Liouvilleschen Eigenwertaufgabe. Bestimmung der Differentialgleichung durch die Eigenwerte,* Acta Math. 78 (1946), 1–96. MR **7**, 382.

15. N. Levinson, *On the uniqueness of the potential in a Schrödinger equation for a given asymptotic phase,* Bull. Amer. Math. Soc. **55** (1949), 517.

16. V. Bargmann, *Remarks on the determination of a central field of force from the elastic scattering phase shifts,* Phys. Rev. (2) **75** (1949), 301–303. MR **11**, 248.

17. M. G. Kreĭn, *On a general method of decomposing Hermite-positive nuclei into elementary products,* C. R. (Dokl.) Acad. Sci. URSS **53** (1946), 3–6. MR **8**, 277.

18. ———, *Solution of the inverse Sturm-Liouville problem*, Dokl. Akad. Nauk SSSR **76** (1951), 21–24. (Russian) MR **12**, 613.

19. N. I. Ahiezer, *Lectures on the theory of approximation,* OGIZ, Moscow, 1947; English transl., Ungar, New York, 1956. MR **10**, 33; **20** #1872.

20. Z. S. Agranovič, *On some questions connected with equations of Sturm-Liouville type on a semi-axis,* Dokl. Akad. Nauk SSSR **66** (1949), 1025–1028. (Russian) MR **11**, 28.

21. I. M. Gel'fand, D. A. Raĭkov and G. E. Šilov, *Commutative normed rings,* Uspehi Mat. Nauk **1** (1946), no. 2 (12), 48–146; English transl., Amer. Math. Soc. Transl. (2) **5** (1957), 115–220. MR **10**, 258; **18**, 714.

22. L. A. Ljusternik, *Basic concepts in functional analysis,* Uspehi Mat. Nauk **1** (1936), 77–140. (Russian)

23. V. V. Nemyckiĭ, *The fixed-point method in analysis,* Usephi Mat. Nauk **1** (1936), 141–174; English transl., Amer. Math. Soc. Transl. (2) **34** (1963), 1–37.

24. I. M. Gel'fand and B. M. Levitan, *On the determination of a differential equation from its spectral function*, Izv. Akad. Nauk SSSR Ser. Mat. **15** (1951), 309–360; English transl., Amer. Math. Soc. Transl. (2) **1** (1955), 253–304. MR **13**, 558; **17**, 489.

25. V. A. Marčenko, *Concerning the theory of a differential operator of the second order,* Dokl. Nauk SSSR **72** (1950), 457–460. (Russian) MR **12**, 183.

26. ———, *Transformation of operators,* Dokl. Akad. Nauk SSSR **74** (1950), 185–188. (Russian) MR **12**, 502.

27. V. A. Marčenko, *On the transformation formulas generated by a linear differential operator of the second order,* Dokl. Akad. Nauk SSSR **74** (1950), 657–660. (Russian) MR **12**, 707.

28. ———, *Generalized almost periodic functions,* Dokl. Akad. Nauk SSSR **74** (1950), 893–895. (Russian) MR **12**, 698.

Translated by:
F. A. Cezus

Amer. Math. Soc. Transl.
(2) Vol. 101, 1973

REMARK ON A THEOREM OF V. A. MARČENKO

B. M. LEVITAN

In the preceding paper (Chapter 2, §2) V. A. Marčenko proved a very interesting asymptotic formula:

$$\rho(\lambda) = \frac{2}{\pi}\sqrt{\lambda} + o(\sqrt{\lambda}). \tag{1}$$

The aim of this note is to show that formula (1) can essentially be obtained from the following special Tauberian theorem of N. Wiener (cf. [1], page 138).

If $\sigma(\mu)$ $(\mu \geqslant 0)$ has bounded variation in each finite interval and

$$\operatorname{Var}_{\mu}^{\mu+1} \sigma(\mu) < C, \tag{2}$$

where C does not depend on μ, then

$$\lim_{\varepsilon\to 0} \int_0^\infty \frac{\sin^2 \varepsilon\mu}{\varepsilon\mu^2}\, d\sigma(\mu) = \frac{\pi}{2} \lim_{\mu\to+\infty} \frac{\sigma(\mu)}{\mu}, \tag{3}$$

where the limits on the left and the right sides exist or do not exist simultaneously.

Suppose

$$\cos\sqrt{\lambda}\, x = \varphi(x, \lambda) + \int_0^x K(x, t)\, \varphi(t, \lambda)\, dt.$$

Integrating the last equality over $(0, \epsilon)$, we get

$$\frac{\sin\sqrt{\lambda}\,\varepsilon}{\sqrt{\lambda}} = \int_0^\varepsilon \left\{1 + \int_t^\varepsilon K(x, t)\, dx\right\} \varphi(t, \lambda)\, dt.$$

Using Parseval's equality, we get

---

* Translation of Trudy Moskov. Mat. Obšč. 1 (1952), 421–422.

AMS (MOS) subject classifications (1970). Primary 34B25; Secondary 40E05.

$$\int_{-\infty}^{\infty}\left(\frac{\sin\sqrt{\lambda}\,\varepsilon}{\sqrt{\lambda}}\right)^2 d\rho(\lambda)=\int_0^{\varepsilon}\left\{1+\int_t^{\varepsilon}K(x,t)\,dx\right\}^2 dt$$

$$=\varepsilon+\int_0^{\varepsilon}O(\varepsilon-t)\,dt=\varepsilon+O(\varepsilon^2). \tag{4}$$

In particular, it follows from the last equality that for all $\epsilon > 0$

$$\int_{-\infty}^{0}\left(\frac{\sin\sqrt{\lambda}\varepsilon}{\sqrt{\lambda}}\right)^2 d\rho(\lambda)<\infty,$$

whence as one of our results we have

$$\int_{-\infty}^{0} e^{\sqrt{|\lambda|}\,\varepsilon}\,d\rho(\lambda)<\infty; \tag{5}$$

that is, Lemma 2.2.1 of the preceding paper.

From inequality (5) and the theorem on the mean it is easy to obtain

$$\int_{-\infty}^{0}\left(\frac{\sin\sqrt{\lambda}\,\varepsilon}{\sqrt{\lambda}}\right)^2 d\rho(\lambda)=\varepsilon^2\int_{-\infty}^{0}\left(\frac{\sin\sqrt{\lambda}\varepsilon}{\varepsilon\sqrt{\lambda}}\right)^2 d\rho(\lambda)=\varepsilon^2\int_{-\infty}^{0}\cosh^2\theta\epsilon\sqrt{|\lambda|}$$

$$=O(\varepsilon^2)\qquad(0<\theta<1). \tag{6}$$

Therefore from (4) and (6) it follows that

$$\int_0^{\infty}\frac{\sin^2\sqrt{\lambda}\,\varepsilon}{\varepsilon\lambda}\,d\rho(\lambda)=1+O(\varepsilon).$$

Or, by setting $\sqrt{\lambda}=\mu$ and $\rho(\lambda)=\sigma(\mu)$, we have

$$\int_0^{\infty}\frac{\sin^2\varepsilon\mu}{\varepsilon\mu^2}\,d\sigma(\mu)=1+O(\varepsilon). \tag{7}$$

Lemma 2.2.3 of the preceding paper implies that condition (2) is satisfied for $\sigma(\mu)$. Therefore from the cited theorem of Wiener and (7) it follows that

$$\frac{\pi}{2}\lim_{\mu\to\infty}\frac{\sigma(\mu)}{\mu}=1;$$

that is, formula (1). It may be that if an estimate of the remainder is used in (7), a more precise estimate of the remainder in (1) could be obtained.

BIBLIOGRAPHY

1. N. Wiener, *The Fourier integral and certain of its applications,* Cambridge Univ. Press, Cambridge, 1933; reprint, Dover, New York, 1959. MR 20 #6634.

Translated by A. COSMIDES

Amer. Math. Soc. Transl.
(2) Vol. 101, 1973

SOME QUESTIONS IN THE THEORY OF ONE-DIMENSIONAL LINEAR DIFFERENTIAL OPERATORS OF THE SECOND ORDER. II

V. A. MARČENKO

CONTENTS

* Translation of Trudy Moskov. Mat. Obšč. 2 (1953), 3–83.
AMS (MOS) subject classifications (1970). Primary 34C20, 42A84; Secondary 34B25, 35F15, 47A70.

CHAPTER 4

A CLOSER INVESTIGATION OF THE PROPERTIES OF A CLASS OF TRANSFORMATION OPERATORS

§ 1. Structure of the operators $V_{\{D^2 L 0 h\}}$ and $V_{\{D^2 L \infty\infty\}}$

Up to this point, in order to apply transformation operators to various problems it was sufficient to know only their simplest properties (essentially just their form and the consequent property of weak boundedness). This situation resulted from the decision to avoid problems connected with uniform metrics on the real line or on a half-line.

In order to apply transformation operators to questions which are inextricably involved with uniform metrics on the real line or on a half-line (in the first case, to the theory of generalized almost periodic functions) it is necessary to investigate the properties of transformation operators which are related to such metrics.

In the present chapter we will study in detail the class of transformation operators $V_{\{L_1 L_2 h_1 h_2\}}$ associated with the differential operators

$$L_1[u] = \frac{d^2}{dx^2} u(x) - q_1(x) u(x), \qquad L_2[u] = \frac{d^2}{dx^2} u(x) - q_2(x) u(x)$$
$$(0 \leqslant x < \infty),$$

which satisfy the restrictions

$$\int_0^\infty (1+x^2)\,|q_1(x)|\,dx < \infty, \qquad \int_0^\infty (1+x^2)\,|q_2(x)|\,dx < \infty, \qquad (4.1.1)$$

which we hereafter assume in all subsequent arguments.

Since an arbitrary transformation operator $V_{\{L_1 L_2 h_1 h_2\}}$ can be written in the form

$$V_{\{L_1 L_2 h_1 h_2\}} = V_{\{D^2 L_2 0 h_2\}} V_{\{L_1 D^2 h_1 0\}} \qquad (h_1, h_2 \neq \infty),$$
$$V_{\{L_1 L_2 \infty\infty\}} = V_{\{D^2 L_2 \infty\infty\}} V_{\{L_1 D^2 \infty\infty\}} \qquad (h_1 = h_2 = \infty)$$

(by D^2 we always mean the operator d^2/dx^2), the study of the structure of the operators $V_{\{L_1 L_2 h_1 h_2\}}$ reduces to finding the solution of the following problems:

1) Determine the structure of $V_{\{D^2 L 0 h\}}$ and $V_{\{D^2 L \infty\infty\}}$.

2) Determine the structure of $V_{\{L D^2 0 h\}} = V^{-1}_{\{D^2 0 h\}}$ and $V_{\{L D^2 \infty\infty\}} = V^{-1}_{\{D^2 L \infty\infty\}}$.

The first of these problems is significantly easier than the second, and we will solve it right now.

To this end, let L be a differential operator given on $0 \leqslant x < \infty$ and satisfying (4.1.1). As in §1 of Chapter 1 we let $\omega_h(\lambda, x)$ denote the solution of the equation $L[u] + \lambda u = 0$ with initial conditions

$$u(0) = 1, \quad u'(0) = h, \quad \text{if} \quad h \neq \infty;$$
$$u(0) = 0, \quad u'(0) = 1, \quad \text{if} \quad h = \infty.$$

By Lemma 1.2.1, for $x \geqslant 0$ we have

$$\omega_h(\lambda, x) = \cos\sqrt{\lambda}\, x + \int_{-|x|}^{|x|} [\mathfrak{A}(x, y) + h\mathfrak{B}(x, y)] \cos\sqrt{\lambda}\, y\, dy,$$

$$\omega_\infty(\lambda, x) = \frac{\sin\sqrt{\lambda}\, x}{\sqrt{\lambda}} + \int_{-|x|}^{|x|} \mathfrak{A}(x, y) \frac{\sin\sqrt{\lambda}\, y}{\sqrt{\lambda}}\, dy.$$

Therefore the operators $V_{\{D^2 L 0 h\}}$ and $V_{\{D^2 L \infty\infty\}}$ are of the form

$$V_{\{D^2L0h\}}[f] = f(x) + \int_0^x A_h(x, y) f(y)\, dy,$$

$$V_{\{D^2L\infty\infty\}}[f] = f(x) + \int_0^x K_\infty(x, y) f(y)\, dy,$$

where

$$A_h(x, y) = \mathfrak{A}(x, y) + \mathfrak{A}(x, -y) + h[\mathfrak{B}(x, y) + \mathfrak{B}(x, -y)], \tag{4.1.2}$$

$$K_\infty(x, y) = \mathfrak{A}(x, y) - \mathfrak{A}(x, -y). \tag{4.1.2'}$$

According to the Corollary of Lemma 1.1.1, in order to establish uniform boundedness of the kernels $\mathfrak{A}(x, y)$ and $\mathfrak{B}(x, y)$ (and therefore $A_h(x, y)$ and $K_\infty(x, y)$) in the half-plane $x \geqslant 0$ it suffices to show that

$$\int_0^\infty (1 + x)\,|q(x)|\, dx < \infty.$$

By our assumption (4.1.1) this inequality is satisfied. Therefore

$$|\mathfrak{A}(x, y)| \leqslant \tilde{\mathfrak{A}}(x, y) < M; \quad |\mathfrak{B}(x, y)| \leqslant \tilde{\mathfrak{B}}(x, y) < M, \quad M < \infty$$

uniformly for all $y \in (-\infty, \infty)$ and $x \geqslant 0$.[1]

The uniform boundedness of these kernels for $x \geqslant 0$ implies that

$$|X^0(x)| \leqslant \tilde{X}^0(x) \leqslant 1 + \int_{-x}^{x} \tilde{\mathfrak{A}}(x, y)\, dy \leqslant 1 + Mx, \tag{4.1.3}$$

---

[1] The symbol $\sim$ indicates the change from the operator $L[u] = u''(x) - q(x)u(x)$ to the operator $\tilde{L}[u] = u''(x) - |q(x)|u(x)$ (see §1 of Chapter 1).

$$|X^1(x)| \leqslant \tilde{X}^1(x) \leqslant \int_{-x}^{x} \tilde{\mathfrak{B}}(x, y)\,dy \leqslant Mx, \tag{4.1.3'}$$

where $X^0(x) = \omega_0(0, x)$ and $X^1(x) = \omega_\infty(0, x)$.

Therefore the integrals

$$\int_0^\infty q(t)\,X^0(t)\,dt = a, \qquad \int_0^\infty q(t)\,X^1(t)\,dt = b \tag{4.1.4}$$

converge absolutely, and

$$\int_0^x q(t)\,X^0(t)\,dt = a - \int_x^\infty q(t)\,X^0(t)\,dt,$$

$$\int_0^x q(t)\,X^1(t)\,dt = b - \int_x^\infty q(t)\,X^1(t)\,dt.$$

By virtue of Lemma 1.1.2, for $x \geqslant 0$ we have

$$\mathfrak{A}(x, y) = \frac{1}{2}[a - \mathfrak{A}_1(x, y)], \quad \mathfrak{B}(x, y) = \frac{1}{2}[1 + b - \mathfrak{B}_1(x, y)], \tag{4.1.5}$$

where

$$\mathfrak{A}_1(x, y) = \mathfrak{A}'(x, y) + \int_x^\infty q(t)\,X^0(t)\,dt,$$

$$\mathfrak{B}_1(x, y) = \mathfrak{B}'(x, y) + \int_x^\infty q(t)\,X^1(t)\,dt$$

(for the definition of $\mathfrak{A}'(x, y)$ and $\mathfrak{B}'(x, y)$ see Lemma 1.2.1).

Furthermore,

$$|\mathfrak{A}_1(x, y)| \leqslant |\mathfrak{A}'(x, y)| + \int_x^\infty |q(t)|\,|X^0(t)|\,dt$$

$$\leqslant \tilde{\mathfrak{A}}'(x, y) + \int_x^\infty |q(t)|\,\tilde{X}^0(t)\,dt = \tilde{\mathfrak{A}}_1(x, y),$$

$$|\mathfrak{B}_1(x, y)| \leqslant |\mathfrak{B}'(x, y)| + \int_x^\infty |q(t)\,x'(t)|\,dt$$

$$\leqslant \tilde{\mathfrak{B}}'(x, y) + \int_x^\infty |q(t)|\,\tilde{X}^1(t)\,dt = \tilde{\mathfrak{B}}_1(x, y)$$

and

$$\int_{-x}^{x} \tilde{\mathfrak{A}}_1(x, y)\,dy = \int_{-x}^{x} \tilde{\mathfrak{A}}'(x, y)\,dy + 2x \int_x^\infty |q(t)|\,\tilde{X}^0(t)\,dt,$$

$$\int_{-x}^{x} \tilde{\mathfrak{B}}_1(x, y)\,dy = \int_{-x}^{x} \tilde{\mathfrak{B}}'(x, y)\,dy + 2x \int_{x}^{\infty} |q(t)|\,\tilde{X}^1(t)\,dt,$$

which by Lemma 1.1.2 implies that

$$\int_{-x}^{x} \tilde{\mathfrak{A}}_1(x, y)\,dy = 2\left[\int_{0}^{x} t\,|q(t)|\,\tilde{X}^0(t)\,dt + x\int_{x}^{\infty} |q(t)|\,\tilde{X}^0(t)\,dt\right],$$

$$\int_{-x}^{x} \tilde{\mathfrak{B}}_1(x, y)\,dy = 2\left[\int_{0}^{x} t\,|q(t)|\,\tilde{X}^1(t)\,dt + x\int_{x}^{\infty} |q(t)|\,\tilde{X}^1(t)\,dt\right].$$

Therefore

$$\int_{-x}^{x} |\mathfrak{A}_1(x, y)|\,dy \leqslant 2\int_{0}^{\infty} t\,|q(t)|\,\tilde{X}^0(t)\,dt < \infty, \tag{4.1.6}$$

$$\int_{-x}^{x} |\mathfrak{B}_1(x, y)|\,dy \leqslant 2\int_{0}^{\infty} t\,|q(t)|\,\tilde{X}^1(t)\,dt < \infty, \tag{4.1.7}$$

since $q(t)$ satisfies (4.1.1), and by (4.1.3) and (4.1.3′) we have $\tilde{X}^0(x) \leqslant 1 + Mx$ and $\tilde{X}^1(x) \leqslant Mx$.

From (4.1.2), (4.1.2′) and (4.1.5) we obtain

$$A_h(x, y) = a + h(1 + b) + K_h(x, y),$$
$$K_\infty(x, y) = \frac{1}{2}[\mathfrak{A}_1(x, -y) - \mathfrak{A}_1(x, y)],$$

where

$$K_h(x, y) = -\frac{1}{2}[\mathfrak{A}_1(x, y) + \mathfrak{A}_1(x, -y) + h\mathfrak{B}_1(x, y) + h\mathfrak{B}_1(x, -y)],$$

and according to (4.1.6) and (4.1.7)

$$\sup_{0\leqslant x<\infty} \int_{0}^{x} |K_h(x, y)|\,dy \leqslant \int_{0}^{\infty} t\,|q(t)|\,\tilde{X}^0(t)\,dt < \infty,$$

$$\sup_{0\leqslant x<\infty} \int_{0}^{x} |K_\infty(x, y)|\,dy \leqslant \int_{0}^{\infty} t\,|q(t)|\,\tilde{X}^1(t)\,dt < \infty.$$

By virtue of (4.1.4) the number $a + hb$ equals

$$a + hb = \int_{0}^{\infty} q(t)\,[X^0(t) + hX^1(t)]\,dt = \int_{0}^{\infty} q(t)\,\omega_h(0, t)\,dt$$

and since

$$\frac{d^2}{dt^2}\omega_h(0, t) - q(t)\,\omega_h(0, t) = 0,$$

we have

$$a + hb = \lim_{N\to\infty} \int_{0}^{N} q(t)\,\omega_h(0, t)\,dt = \lim_{N\to\infty} \frac{d}{dt}\omega_h(0, t)\Big|_0^N,$$

from which, considering the initial conditions of the solution $\omega_h(0, t)$, we find that

$$a + hb = \lim_{N\to\infty} (\omega_h'(0, N) - \omega_h'(0, 0)) = \omega_h'(0, \infty) - h;$$

that is,

$$a + h(1 + b) = \omega_h'(0, \infty) = \lim_{x\to\infty} \omega_h'(0, x).$$

We have proved the following theorem.

THEOREM 4.1.1. *The transformation operators* $V_{\{D^2L0h\}}$ *and* $V_{\{D^2L\infty\infty\}}$ *are of the form*

$$V_{\{D^2L0h\}}[f] = f(x) + c\int_0^x f(t)\,dt + \int_0^x K_h(x, t) f(t)\,dt,$$

$$V_{\{D^2L\infty\infty\}}[f] = f(x) + \int_0^x K_\infty(x, t) f(t)\,dt,$$

where

$$\sup_{0\le x<\infty} \int_0^x |K_h(x, t)|\,dt = |K_h| < \infty,$$

$$\sup_{0\le x<\infty} \int_0^x |K_\infty(x, t)|\,dt = |K_\infty| < \infty$$

$$c = \lim_{N\to\infty} \omega_h'(0, N) = \int_0^\infty q(t)\,\omega_h(0, t)\,dt + h.$$

This theorem solves the first of the two problems that we stated. And from this theorem we can deduce certain properties of $\omega_h(s^2, x)$ which will be useful in solving the second problem. We have

$$\omega_h(s^2, x) = V_{\{D^2L0h\}}[\cos sx], \qquad \text{if} \quad h \ne \infty,$$
$$\omega_\infty(s^2, x) = V_{\{D^2L\infty\infty\}}\left[\frac{\sin sx}{s}\right], \qquad \text{if} \quad h = \infty,$$

from which by Theorem 4.1.1 we obtain

$$\omega_h(s^2, x) = \cos sx + c\,\frac{\sin sx}{s} + \int_0^x K_h(x, t)\cos st\,dt, \tag{4.1.8}$$

$$\omega_\infty(s^2, x) = \frac{\sin sx}{s} + \int_0^x K_\infty(x, t)\,\frac{\sin st}{s}\,dt, \tag{4.1.8'}$$

where $c = \int_0^\infty q(t)\omega_h(0, t)\,dt + h$. From these equations it is clear that for fixed x the functions $\omega_h(s^2, x)$ are even, entire functions of exponential type [19] with respect to s of order $\sigma = x$.

Furthermore, for real s we have

$$|\omega_h(s^2, x)| \leqslant 1 + |c|x + |K_h|, \quad |\omega_\infty(s^2, x)| \leqslant x + |K_\infty|x, \tag{4.1.9}$$

where

$$|K_h| = \sup_{0\leqslant x<\infty} \int_0^x |K_h(x, t)|\,dt < \infty, \quad |K_\infty| = \sup_{0\leqslant x<\infty} \int_0^x |K_\infty(x, t)|\,dt < \infty.$$

§2. Auxiliary asymptotic formulas

In this section we will introduce asymptotic formulas for the solutions $\omega_h(s^2, x)$ (where s is real) analogous to the formulas in Lemma 2.1.1 but more precise.

Let $h \neq \infty$ be arbitrary but fixed from here on. Then

$$\omega_h(s^2, x) = \cos sx + h\frac{\sin sx}{s} + \int_0^x \frac{\sin s(x-t)}{s} q(t)\,\omega_h(s^2, t)\,dt,$$

which by (4.1.1) and (4.1.9) implies that for real s

$$\omega_h(s^2, x) = \cos sx + h\frac{\sin sx}{s}$$
$$+ \int_0^\infty \frac{\sin s(x-t)}{s} q(t)\,\omega_h(s^2, t)\,dt - \int_x^\infty \frac{\sin s(x-t)}{s} q(t)\,\omega_h(s^2, t)\,dt,$$

or

$$\omega_h(s^2, x) = m(s)\cos sx + n(s)\frac{\sin sx}{s} + r(s, x), \tag{4.2.1}$$

where

$$m(s) = 1 - \frac{1}{s}\int_0^\infty \sin st\,\omega_h(s^2, t)\,q(t)\,dt, \tag{4.2.2}$$

$$n(s) = h + \int_0^\infty \cos st\,\omega_h(s^2, t)\,q(t)\,dt, \tag{4.2.3}$$

$$r(s, x) = \int_x^\infty \frac{\sin s(x-t)}{s}\,\omega_h(s^2, t)\,q(t)\,dt. \tag{4.2.4}$$

We note that from the formula for $n(s)$ we have

$$n(0) = h + \int_0^\infty q(t)\,\omega_h(0, t)\,dt = \lim_{N\to\infty} \omega_h'(0, N) = c, \tag{4.2.5}$$

where c is the constant which occurs in Theorem 4.1.1. The equation $n(0) = c$ is crucial in later arguments.

By analogous methods we obtain for solutions $\omega_\infty(s^2, x)$, where s is real, the formula

$$\omega_\infty(s^2, x) = m_1(s)\cos sx + n_1(s)\frac{\sin sx}{s} + r_1(s, x), \qquad (4.2.1')$$

where

$$m_1(s) = -\frac{1}{s}\int_0^\infty \sin st\, \omega_\infty(s^2, t)\, q(t)\, dt, \qquad (4.2.2')$$

$$n_1(s) = 1 + \int_0^\infty \cos st\, \omega_\infty(s^2, t)\, q(t)\, dt, \qquad (4.2.3')$$

$$r_1(s, x) = \int_x^\infty \frac{\sin s(x-t)}{s}\, \omega_\infty(s^2, t)\, q(t)\, dt, \qquad (4.2.4')$$

and thus

$$n_1(0) = 1 + \int_0^\infty q(t)\, \omega_\infty(0, t)\, dt = \lim_{N\to\infty} \omega'_\infty(0, N). \qquad (4.2.5')$$

LEMMA 4.2.1. *The functions* $m(s)$ *and* $n(s)$ (*or* $m_1(s)$ *and* $n_1(s)$) *cannot vanish simultaneously for any real* s.

PROOF. Suppose that $m(s_0) = n(s_0) = 0$. From (4.2.1) we then have

$$\omega_h(s_0, x) = r(s_0, x) = \int_x^\infty \frac{\sin s_0(x-t)}{s_0}\, \omega_h(s_0^2, t)\, q(t)\, dt.$$

Since for $t \geqslant x \geqslant 0$ we have

$$\left|\frac{\sin s_0(x-t)}{s_0}\right| \leqslant t,$$

by (4.1.9) we see that

$$|\omega_h(s_0^2, x)| \leqslant \int_x^\infty t(1 + |c|t + |K_h|)\,|q(t)|\, dt.$$

Condition (4.1.1) implies that the right side of this inequality goes to zero as $x \to \infty$. Hence there exists a number N such that $x \geqslant N$ implies

$$|\omega_h(s_0^2, x)| \leqslant \int_x^\infty t(1 + |c|t + |K_h|)\,|q(t)|\, dt < \frac{1}{2}$$

and therefore

$$M = \sup_{x \geqslant N} |\omega_h(s_0^2, x)| < \frac{1}{2},$$

which implies that

$$M \leqslant \int_N^\infty t\,|\omega_h(s_0^2, t)|\,|q(t)|\,dt \leqslant M\cdot\frac{1}{2},$$

which is possible only if $M = 0$. Thus $\omega_h(s_0^2, x) = 0$ for $x \geqslant N$, which is impossible since $\omega_h(s_0^2, x)$ is not the trivial solution of $L[u] + s_0^2 u = 0$. Therefore our assumption has been contradicted, and the functions $m(s)$ and $n(s)$ cannot vanish simultaneously. The proof for $m_1(s)$ and $n_1(s)$ is analogous.

The lemma is proved.

From the lemma we see that the functions

$$m^2(s) + \frac{n^2(s)}{s^2}, \quad m_1^2(s) + \frac{n_1^2(s)}{s^2}$$

are different from zero for all real s. If we divide both sides of (4.2.1) and (4.2.1′) by these functions and express the functions $\sin sx$ and $\cos sx$ by Euler's formula in terms of e^{isx} and e^{-isx}, we obtain

$$\frac{\omega_h(s^2, x)}{m^2(s) + \frac{n^2(s)}{s^2}} = \frac{e^{isx}}{2\left[m(s) + i\frac{n(s)}{s}\right]} + \frac{e^{-isx}}{2\left[m(s) - i\frac{n(s)}{s}\right]} + \frac{r(s, x)}{m^2(s) + \frac{n^2(s)}{s^2}},$$

$$\frac{\omega_\infty(s^2, x)}{m_1^2(s) + \frac{n_1^2(s)}{s^2}} = \frac{e^{isx}}{2\left[m_1(s) + i\frac{n_1(s)}{s}\right]} + \frac{e^{-isx}}{2\left[m_1(s) - i\frac{n_1(s)}{s}\right]} + \frac{r_1(s, x)}{m_1^2(s) + \frac{n_1^2(s)}{s^2}}.$$

Then if we introduce the notation

$$k_1(s) = \frac{s}{sm(s) + in(s)} - 1, \quad p_1(s) = \frac{1}{sm_1(s) + in_1(s)} + i$$

and notice that the functions $m(s)$, $n(s)$, $m_1(s)$ and $n_1(s)$ are even in s, we obtain

$$\frac{s^2\omega_h(s^2, x)}{s^2 m^2(s) + n^2(s)} = \cos sx + \frac{1}{2}\left[k_1(s)e^{isx} + k_1(-s)e^{-isx}\right] + \frac{s^2 r(s, x)}{s^2 m^2(s) + n^2(s)}, \tag{4.2.6}$$

$$\frac{s^2\omega_\infty(s^2, x)}{s^2 m_1^2(s) + n_1^2(s)} = s\left\{\sin sx + \frac{1}{2}\left[p_1(s)e^{isx} + p_1(-s)e^{-isx}\right] + \frac{s r_1(s, x)}{s^2 m_1^2(s) + n_1^2(s)}\right\}. \tag{4.2.6′}$$

Hereafter, instead of studying the functions $\omega_h(s^2, x)$ and $\omega_\infty(s^2, x)$, we will study the functions in the above equations, since they play an essential role in the solution of the second problem stated in the preceding section.

The investigation of either of these two functions is analogous to the investigation of the other. We will consider the first function $(h \neq \infty)$ in detail, and for the second function $(h = \infty)$ we will only formulate the corresponding propositions.

First we consider the functions $r(s, x)$ and $r_1(s, x)$.

LEMMA 4.2.2. *The functions* $r(s, x)$ *and* $r_1(s, x)$ *can be written in the form*

$$r(s, x) = \frac{c}{s} \int_{-\infty}^{\infty} A(x, u) \sin su \, du + \int_{-\infty}^{\infty} B(x, u) \cos su \, du,$$

$$r_1(s, x) = \frac{1}{s} \int_{-\infty}^{\infty} A_1(x, u) \sin su \, du,$$

where all of the integrals are absolutely convergent and

$$|A(x, u)| \leqslant \frac{1}{2} \int_x^{\infty} |q(t)| \, dt, \quad |B(x, u)| \leqslant \frac{1}{2} (1 + |K_h|) \int_x^{\infty} |q(t)| \, dt,$$

$$|A_1(x, u)| \leqslant \frac{1}{2} (1 + |K_\infty|) \int_x^{\infty} |q(t)| \, dt.$$

The constants $c = n(0)$, $|K_h|$ *and* $|K_\infty|$ *are the same as in Theorem* 4.1.1.

PROOF. If we substitute for $\omega_h(s^2, t)$ in (4.2.4) its expansion (4.1.8), we obtain

$$r(s, x) = \frac{c}{s} \int_x^{\infty} q(t) \frac{\sin s(x-t)}{s} \sin st \, dt$$

$$+ \int_x^{\infty} q(t) \left[\frac{\sin s(x-t) \cos st}{s} + \int_0^t K_h(t, y) \frac{\sin s(x-t) \cos sy}{s} dy \right] dt.$$

From the identity

$$\frac{\sin s(x-t)}{s} e^{isz} = -\frac{1}{2} \int_{x-t+z}^{-x+t+z} e^{isu} \, du$$

for $t \geqslant x \geqslant 0$ we see that

$$\frac{\sin s(x-t) \sin st}{s} = \int_{-\infty}^{\infty} E_1(x, t, u) \sin su \, du,$$

$$\frac{\sin s(x-t) \cos t}{s} = \int_{-\infty}^{\infty} E_1(x, t, u) \cos su \, du,$$

$$\frac{\sin s(x-t) \cos sy}{s} = \int_{-\infty}^{\infty} E_2(x, t, y, u) \cos su \, du,$$

where

$$E_1(x, t, u) = \begin{cases} -\frac{1}{2} & \text{for} \quad |u-t| \leqslant t-x, \\ 0 & \text{for} \quad |u-t| > t-x, \end{cases}$$

$$E_2(x, t, y, u) = \begin{cases} -\frac{1}{2} & \text{for} \quad |u-y| \leqslant t-x, \\ 0 & \text{for} \quad |u-y| > t-x. \end{cases}$$

Furthermore,

$$\int_0^t K_h(t, y) \frac{\sin s(x-t)}{s} \cos sy\, dy = \int_0^t K_h(t, y) \left[\int_{-\infty}^{\infty} E_2(x, t, y, u) \cos su\, du \right] dy$$

$$= \int_{-\infty}^{\infty} \left[\int_0^t K_h(t, y) E_2(x, t, y, u)\, dy \right] \cos su\, du,$$

and if we set

$$E_3(x, t, u) = E_1(x, t, u) + \int_0^t K_h(t, y) E_2(x, t, y, u)\, dy,$$

we obtain

$$\frac{\sin s(x-t)}{s} \cos st + \int_0^t K_h(t, y) \frac{\sin s(x-t)}{s} \cos sy\, dy = \int_{-\infty}^{\infty} E_3(x, t, u) \cos su\, du,$$

where we can easily show that $E_3(x, t, u) = 0$ for $|u-t| > t-x$, and everywhere

$$|E_3(x, t, u)| \leqslant |E_1(x, t, u)| + \int_0^t |K_h(t, y)|\, |E_2(x, t, y, u)|\, dy \leqslant \frac{1}{2}(1 + |K_h|),$$

and by Theorem 4.1.1

$$|K_h| = \sup_{0 \leqslant t < \infty} \int_0^t |K_h(t, y)|\, dy < \infty.$$

For the function $r(s, x)$ we now have

$$r(s, x) = \frac{c}{s} \int_x^{\infty} q(t) \int_{-\infty}^{\infty} E_1(x, t, u) \sin su\, du\, dt$$

$$+ \int_x^{\infty} q(t) \int_{-\infty}^{\infty} E_3(x, t, u) \cos su\, du\, dt,$$

and if we change the order of integration[2] and use the notation

$$A(x, u) = \int_x^\infty q(t) E_1(x, t, u)\, dt, \quad B(x, u) = \int_x^\infty q(t) E_3(x, t, u)\, dt,$$

we obtain

$$r(s, x) = \frac{c}{s} \int_{-\infty}^{\infty} A(x, u) \sin su\, du + \int_{-\infty}^{\infty} B(x, u) \cos su\, du,$$

where

$$|A(x, u)| \leqslant \int_x^\infty |q(t)|\, |E_1(x, t, u)|\, dt \leqslant \frac{1}{2} \int_x^\infty |q(t)|\, dt,$$

$$|B(x, u)| \leqslant \int_x^\infty |q(t)|\, |E_3(x, t, u)|\, dt \leqslant \frac{1}{2} (1 + |K_h|) \int_x^\infty |q(t)|\, dt,$$

which is what we wanted to show.

Now we consider the functions $k_1(s)$ and $p_1(s)$.

If we substitute for $\omega_h(s^2, t)$ in (4.2.2) and (4.2.3) its expansion (4.1.8), we obtain

$$m(s) = 1 - \frac{c}{s^2} \alpha_{11}(s) - \frac{\alpha_{12}(s)}{s}, \tag{4.2.7}$$

$$n(s) = h + \frac{c}{s} \alpha_{21}(s) + \alpha_{22}(s), \tag{4.2.8}$$

where we use the notation

$$\int_x^\infty |q(t)| \left[\int_{-\infty}^{\infty} |E_1(x, t, u)|\, du \right] dt \leqslant \int_x^\infty |q(t)|\, (t - x)\, dt < \infty,$$

$$\int_x^\infty |q(t)| \left[\int_{-\infty}^{\infty} |E_3(x, t, u)|\, du \right] dt \leqslant (1 + |K_h|) \int_x^\infty |q(t)|\, (t - x)\, dt < \infty,$$

---

(2) Since $q(t)$ satisfies (4.1.1) and the functions $E_1(x, t, u)$ and $E_2(x, t, u)$ equal zero outside invervals of length $2(t - x)$ and are uniformly bounded, we know that

$$\alpha_{ik}(s) = \int_0^\infty \varphi_{ik}(s, t)\, q(t)\, dt, \qquad (i, k = 1, 2)$$

$$\varphi_{11}(s, t) = \sin^2 st, \qquad \varphi_{12}(s, t) = \sin st\, v_h(s, t),$$

$$\varphi_{21}(s, t) = \cos st \sin st, \qquad \varphi_{22}(s, t) = \cos st\, v_h(s, t),$$

and by virtue of Fubini's theorem we can change the order of integration.

$$v_h(s, t) = \cos st + \int_0^t K_h(t, y) \cos sy \, dy.$$

From the last equation it is clear that for a fixed t the function $v_h(s, t)$ is an even entire function of exponential type of order $\sigma = t$ relative to s. And by Theorem 4.1.1 we have

$$|v_h(s, t)| \leqslant 1 + \int_0^t |K_h(t, y)| \, dy \leqslant 1 + |K_h| < \infty$$

uniformly for all real s and $t \geqslant 0$.

Hence for fixed t the functions $\phi_{ik}(s, t)$ as functions of s are entire functions of exponential type of order $\sigma = 2t$, and are even if $i + k$ is even, and odd if $i + k$ is odd; moreover,

$$|\varphi_{ik}(s, t)| \leqslant 1 + |K_h| < \infty \qquad (i, k = 1 \text{ or } 2)$$

uniformly for all real s and $t \geqslant 0$.

Therefore by Bernstein's inequality we have

$$\left| \frac{\partial^n}{\partial s^n} \varphi_{ik}(s, t) \right| \leqslant (2t)^n (1 + |K_h|)$$

uniformly for all real s and $t \geqslant 0$.

From these inequalities and (4.1.1) it follows that

$$\frac{d^n}{ds^n} \alpha_{ik}(s) = \int_0^\infty \frac{\partial^n}{\partial s^n} \varphi_{ik}(s, t) \, q(t) \, dt,$$

$$\sup_{-\infty < s < \infty} \left| \frac{d^n}{ds^n} \alpha_{ik}(s) \right| \leqslant \int_0^\infty (1 + |K_h|)(2t)^n |q(t)| \, dt < \infty,$$

if n equals 0, 1, or 2.

Thus the functions $\alpha_{ik}(s)$ $(i, k = 1$ or $2)$ are twice continuously differentiable and are bounded together with their first two derivatives on the whole real line $-\infty < s < \infty$.

LEMMA 4.2.3. *The functions* $m(s)$ *and* $n(s)/s$ *are twice continuously differentiable at all points* $s \neq 0$, *and as* $|s| \to \infty$

$$m(s) = 1 + O\left(\frac{1}{s}\right), \quad m'(s) = O\left(\frac{1}{s}\right), \quad m''(s) = O\left(\frac{1}{s}\right),$$

$$\frac{n(s)}{s} = O\left(\frac{1}{s}\right), \quad \left(\frac{n(s)}{s}\right)' = O\left(\frac{1}{s}\right), \quad \left(\frac{n(s)}{s}\right)'' = O\left(\frac{1}{s}\right).$$

The functions $sm_1(s)$ *and* $n_1(s)$ *are twice continuously differentiable for all* $s \neq 0$, *and as* $|s| \to \infty$

$$sm_1(s)=O\left(\frac{1}{s}\right),\quad (sm_1(s))'=O\left(\frac{1}{s}\right),\quad (sm_1(s))''=O\left(\frac{1}{s}\right),$$
$$n_1(s)=1+O\left(\frac{1}{s}\right),\quad n_1'(s)=O\left(\frac{1}{s}\right),\quad n''(s)=O\left(\frac{1}{s}\right).$$

PROOF. The validity of the lemma for the functions $m(s)$ and $n(s)/s$ follows immediately from (4.2.7) and (4.2.8), since the functions $\alpha_{ik}(s)$ are twice continuously differentiable and are bounded together with their first two derivatives on $-\infty<s<\infty$. For the functions $sm_1(s)$ and $n_1(s)$ the lemma can be proved in the same way, using the expression (4.1.8′) for $\omega_\infty(s^2, t)$.

LEMMA 4.2.4. *The function* $m(s)+in(s)/s$ $(s\to 0)$ *can be written in the form*
$$m(s)+i\frac{n(s)}{s}=m(0)+i\frac{n(0)}{s}+As+\frac{\gamma(s)}{s}+\frac{n(0)}{s^2}\delta(s),$$
where A *is a constant, the functions* $\gamma(s)$ *and* $\delta(s)$ *are twice continuously differentiable, and*
$$\gamma(0)=\gamma'(0)=\gamma''(0)=\delta(0)=\delta'(0)=\delta''(0)=0.$$
The function $sm_1(s)+in_1(s)$ *can be written in the form*
$$sm_1(s)+in_1(s)=in_1(0)+m_1(0)s+\frac{\gamma_1(s)}{s},$$
where $\gamma_1(s)$ *is twice continuously differentiable and*
$$\gamma_1(0)=\gamma_1'(0)=\gamma_1''(0)=0.$$

PROOF. We have shown that in (4.2.7) and (4.2.8) the functions $\alpha_{ik}(s)$ are twice continuously differentiable, are even if $i+k$ is even, and are odd if $i+k$ is odd. Therefore
$$\alpha_{11}'(0)=\alpha_{22}'(0)=\alpha_{12}(0)=\alpha_{12}''(0)=\alpha_{21}(0)=\alpha_{21}''(0)=0.$$
Also
$$\alpha_{11}(s)=\int_0^\infty \sin^2 st\, q(t)\,dt$$
and thus $\alpha_{11}(0)=0$.

We can now write the functions $\alpha_{ik}(s)$ in the form
$$\alpha_{11}(s)=\frac{s^2}{2}\alpha_{11}''(0)+\beta_{11}(s),\quad \alpha_{12}(s)=s\alpha_{12}'(0)+\beta_{12}(s),$$
$$\alpha_{21}(s)=s\alpha_{21}'(0)+\beta_{21}(s),\quad \alpha_{22}(s)=\alpha_{22}(0)+\frac{s^2}{2}\alpha''_{2}(0)+\beta_{2\circ}(s),$$
where the functions $\beta_{ik}(s)$ are twice continuously differentiable and
$$\beta_{ik}(0)=\beta_{ik}'(0)=\beta_{ik}''(0)=0\qquad (i,\ k=1\ \text{or}\ 2).$$

If we substitute these expressions into (4.2.7) and (4.2.8), we find that

$$m(s)=1-\frac{c}{2}\alpha''_{11}(0)-\frac{c}{s^2}\beta_{11}(s)-\alpha'_{12}(0)-\frac{\beta_{12}(s)}{s},$$

$$n(s)=h+c\alpha'_{21}(0)+\frac{c}{s}\beta_{21}(s)+\alpha_{22}(0)+\frac{s^2}{2}\alpha''_{22}(0)+\beta_{22}(s),$$

and since $s^{-2}\beta_{ik}(s)\to 0$ as $s\to 0$, and $c=n(0)$ by (4.2.5), we have

$$m(s)=m(0)-\frac{\beta_{12}(s)}{s}-\frac{n(0)}{s^2}\beta_{11}(s),$$

$$n(s)=n(0)+\frac{s^2}{2}\alpha''_{22}(0)+\beta_{22}(s)+\frac{n(0)}{s}\beta_{21}(s).$$

Consequently

$$m(s)+i\frac{n(s)}{s}=m(0)+i\frac{n(0)}{s}+As+\frac{\gamma(s)}{s}+\frac{n(0)}{s^2}\delta(s),$$

where $A=\tfrac{1}{2}\alpha''_{22}(0)$ and $\gamma(s)=-\beta_{12}(s)+i\beta_{22}(s)$, the functions $\delta(s)=-\beta_{11}(s)+i\beta_{21}(s)$ are twice continuously differentiable, and

$$\gamma(0)=\gamma'(0)=\gamma''(0)=\delta(0)=\delta'(0)=\delta''(0)=0,$$

which is what we wanted to show.

LEMMA 4.2.5. *The function*

$$k_1(s)=\frac{s}{sm(s)+in(s)}-1$$

is continuously differentiable on the whole real line $-\infty<s<\infty$ *at least once if* $n(0)=0$ *and at least twice if* $n(0)\neq 0$, *and as* $|s|\to\infty$

$$k_1(s)=O\left(\frac{1}{s}\right),\quad k'_1(s)=O\left(\frac{1}{s}\right),\quad k''_1(s)=O\left(\frac{1}{s}\right).$$

If $n_1(0)\neq 0$, *then the function*

$$p_1(s)=\frac{1}{sm_1(s)+in_1(s)}+i$$

is continuously differentiable on the whole real line $-\infty<s<\infty$, *and as* $|s|\to\infty$

$$p_1(s)=O\left(\frac{1}{s}\right),\quad p'_1(s)=O\left(\frac{1}{s}\right).$$

PROOF. Since $m(s)+in(s)/s$ is not equal to zero for any real s (Lemma 4.2.1) and is twice continuously differentiable at all points $s\neq 0$ (Lemma 4.2.3), the function $k_1(s)$ is also twice continuously differentiable at all points $s\neq 0$. By Lemma 4.2.3 we know that as $|s|\to\infty$

$$k_1(s=\frac{s}{sm(s)+in(s)}-1=\frac{1-m(s)-i\frac{n(s)}{s}}{m(s)+i\frac{n(s)}{s}}=\frac{O\left(\frac{1}{s}\right)}{1+O\left(\frac{1}{s}\right)}=O\left(\frac{1}{s}\right),$$

$$k'_1(s)=-\frac{m'(s)+i\left[\frac{n(s)}{s}\right]'}{\left(m(s)+i\frac{n(s)}{s}\right)^2}=\frac{O\left(\frac{1}{s}\right)}{1+O\left(\frac{1}{s}\right)}=O\left(\frac{1}{s}\right),$$

$$k_1''(s) = -\frac{m''(s) + i\left[\frac{n(s)}{s}\right]''}{\left(m(s) + i\frac{n(s)}{s}\right)^2} + 2\frac{\left\{m'(s) + i\left[\frac{n(s)}{s}\right]'\right\}^2}{\left(m(s) + i\frac{n(s)}{s}\right)^3}$$

$$= \frac{o\left(\frac{1}{s}\right)}{1 + o\left(\frac{1}{s}\right)} + \frac{o\left(\frac{1}{s^2}\right)}{1 + o\left(\frac{1}{s}\right)} = o\left(\frac{1}{s}\right).$$

In order to complete the proof, it remains to consider $k_1(s)$ in a neighborhood of zero.

We note that if a continuous function $\phi(s)$ has n continuous derivatives at all points except perhaps at $s = 0$ and if these derivatives converge to finite limits as $s \to 0$, then $\phi(s)$ is n times continuously differentiable at all points (including $s = 0$).

We consider two cases.

1) Let $n(0) = 0$. For $s \neq 0$ we have by Lemma 4.2.4

$$k'(s) = -\frac{m'(s) + i\left[\frac{n(s)}{s}\right]'}{\left(m(s) + i\frac{n(s)}{s}\right)^2} = -\frac{A + \frac{\gamma'(s)}{s} - \frac{\gamma(s)}{s^2}}{\left(m(0) + As + \frac{\gamma(s)}{s}\right)^2},$$

where $m(0) \neq 0$ (this follows from Lemma 4.2.1 since $n(0) = 0$). From the properties of $\gamma(s)$ we know that

$$\lim_{s\to 0} \frac{\gamma(s)}{s^2} = \lim_{s\to 0} \frac{\gamma'(s)}{s} = 0.$$

Therefore the limit

$$\lim_{s\to 0} k_1'(s) = -\frac{A}{m^2(0)}$$

exists and is finite, and by the remarks above we know that $k_1(s)$ is continuously differentiable on the whole real line $-\infty < s < \infty$.

2) Let $n(0) \neq 0$. We have

$$k_1(s) = \frac{s}{sm(s) + in(s)} - 1$$

and if $s \neq 0$, then

$$k_1'(s) = \frac{1}{sm(s) + in(s)} - \frac{s(sm(s) + in(s))'}{(sm(s) + in(s))^2},$$

$$k_1''(s) = -\frac{2[sm(s) + in(s)]'}{[sm(s) + in(s)]^2} - \frac{s[sm(s) + in(s)]''}{[sm(s) + in(s)]^2} + \frac{2s[(sm(s) + in(s))']^2}{[sm(s) + in(s)]^3}.$$

As $s \to 0$ the function $sm(s) + in(s)$ converges to the nonzero limit $in(0)$. Hence for the proof of the existence and finiteness of the limits of $k_1'(s)$ and $k_1''(s)$ as $s \to 0$ it suffices to prove the existence and finiteness of the limits

$$\lim_{s\to 0}[sm(s) + in(s)]'; \quad \lim_{s\to 0} s[sm(s) + in(s)]''.$$

By Lemma 4.2.4 we have

$$[sm(s)+in(s)]' = m(0) + 2As + \gamma'(s) + \frac{n(0)}{s}\delta'(s) - \frac{n(0)}{s^2}\delta(s),$$

$$s[sm(s)+in(s)]'' = 2As + s\gamma''(s) + n(0)\delta''(s) - \frac{2n(0)}{s}\delta'(s) + \frac{2n(0)}{s^2}\delta(s).$$

From the properties of $\gamma(s)$ and $\delta(s)$ we can see that

$$\lim_{s\to 0}\gamma'(s) = \lim_{s\to 0}\gamma''(s) = \lim_{s\to 0}\frac{\delta(s)}{s^2} = \lim_{s\to 0}\frac{\delta'(s)}{s} = \lim_{s\to 0}\delta''(s) = 0.$$

Therefore the following limits exist:

$$\lim_{s\to 0}[sm(s)+in(s)]' = m(0), \qquad \lim_{s\to 0} s[sm(s)+in(s)]'' = 0,$$

and along with them the limits

$$\lim_{s\to 0} k_1'(s) = \frac{1}{in(0)}, \qquad \lim_{s\to 0} k_1''(s) = \frac{2m(0)}{n(0)}$$

also exist; from the remarks above we now know that $k_1(s)$ is twice continuously differentiable on $-\infty < s < \infty$.

The lemma is proved.

LEMMA 4.2.6. *The functions*

$$k_2(s) = \frac{cs}{s^2m^2(s)+n^2(s)}, \qquad k_3(s) = \frac{s^2}{s^2m^2(s)+n^2(s)} - 1 \qquad (c = n(0))$$

are continuously differentiable on $-\infty < s < \infty$, *and*

$$k_2(s) = O\left(\frac{1}{s}\right), \quad k_2'(s) = O\left(\frac{1}{s}\right), \quad k_3(s) = O\left(\frac{1}{s}\right), \quad k_3'(s) = O\left(\frac{1}{s}\right)$$

as $|s| \to \infty$. *If* $n_1(0) \neq 0$, *the function*

$$p_2(s) = \frac{1}{s^2m_1^2(s)+n_1^2(s)} - 1$$

is continuously differentiable on $-\infty < s < \infty$, *and*

$$p_2(s) = O\left(\frac{1}{s}\right), \qquad p_2'(s) = O\left(\frac{1}{s}\right)$$

as $|s| \to \infty$.

PROOF. From the definition of $k_1(s)$ (see Lemma 4.2.5) and $k_3(s)$ it follows that

$$(k_1(s)-1)(k_1(-s)-1) = k_3(s)+1;$$

that is,

$$k_3(s) = k_1(s)k_1(-s) - k_1(s) - k_1(-s),$$

from which it is clear that the assertions concerning $k_3(s)$ follow immediately from Lemma 4.2.5. Furthermore

$$k_2(s) = \frac{c}{s}[k_3(s)+1] = \frac{c}{s}[k_1(s)k_1(-s) - k_1(s) - k_1(-s) + 1] = \frac{cf(s)}{s},$$

where

$$f(s) = k_1(s)k_1(-s) - k_1(s) - k_1(-s) + 1.$$

If $n(0)=0$, then $k_2(s)\equiv 0$ (since $c=n(0)$), and the lemma is trivially true.

If $n(0)\neq 0$, then by Lemma 4.2.5 the function $k_1(s)$, and hence $f(s)$, is everywhere twice continuously differentiable, and thus $f'(0)=0$, since $f(s)$ is even. Also

$$f(0) = \lim_{s\to 0}\frac{s}{c}k_2(s) = \lim_{s\to 0}\frac{s^2}{s^2m^2(s)+n^2(s)} = 0.$$

Therefore $f(s)$ can be written in the form

$$f(s) = \frac{s^2}{2}f''(0) + f_1(s),$$

where $f_1(s)$ is twice continuously differentiable and

$$f_1(0) = f_1'(0) = f_1''(0) = 0.$$

The remainder of the proof is completely analogous to the proof of the preceding lemma.

REMARK. Since $m(s)$ and $n(s)$ (or $m_1(s)$ and $n_1(s)$) are even, $k_2(s)$ is odd and $k_3(s)$ ($p_2(s)$) is even.

THEOREM 4.2.1. *The functions* $s^2\omega_h(s^2,x)/(s^2m^2(s)+n^2(s))$ *can be written in the form*

$$\frac{s^2\omega_h(s^2,x)}{s^2m^2(s)+n^2(s)} = \cos sx + \frac{1}{2}[k_1(s)e^{isx} + k_1(-s)e^{-isx}]$$
$$+ k_2(s)\int_{-\infty}^{\infty} A(x,u)\sin su\,du + [k_3(s)+1]\int_{-\infty}^{\infty} B(x,u)\cos su\,du,$$

where the integrals are absolutely convergent and the following conditions are satisfied.

1) *For* $A(x,u)$ *and* $B(x,u)$ *we have the bounds*

$$|A(x,u)| \leqslant \frac{1}{2}\int_x^{\infty}|q(t)|\,dt, \quad |B(x,u)| \leqslant \frac{1}{2}(1+|K_h|)\int_x^{\infty}|q(t)|\,dt.$$

2) *The functions* $k_1(s)$, $k_2(s)$ *and* $k_3(s)$ *are the Fourier transforms of the functions* $h_1(x)$, $h_2(x)$ *and* $h_3(x)$ *which are integrable on the whole real line; that is,*

$$k_j(s) = \int_{-\infty}^{\infty} h_j(x)\, e^{-isx}\, dx; \qquad \int_{-\infty}^{\infty} |h_j(x)|\, dx < \infty \qquad (j = 1, 2, 3).$$

3) *If* $n(0) \neq 0$, *then for* $h_1(x)$ *we have the stronger inequality*

$$\int_{-\infty}^{\infty} (1 + |x|)\, |h_1(x)|\, dx < \infty.$$

PROOF. If we substitute for $r(s, x)$ in (4.2.6) its representation given in Lemma 4.2.2, we obtain

$$\frac{s^2 \omega_h(s^2, x)}{s^2 m^2(s) + n^2(s)} = \cos sx + \frac{1}{2}[k_1(s) e^{isx} + k_1(-s) e^{-isx}]$$

$$+ \frac{cs}{s^2 m^2(s) + n^2(s)} \int_{-\infty}^{\infty} A(x, u) \sin su\, du + \frac{s^2}{s^2 m^2(s) + n^2(s)} \int_{-\infty}^{\infty} B(x, u) \cos su\, du,$$

and then by the definition of $k_2(s)$ and $k_3(s)$ (see Lemma 4.2.6)

$$\frac{s^2 \omega_h(s^2, x)}{s^2 m^2(s) + n^2(s)} = \cos sx + \frac{1}{2}[k_1(s) e^{isx} + k_1(-s) e^{-isx}]$$

$$+ k_2(s) \int_{-\infty}^{\infty} A(x, u) \sin su\, du + [k_3(s) + 1] \int_{-\infty}^{\infty} B(x, u) \cos su\, du.$$

The desired representation of $s^2 \omega_h(s^2, x)/(s^2 m^2(s) + n^2(s))$ has been achieved.

Assertion 1) was proved in Lemma 4.2.2.

Furthermore, from the bounds obtained in Lemmas 4.2.5 and 4.2.6 it follows that the functions $k_j(s)$, $j = 1, 2, 3$, and their first derivatives are square-integrable on $-\infty < s < \infty$. Hence by classical Fourier integral theory the sequences of functions

$$\frac{1}{2\pi} \int_{-N}^{N} k_j(s)\, e^{isx}\, ds \qquad (j = 1, 2, 3)$$

as $N \to \infty$ converge in the metric of $L^2(-\infty, \infty)$ to the functions $h_j(x)$, $j = 1, 2, 3$, whose Fourier transforms equal $k_j(s)$:

$$h_j(x) = \frac{1}{2\pi} \int_{-\infty}^{\infty} k_j(s)\, e^{isx}\, ds, \quad k_j(s) = \int_{-\infty}^{\infty} h_j(x)\, e^{-isx}\, dx \qquad (j = 1, 2, 3)$$

(here the integrals converge in the metric of $L^2(-\infty, \infty)$).

For the proof of assertion 2) it remains to show that

$$\int_{-\infty}^{\infty} |h_j(x)|\, dx < \infty \qquad (j = 1, 2, 3).$$

Integrating by parts, we find that

$$\int_{-N}^{N} k_j(s)\, e^{isx}\, ds = \frac{k_j(s) e^{isx}}{ix}\Bigg|_{-N}^{N} - \frac{1}{ix}\int_{-N}^{N} k_j'(s)\, e^{isx}\, ds$$

and since $k_j(s) = O(1/s)$ and $k_j'(s) = O(1/s)$ as $|s| \to \infty$, we have

$$h_j(x) = -\frac{1}{2ix\pi}\int_{-\infty}^{\infty} k_j'(s)\, e^{isx}\, ds \qquad (j = 1, 2, 3),$$

where the integral converges in the metric of $L^2(-\infty, \infty)$ to some function $g_j(x) \in L^2(-\infty, \infty)$. Therefore

$$\int_{-\infty}^{\infty} |h_j(x)|\, dx = \int_{-1}^{1} |h_j(x)|\, dx + \int_{1}^{\infty} \left|\frac{g_j(x)}{x}\right| dx + \int_{-\infty}^{-1} \left|\frac{g_j(x)}{x}\right| dx$$

and by Bunjakovskiĭ's inequality

$$\int_{-8}^{\infty} h_j(x)|\, dx \leqslant \int_{-1}^{1} |h_j(x)|\, dx + \left\{\int_{1}^{\infty} |g_j(x)|^2\, dx \int_{1}^{\infty} \frac{dx}{x^2}\right\}^{\frac{1}{2}}$$

$$+ \left\{\int_{-\infty}^{-1} |g_j(x)|^2\, dx \int_{-\infty}^{-1} \frac{dx}{x^2}\right\}^{\frac{1}{2}} < \infty \qquad (j = 1, 2, 3),$$

which proves assertion 2).

The proof of assertion 3) is similar. We note only that by Lemma 4.2.5 with $n(0) \neq 0$ the function $k_1(s)$ is twice continuously differentiable and

$$k_1(s) = O\left(\frac{1}{s}\right), \quad k_1'(s) = O\left(\frac{1}{s}\right), \quad k_1''(s) = O\left(\frac{1}{s}\right) \qquad ((s) \to \infty).$$

The theorem is proved.

Similarly we can prove

THEOREM 4.2.1′. *The functions* $s^2\omega_\infty(s^2, x)/(s^2m_1^2(s) + n_1^2(s))$ *can be written in the form*

$$\frac{s^2\omega_\infty(s^2, x)}{s^2m_1^2(s) + n_1^2(s)} = s\left\{\sin sx + \frac{1}{2}\left[p_1(s)\, e^{isx} - p_1(-s)\, e^{-isx}\right]\right.$$

$$\left. + [p_2(s) + 1]\int_{-\infty}^{\infty} A_1(x, u) \sin su\, du\right\},$$

where the integrals are absolutely convergent and

1) *for* $A_1(x, u)$ *we have the bound*

$$|A_1(x, u)| \leqslant \frac{1}{2}(1 + |K_\infty|)\int_{x}^{\infty} |q(t)|\, dt;$$

2) *if* $n_1(0) \neq 0$, *then* $p_1(s)$ *and* $p_2(s)$ *are the Fourier transforms of*

the functions $g_1(x)$ *and* $g_2(x)$ *which are integrable on the whole real line; that is,*

$$p_j(s)=\int_{-\infty}^{\infty} g_j(x)\,e^{-isx}\,dx,\qquad \int_{-\infty}^{\infty}|g_j(x)|\,dx<\infty\qquad (j=1,\,2).$$

REMARK ON THEOREMS 4.2.1 AND 4.2.1′. From (4.2.5) and (4.2.5′) we have

$$n(0)=\lim_{x\to\infty}\omega_h'(0,\,x),\qquad n_1(0)=\lim_{x\to\infty}\omega_\infty'(0,\,x).$$

Hence if $n(0)=0$ ($n_1(0)=0$), then $\omega_h'(0,x)=o(1)$ ($\omega_\infty'(0,x)=o(1)$) as $x\to\infty$. By Lemma 2.1.2 (see part I of this article) this means that the solution $\omega_h(0,x)$ ($\omega_\infty(0,x)$) is bounded on $0\leqslant x<\infty$. Consequently $n(0)=0$ ($n_1(0)=0$) if and only if the solution $\omega_h(0,x)$ ($\omega_\infty(0,x)$) is bounded on $0\leqslant x<\infty$.

§3. Structure of the operators $V_{\{LD^2h0\}}$ and $V_{\{LD^2\infty\infty\}}$

We introduce the following normed linear spaces of functions:

$L^2[0,\infty)$ with norm $\|f\|_{L^2}=\left[\int_0^\infty |f(t)|^2\,dt\right]^{\frac{1}{2}}$,

$L^1[0,\infty)$ with norm $\|f\|_{L^1}=\int_0^\infty |f(t)|\,dt$,

$L_x[0,\infty)$ with norm $\|f\|_{L_x}=\int_0^\infty (1+t)\,|f(t)|\,dt$,

$M_1[0,\infty)$ with norm $\|F\|_{M_1}=\operatorname*{ess\,sup}_{0\leqslant x<\infty}|F(x)|$,

$M_x[0,\infty)$ with norm $\|F\|_{M_x}=\operatorname*{ess\,sup}_{0\leqslant x<\infty}|(1+x)^{-1}F(x)|$.

For the sake of brevity we will often denote these spaces by L^2, L^1, L_x, M_1 and M_x. As a rule we will denote elements of the spaces M_1 and M_x by upper-case letters F, $F(x)$ and we will denote elements of L^2, L^1, and L_x by lower-case letters f, $f(x)$, and so forth.

We will let $\mathfrak{A}$ denote the set of twice continuously differentiable functions on $0\leqslant x<\infty$ which are zero outside of some finite interval. The set $\mathfrak{A}$ is contained in each of the spaces just listed and is dense in L^2, L^1 and L_x.

In functional analysis it is shown that a function $\Phi(x)$ which is integrable in each finite subinterval of $0\leqslant x<\infty$ satisfies the equations

$$\|\Phi\|_{L^1}=\lim_{N\to\infty}\int_0^N|\Phi(t)|\,dt=\sup_{h\in\mathfrak{A}}\frac{1}{\|h\|_{M_1}}\left|\int_0^\infty \Phi(t)\,h(t)\,dt\right|,\qquad(4.3.1)$$

$$\|\Phi\|_{L_x} = \lim_{N\to\infty} \int_0^N (1+t)|\Phi(t)|\,dt = \sup_{h\in\mathfrak{A}} \frac{1}{\|h\|_{M_x}} \left| \int_0^\infty \Phi(t)\,h(t)\,dt \right|, \tag{4.3.1'}$$

$$\|\Phi\|_{M_1} = \operatorname*{ess\,sup}_{0\leqslant x<\infty} |\Phi(x)| = \sup_{h\in\mathfrak{A}} \frac{1}{\|h\|_{L^1}} \left| \int_0^\infty \Phi(t)\,h(t)\,dt \right|, \tag{4.3.2}$$

$$\|\Phi\|_{M_x} = \operatorname*{ess\,sup}_{0\leqslant x<\infty} |(1+x)^{-1}\Phi(x)| = \sup_{h\in\mathfrak{A}} \frac{1}{\|h\|_{L_x}} \left| \int_0^\infty \Phi(t)\,h(t)\,dt \right|. \tag{4.3.2'}$$

If the right side of any of these equations does not go to infinity (does go to infinity), then $\Phi(x)$ belongs (does not belong) to the corresponding space.

By an operator acting on one of these spaces we mean an additive homogeneous operation which associates with each function in the space (or with each function in a dense subset of the space) some function which is integrable in each finite subinterval of $0 \leqslant x < \infty$ but which need not belong to the space.

If an operator A acts on X (here X is one of the spaces L^2, L^1, L_x, M_1, M_x), then the norm of A in this space is defined to be

$$|A|_X = \sup_{f\in O_A} \frac{\|A[f]\|_X}{\|f\|_X},$$

where $O_A \subset X$ is the domain of definition of A.

This section is devoted to studying the structure of the transformation operators $V_{\{LD^2h0\}}$ and $V_{\{LD^2\infty\infty\}}$ when the differential operator L satisfies (4.1.1). Of prime interest is the behavior of these operators on functions in M_1 and M_x. Since the investigations of the two operators are analogous, we will usually consider one operator $V_{\{LD^2h0\}}$ in detail, and for $V_{\{LD^2\infty\infty\}}$ we will only formulate the final result (Theorem 4.3.1′).

We recall some facts from spectral analysis for differential operators L which satisfy (4.1.1).

As we remarked earlier (Chapter 2, §1, part F), in the case we are considering we know that a boundary condition at zero

$$y'(0) - hy(0) = 0$$

uniquely determines a boundary value problem for L. The spectrum of this boundary value problem consists of a finite number of negative eigenvalues $\lambda_1 < \lambda_2 < \cdots < \lambda_n < 0$ and the whole nonnegative semi-axis $0 \leqslant \lambda < \infty$.

We set

$$s_1^2 = \lambda_1, \quad s_2^2 = \lambda_2, \quad \ldots, \quad s_n^2 = \lambda_n.$$

Then the inversion formulas generated by the boundary value problem can be written in the form

$$f(x)=\sum_{\nu=1}^{n} E_f(s_\nu^2)\frac{\omega_h(s_\nu^2, x)}{\|\omega_h(s_\nu^2, x)\|_{L^2}^2}+\frac{2}{\pi}\int_0^\infty E_f(s^2)\frac{s^2\omega_h(s^2, x)}{s^2m^2(s)+n^2(s)}\,ds, \tag{4.3.3}$$

$$E_f(s^2)=\int_0^\infty f(x)\,\omega_h(s^2, x)\,dx, \tag{4.3.3'}$$

where, as in the preceding section, $\omega_h(s^2, x)$ denotes the solution of $L[u]+s^2u=0$ with initial conditions $u(0)=1$ and $u'(0)=h$; and

$$m(s)=1-\frac{1}{s}\int_0^\infty \sin st\,\omega_h(s^2, t)\,q(t)\,dt,$$

$$n(s)=h+\int_0^\infty \cos st\,\omega_h(s^2, t)\,q(t)\,dt.$$

By Corollary 1 of Lemma 2.1.2 (see part I) we have the following asymptotic formulas for the eigenfunctions $\omega_h(s_\nu^2, x)$ corresponding to the negative eigenvalues:

$$\omega_h(s_\nu^2, x)=O(e^{-|s_\nu|x})\qquad (x\to\infty),$$

and therefore the set of linear combinations of the eigenfunctions $\omega_h(s_\nu^2, x)$, $\nu=1,\cdots,n$, is a finite-dimensional subspace of each of the spaces at the start of this section.

We introduce the projection operator $P_{\{Lh\}}=P$ onto this subspace by means of the formula

$$P[f]=\int_0^\infty\sum_{\nu=1}^{n}\frac{\omega_h(s_\nu^2, x)\,\omega_h(s_\nu^2, t)}{\|\omega_h(s_\nu^2, x)\|_{L^2}^2}f(t)\,dt \tag{4.3.4}$$

for all functions $f(t)$ satisfying

$$\int_0^\infty e^{-|s_n|t}|f(t)|\,dt<\infty, \tag{4.3.4'}$$

where $|s_n|=\min_{1\leqslant\nu\leqslant n}|s_\nu|$. Since this condition is satisfied by all of the functions in each of our spaces, the operator $P=P_{\{Lh\}}$ is defined on all of these spaces.

We turn now to the study of the structure of the transformation operator $V=V_{\{LD^2h0\}}$. We have

$$V[F]=F(x)+\int_0^x H(x, t)\,F(t)\,dt,$$

where the kernel $H(x, t)$ is bounded in each finite portion of the (x, t)-plane.

On all functions satisfying (4.3.4′) (and thus on all functions in our spaces) the operator V can be written in the form

$$V = VP + V(E - P),$$

where E is the identity operator. From the definition of P we have

$$VP[F] = V\left[\int_0^\infty \sum_{\nu=1}^n \frac{\omega_h(s_\nu^2, x)\,\omega_h(s_\nu^2, t)}{\|\omega_h(s_\nu^2, x)\|_{L2}^2} F(t)\,dt\right]$$

$$= \int_0^\infty \sum_{\nu=1}^n \frac{V[\omega_h(s_\nu^2, x)]}{\|\omega_h(s_\nu^2, x)\|_{L2}^2}\,\omega_h(s_\nu^2, t)\,F(t)\,dt$$

and, since $s_\nu^2 = \lambda_\nu < 0$ $(\nu = 1, \cdots, n)$, then

$$V[\omega_h(s_\nu^2, x)] = \cos s_\nu x = \operatorname{ch}|s_\nu|\,x,$$

which implies that

$$VP[F] = \int_0^\infty \sum_{\nu=1}^n \frac{\cosh|s_\nu|x}{\|\omega_h(s_\nu^2, x)\|_{L2}^2}\,\omega_h(s_\nu^2, t)\,F(t)\,dt.$$

But then

$$V(E - P)[F] = F(x) + \int_0^\infty \overset{+}{H}(x, t)\,F(t)\,dt, \tag{4.3.5}$$

where

$$\overset{+}{H}(x, t) = H(x, t) - \sum_{\nu=1}^n \frac{\cosh|s_\nu|x\,\omega_h(s_\nu^2, t)}{\|\omega_h(s_\nu^2, x)\|_{L2}^2} \tag{4.3.5'}$$

and $H(x, t) = 0$ for $t > x$.

We have proved the following lemma.

LEMMA 4.3.1. *On all functions $F(x)$ satisfying (4.3.4′) the transformation operator $V_{\{LD^2h0\}}$ can be written in the form*

$$V_{\{LD^2h0\}} = \overline{V}_{\{LD^2h0\}} + \overset{+}{V}_{\{LD^2h0\}},$$

where

$$\overline{V}_{\{LD^2h0\}}[F] = \int_0^\infty \sum_{\nu=1}^n \frac{\cosh|s_\nu|x\,\omega_h(s_\nu^2, t)}{\|\omega_h(s_\nu^2, x)\|_{L2}^2} F(t)\,dt,$$

$$\overset{+}{V}_{\{LD^2h0\}}[F] = F(x) + \int_0^\infty \overset{+}{H}(x, t)\,F(t)\,dt,$$

and $s_1^2 < s_2^2 < \cdots < s_n^2 < 0$ *are the negative eigenvalues of the boundary value problem for* L *with boundary condition*

$$y'(0) - hy(0) = 0.$$

The significance of this lemma will become clear when we prove the boundedness of the operator $\overset{+}{V} = \overset{+}{V}_{\{LD^2h0\}}$ in the space M_1. In that proof we will use a rather roundabout but natural method.(3)

We introduce the operator $\overset{+}{U} = \overset{+}{U}_{\{LD^2h0\}}$ defined by the formula

$$\overset{+}{U}[f] = f(x) + \int_0^\infty f(t)\overset{+}{H}(t, x)\,dt \tag{4.3.6}$$

on all functions $f(x)$ in $\mathfrak{A}$. If $f(x) \in \mathfrak{A}$ then $\overset{+}{U}[f]$ belongs to each of the spaces L^1, L_x and L^2. A function $f(x) \in \mathfrak{A}$ is bounded and equals zero outside some interval $(0, N)$.

By (4.3.5′) we have

$$\int_0^\infty f(t)\overset{+}{H}(t, x)\,dt = \int_0^\infty f(t) H(t, x)\,dt - \int_0^\infty \sum_{\nu=1}^n \frac{\omega_h(s_\nu^2, x)}{\|\omega_h(s_\nu^2, x)\|_{L^2}^2} \cosh|s_\nu|\, tf(t)\,dt$$

and since $H(t, x) = 0$ for $x > t$, the function $\int_0^\infty f(t)H(t, x)\,dt$ is also bounded and equal to zero outside $(0, N)$. Furthermore,

$$\int_0^\infty \sum_{\nu=1}^n \frac{\omega_h(s_\nu^2, x)}{\|\omega_h(s_\nu^2, x)\|_{L^2}^2} \cosh|s_\nu|\, tf(t)\,dt = \sum_{\nu=1}^n a_\nu \omega_h(s_\nu^2, x),$$

where

$$a_\nu = \int_0^N \frac{\cosh|s_\nu|t}{\|\omega_h(s_\nu^2, x)\|_{L^2}^2} f(t)\,dt. \qquad (\nu = 1, 2, \ldots, n).$$

From the asymptotic formulas for the eigenfunctions $\omega_h(s_\nu^2, x)$ it follows that

$$\sum_{\nu=1}^n a_\nu \omega_h(s_\nu^2, x) = O(e^{-|s_n| x}),$$

where

$$|s_n| = \min_{1 \leq \nu \leq n} |s_\nu|.$$

Finally we have

(3) In the terminology of functional analysis the method consists of switching from the operator $\overset{+}{V}$ mapping M_1 into itself to the adjoint operator $\overset{+}{U}$ mapping L^1 (the dual space of M_1) into itself. Since the norms of adjoint operators are equal, if we establish the fact that $\overset{+}{U}$ is bounded, we then know that $\overset{+}{V}$ is bounded.

$$|\overset{+}{U}[f]| \leqslant Ce^{-|s_n| x};$$

where C is some constant. This inequality shows that $\overset{+}{U}[f]$ belongs to L^1, L_x and L^2 if $f(x) \in \mathfrak{A}$, which is what we wanted to show.

LEMMA 4.3.2. *If* $F(x)$ *satisfies* (4.3.4′) *and* $f(x) \in \mathfrak{A}$, *then*

$$\int_0^\infty f(x)\, \overset{+}{V}[F(x)]\, dx = \int_0^\infty \overset{+}{U}[f(x)]\, F(x)\, dx.$$

PROOF. We have

$$\int_0^\infty f(x) \overset{+}{V}[F(x)]\, dx = \int_0^\infty f(x)\Big[F(x) + \int_0^\infty \overset{+}{H}(x, t)\, F(t)\, dt\Big]\, dx$$

$$= \int_0^\infty f(x)\, F(x)\, dx + \int_0^\infty f(x) \int_0^\infty \overset{+}{H}(x, t)\, F(t)\, dt\, dx.$$

Since the function $f(x) \in \mathfrak{A}$ equals zero outside some finite interval, we can interchange the order of integration in the last integral, and we find that

$$\int_0^\infty f(x)\overset{+}{V}[F(x)]\, dx = \int_0^\infty f(x)\, F(x)\, dx + \int_0^\infty F(t) \int_0^\infty \overset{+}{H}(x, t)\, f(x)\, dx\, dt,$$

or

$$\int_0^\infty f(x)\overset{+}{V}[F(x)]\, dx = \int_0^\infty F(x)\, [f(x) + \int_0^\infty \overset{+}{H}(t, x)\, f(t)\, dt]\, dx = \int_0^\infty F(x)\overset{+}{U}[f(x)]\, dx,$$

and the lemma is proved.

The operator $\overset{+}{U}$ is defined on all of $\mathfrak{A}$, which is dense in L^1 and in L_x. Thus we can consider it as an operator given by (4.3.6) on a dense subspace of L^1 or L_x. We define the usual norms:

$$|\overset{+}{U}|_{L^1} = \sup_{f \in \mathfrak{A}} \frac{\|\overset{+}{U}[f]\|_{L^1}}{\|f\|_{L^1}},$$

if we look at the operator in L^1, and

$$|\overset{+}{U}|_{L_x} = \sup_{f \in \mathfrak{A}} \frac{\|\overset{+}{U}[f]\|_{L_x}}{\|f\|_{L_x}},$$

if we look at the operator in L_x.

LEMMA 4.3.3. *The norm* $|\overset{+}{V}|_{M_1}$ $(|\overset{+}{V}|_{M_x})$ *of* $\overset{+}{V} = \overset{+}{V}_{\{LD^2h0\}}$ *in* M_1 (M_x) *equals the norm* $|\overset{+}{U}|_{L^1}$ $(|\overset{+}{U}|_{L_x})$ *of* $\overset{+}{U} = \overset{+}{U}_{\{LD^2h0\}}$ *in* L^1 (L_x); *that is,*

$$\sup_{F\in M_1} \frac{\| \overset{+}{V}[F] \|_{M_1}}{\| F \|_{M_1}} = \sup_{f\in\mathfrak{A}} \frac{\| \overset{+}{U}[f] \|_{L^1}}{\| f \|_{L^1}},$$

$$\left(\sup_{F\in M_x} \frac{\| \overset{+}{V}[F] \|_{M_x}}{\| F \|_{M_x}} = \sup_{f\in\mathfrak{A}} \frac{\| \overset{+}{U}[f] \|_{L_x}}{\| f \|_{L_x}} \right).$$

PROOF. By (4.3.2)

$$\sup_{F\in M_1} \frac{\| \overset{+}{V}[F] \|_{M_1}}{\| F \|_{M_1}} = \sup_{F\in M_1} \frac{1}{\| F \|_{M_1}} \sup_{f\in\mathfrak{A}} \frac{1}{\| f \|_{L^1}} \left| \int_0^\infty \overset{+}{V}[F(x)]\, f(x)\, dx \right|,$$

and then by Lemma 4.3.2

$$\sup_{F\in M_1} \frac{\| \overset{+}{V}[F] \|_{M_1}}{\| F \|_{M_1}} = \sup_{\substack{F\in M_1\\ f\in\mathfrak{A}}} \frac{1}{\| F \|_{M_1} \| f \|_{L^1}} \left| \int_0^\infty \overset{+}{U}[f(x)]\, F(x)\, dx \right|.$$

On the other hand, from (4.3.1) we have

$$\sup_{F\in M_1} \frac{1}{\| F \|_{M_1}} \left| \int_0^\infty \overset{+}{U}[f(x)]\, F(x)\, dx \right| = \| \overset{+}{U}[f] \|_{L^1}$$

and thus

$$\sup_{F\in M_1} \frac{\| \overset{+}{V}[F] \|_{M_1}}{\| F \|_{M_1}} = \sup_{f\in\mathfrak{A}} \frac{\| \overset{+}{U}[f] \|_{L^1}}{\| f \|_{L^1}},$$

or $|\overset{+}{V}|_{M_1} = |\overset{+}{U}|_{L^1}$, which is what we wanted to show.

The proof for the spaces M_x and L_x is by analogy.

The lemma is proved.

In our investigation of the operator $\overset{+}{U} = \overset{+}{U}_{\{LD^2h0\}}$ the following lemma plays an essential role.

LEMMA 4.3.4. *If* $f(x) \in \mathfrak{A}$, *then*

$$\overset{+}{U}_{\{LD^2h0\}}[f] = \frac{2}{\pi} \int_0^\infty C_f(s) \frac{s^2 \omega_h(s^2, x)}{s^2 m^2(s) + n^2(s)}\, ds,$$

where

$$C_f(s) = \int_0^\infty f(x) \cos sx\, dx,$$

and the integral is absolutely convergent.

PROOF. As we proved earlier, the function $g(x) = \overset{+}{U}_{\{LD^2h0\}}[f(x)]$ belongs to each of the spaces L^2, L^1 and L_x. Thus we can apply the inversion formulas (4.3.3) and (4.3.3′). We have

$$E_g(s^2) = \int_0^\infty g(x)\,\omega_h(s^2, x)\,dx = \int_0^\infty \overset{+}{U}_{\{LD^2h0\}}[f(x)]\,\omega_h(s^2, x)\,dx,$$

which according to Lemma 4.3.2 implies that

$$E_g(s^2) = \int_0^\infty f(x)\,\overset{+}{V}_{\{LD^2h0\}}[\omega_h(s^2, x)]\,dx,$$

where

$$\overset{+}{V}_{\{LD^2h0\}}[\omega_h(s^2, x)] = V_{\{LD^2h0\}}(E - P_{\{Lh\}})[\omega_h(s^2, x)].$$

Let s^2 equal the eigenvalue s_ν^2. From the definition of $P_{\{Lh\}}$ it is clear that $P_{\{Lh\}}[\omega_h(s_\nu^2, x)] = \omega_h(s_\nu^2, x)$; that is, $(E - P_{\{Lh\}})[\omega_h(s_\nu^2, x)] \equiv 0$. Therefore

$$\overset{+}{V}_{\{LD^2h0\}}[\omega_h(s_\nu^2, x)] \equiv 0$$

and

$$E_g(s_\nu^2) = 0 \qquad (\nu = 1, 2, \ldots, n). \tag{4.3.7}$$

Now let $s^2 \geqslant 0$. It is easy to see that $P_{\{Lh\}}[\omega_h(s^2, x)] \equiv 0$. Consequently

$$\overset{+}{V}_{\{LD^2h0\}}[\omega_h(s^2, x)] = V_{\{LD^2h0\}}[\omega_h(s^2, x)] = \cos sx$$

and

$$E_g(s^2) = \int_0^\infty f(x)\cos sx\,dx = C_f(s) \qquad (s^2 \geqslant 0). \tag{4.3.7'}$$

By using (4.3.3) and (4.3.7), (4.3.7′) we obtain

$$g(x) = \overset{+}{U}_{\{LD^2h0\}}[f(x)] = \frac{2}{\pi}\int_0^\infty C_f(s)\,\frac{s^2\omega_h(s^2, x)}{s^2m^2(s) + n^2(s)}\,ds.$$

To conclude the proof we have to show that this integral converges absolutely. Since $f(x) \in \mathfrak{A}$ is twice continuously differentiable and equal to zero outside some finite interval, integration by parts gives us

$$C_f(s) = \int_0^\infty f(x)\cos sx\,dx = \frac{f'(0)}{s^2} - \frac{1}{s^2}\int_0^\infty f''(x)\cos sx\,dx;$$

that is, $C_f(s) = O(s^{-2})$ as $|s| \to \infty$, which implies absolute convergence of the integral.

The lemma is proved.

Now we can give a complete answer to the question of whether $\overset{+}{U}_{\{LD^2h0\}}$ is bounded in L^1 and in L_x.

LEMMA 4.3.5. *The operator* $\overset{+}{U}_{\{LD^2h0\}}$ *is bounded in the space* $L^1[0, \infty)$; *that is,*

$$|\overset{+}{U}_{\{LD^2h0\}}|_{L^1} = \sup_{f\in\mathfrak{A}} \frac{\|\overset{+}{U}_{\{LD^2h0\}}[f]\|_{L^1}}{\|f\|_{L^1}} < \infty.$$

If $\omega_h'(0, \infty) = n(0) \neq 0$, *then* $\overset{+}{U}_{\{LD^2h0\}}$ *is also bounded in* $L_x[0, \infty)$; *that is,*

$$|\overset{+}{U}_{\{LD^2h0\}}|_{L_x} = \sup_{f\in\mathfrak{A}} \frac{\|\overset{+}{U}_{\{LD^2h0\}}[f]\|_{L_x}}{\|f\|_{L_x}} < \infty.$$

PROOF. Let $f(x)$ be an arbitrary function in $\mathfrak{A}$. By Lemma 4.3.4 we have

$$g(x) = \overset{+}{U}_{\{LD^2h0\}}[f(x)] = \frac{2}{\pi}\int_0^\infty C_f(s)\frac{s^2\omega_h(s^2, x)}{s^2m^2(s) + n^2(s)}\,ds,$$

where

$$C_f(s) = \int_0^\infty f(x)\cos sx\,dx.$$

Then by Theorem 4.2.1 we have

$$g(x) = \frac{2}{\pi}\int_0^\infty C_f(s)\cos sx\,ds + \frac{1}{\pi}\int_0^\infty C_f(s)\left[k_1(s)e^{isx} + k_1(-s)e^{-isx}\right]ds$$
$$-\int_{-\infty}^\infty A(x, u)\left[\frac{2}{\pi}\int_0^\infty C_f(s)k_2(s)\sin su\,ds\right]du$$
$$-\int_{-\infty}^\infty B(x, u)\left[\frac{2}{\pi}\int_0^\infty C_f(s)(1 + k_3(s))\cos su\,ds\right]du.$$

We will denote the even extension of $f(x)$ onto the negative semi-axis by the same symbol $f(x)$. Since the functions $C_f(s)$ and $k_3(s)$ are even and $k_2(s)$ is odd, when we apply the classical Fourier inversion formulas to the above equation we obtain

$$g(x) = f(x) + \frac{1}{2\pi}\int_{-\infty}^\infty 2C_f(s)k_1(s)e^{ixs}\,ds$$
$$-\int_{-\infty}^\infty A(x, u)\left[\frac{1}{2\pi}\int_{-\infty}^\infty 2C_f(s)k_2(s)e^{isu}\,ds\right]du$$
$$-\int_{-\infty}^\infty B(x, u)f(u)\,du - \int_{-\infty}^\infty B(x, u)\left[\frac{1}{2\pi}\int_{-\infty}^\infty 2C_f(s)k_3(s)e^{isu}\,ds\right]du.$$

By assertion 2) of Theorem 4.2.1 the functions $k_j(s)$, $j = 1, 2, 3$, are the Fourier transforms of functions $h_j(x)$ which are integrable on $-\infty < x < \infty$. Then by virtue of classical convolution and inversion formulas we have

$$\frac{1}{2\pi} \int_{-\infty}^{\infty} 2C_f(s)\, k_j(s)\, e^{isu}\, ds = f * h_j(u) \qquad (j = 1, 2, 3),$$

where

$$f * h_j(u) = \int_{-\infty}^{\infty} f(u-t)\, h_j(t)\, dt,$$

and

$$\int_{-\infty}^{\infty} |f * h_j(u)|\, du \leqslant \int_{-\infty}^{\infty} |f(u)|\, du \cdot \int_{-\infty}^{\infty} |h_j(u)|\, du = 2\|f\|_{L^1} \int_{-\infty}^{\infty} |h_j(u)|\, du. \tag{4.3.8}$$

Therefore

$$g(x) = f(x) + f * h_1(x) - \int_{-\infty}^{\infty} A(x, u)\, f * h_2(u)\, du - \int_{-\infty}^{\infty} B(x, u)[f(u) - f * h_3(u)]\, du.$$

By assertion 1) of Theorem 4.2.1 and inequality (4.3.8) we have

$$\left| \int_{-\infty}^{\infty} A(x, u)\, f * h_2(u)\, du \right| \leqslant \frac{1}{2} \int_{x}^{\infty} |q(t)|\, dt \int_{-\infty}^{\infty} |f * h_2(u)|\, du$$

$$\leqslant C_1 \|f\|_{L^1} \int_{x}^{\infty} |q(t)|\, dt,$$

where

$$C_1 = \int_{-\infty}^{\infty} |h_2(x)|\, dx < \infty.$$

Similarly

$$\left| \int_{-\infty}^{\infty} B(x, u)[f(u) - f * h_3(u)]\, du \right| \leqslant C_2 \|f\|_{L^1} \int_{x}^{\infty} |q(t)|\, dt,$$

where

$$C_2 = (1 + |K_h|)\left(1 + \int_{-\infty}^{\infty} |h_3(u)|\, du\right) < \infty.$$

It follows that

$$|g(x)| \leqslant |f(x)| + |f * h_1(x)| + C_3 \|f\|_{L^1} \int_{x}^{\infty} |q(t)|\, dt, \tag{4.3.9}$$

where $C_3 = C_1 + C_2 < \infty$. Therefore

$$\|g\|_{L^1} \leqslant \int_0^\infty |f(x)|\,dx + \int_0^\infty |f * h_1(x)|\,dx + C_3\|f\|_{L^1}\int_0^\infty \Big[\int_x^\infty |q(t)|\,dt\Big]\,dx,$$

which by (4.3.8) implies that

$$\|g\|_{L^1} \leqslant \|f\|\Big[1 + 2\int_{-\infty}^\infty |h_1(x)|\,dx + C_3\int_0^\infty \Big[\int_x^\infty |q(t)|\,dt\Big]\,dx\Big].$$

Integration by parts gives us

$$\int_0^N \Big[\int_x^\infty |q(t)|\,dt\Big]\,dx = N\int_N^\infty |q(t)|\,dt + \int_0^N x\,|q(x)|\,dx$$

$$\leqslant \int_N^\infty t\,|q(t)|\,dt + \int_0^N x\,|q(x)|\,dx = \int_0^\infty t\,|q(t)|\,dt < \infty,$$

and then as $N \to \infty$ we find that

$$\int_0^\infty \Big[\int_x^\infty |q(t)|\,dt\Big]\,dx \leqslant \int_0^\infty t\,|q(t)|\,dt < \infty.$$

Consequently

$$1 + 2\int_{-\infty}^\infty |h_1(x)|\,dx + C_3\int_0^\infty \Big[\int_x^\infty |q(t)|\,dt\Big]\,dx = C < \infty$$

and

$$\|g\|_{L^1} = \|\overset{+}{U}_{\{LD^2h0\}}[f]\|_{L^1} \leqslant C\|f\|_{L^1},$$

where the constant C does not depend on $f(x)$. Since $f(x) \in \mathfrak{A}$ was arbitrary, the last inequality shows that

$$|\overset{+}{U}_{\{LD^2h0\}}|_{L^1} = \sup_{f\in\mathfrak{A}} \frac{\|\overset{+}{U}_{\{LD^2h0\}}[f]\|_{L^1}}{\|f\|_{L^1}} < \infty,$$

and the first part of the lemma is proved.

The proof of the second part is analogous. Since $\|f\|_{L^1} \leqslant \|f\|_{L_x}$, inequality (4.3.9) implies that

$$|g(x)| \leqslant |f(x)| + |f * h_1(x)| + C_3\|f\|_{L_x}\int_x^\infty |q(t)|\,dt,$$

$$\|g\|_{L_x} \leqslant \|f\|_{L_x} + \int_{-\infty}^\infty (1+|x|)\,|f * h_1(x)|\,dx + C_3\|f\|_{L_x}\int_0^\infty (1+x)\Big[\int_x^\infty |q(t)|\,dt\Big]\,dx.$$

For $n(0) \neq 0$ we see by assertion 3) of Theorem 4.2.1 that

$$\int_{-\infty}^{\infty} (1+|x|)\,|h_1(x)|\,dx = C_4 < \infty,$$

and since

$$\int_{-\infty}^{\infty} (1+|x|)\,|f * h(x)|\,dx \leqslant \int_{-\infty}^{\infty} (1+|x|)\,|f(x)|\,dx \int_{-\infty}^{\infty} (1+|x|)\,|h(x)|\,dx,$$

we have

$$\int_{-\infty}^{\infty} (1+|x|)\,|f * h_1(x)|\,dx \leqslant 2C_4 \|f\|_{L_x}$$

and

$$\|g\|_{L_x} \leqslant \|f\|_{L_x} \{1 + 2C_4 + C_3 \int_0^{\infty} (1+x)\,[\int_x^{\infty} |q(t)|\,dt]\,dx\}.$$

Integration by parts gives us

$$\int_0^{\infty} (1+x)\,[\int_x^{\infty} |q(t)|\,dt]\,dx \leqslant \int_0^{\infty} t\,|q(t)|\,dt + \frac{1}{2}\int_0^{\infty} t^2\,|q(t)|\,dt < \infty.$$

Consequently

$$1 + 2C_4 + C_3 \int_0^{\infty} (1+x)\,[\int_x^{\infty} |q(t)|\,dt]\,dx = C' < \infty$$

and $\|g\|_{L_x} = \| U_{\{LD^2h0\}}[f]\|_{L_x} \leqslant C' \|f\|_{L_x}$, where the constant C' does not depend on $f(x)$. Therefore

$$|\overset{+}{U}_{\{LD^2h0\}}|_{L_x} = \sup_{f \in \mathfrak{A}} \frac{\|\overset{+}{U}_{\{LD^2h0\}}[f]\|_{L_x}}{\|f\|_{L_x}} < \infty,$$

and the lemma is proved.

The following theorem concludes our study of $V_{\{LD^2h0\}}$.

THEOREM 4.3.1. *On all functions* $F(x)$ *satisfying* (4.3.4′) *the transformation operator* $V_{\{LD^2h0\}}$ *can be written in the form*

$$V_{\{LD^2h0\}} = \overset{-}{V}_{\{LD^2h0\}} + \overset{+}{V}_{\{LD^2h0\}},$$

where

$$\overset{-}{V}_{\{LD^2h0\}}[F] = \int_0^{\infty} \sum_{\nu=1}^{n} \frac{\cosh|s_\nu|x\,\omega_h(s_\nu^2, t)}{\|\omega_h(s_\nu^2, x)\|_{L^2}^2} F(t)\,dt,$$

$$\overset{+}{V}_{\{LD^2h0\}}[F] = F(x) + \int_0^{\infty} \overset{+}{H}_h(x,t)\,F(t)\,dt$$

and $s_1^2 < s_2^2 < \cdots < s_n^2 < 0$ *are the negative eigenvalues of the boundary value problem for* L *with boundary condition*

$$y'(0) - hy(0) = 0.$$

The operator $\overset{+}{V}_{\{LD^2h0\}}$ *is bounded in* M_1:

$$1) \qquad \operatorname*{ess\,sup}_{0 \leq x < \infty} \int_0^\infty |\overset{+}{H}_h(x, t)|\, dt < \infty.$$

If $\omega_h'(0, \infty) = n(0) \neq 0$ *(that is, if* $\omega_h(0, x)$ *is not bounded), then* $\overset{+}{V}_{\{LD^2h0\}}$ *is also bounded in* M_x:

$$2) \qquad \operatorname*{ess\,sup}_{0 \leq x < \infty} \frac{1}{1+x} \int_0^\infty |\overset{+}{H}_h(x, t)|\,(1+t)\, dt < \infty.$$

PROOF. The representation of $V_{\{LD^2h0\}}$ as the sum of $\overline{V}_{\{LD^2h0\}}$ and $\overset{+}{V}_{\{LD^2h0\}}$ was established in Lemma 4.3.1. The boundedness of $\overset{+}{V}_{\{LD^2h0\}}$ in M_1 (and in M_x for $n(0) \neq 0$) follows, according to Lemma 4.3.3, from the boundedness of $\overset{+}{U}_{\{LD^2h0\}}$ in L^1 (and in L_x for $n(0) \neq 0$), which we established in Lemma 4.3.5. Finally, it is a well-known fact of functional analysis that $\overset{+}{V}_{\{LD^2h0\}}$ is bounded in M_1 (respectively, in M_x) if and only if 1) (respectively, 2)) holds.

The theorem is proved.

For the operator $V_{\{LD^2\infty\infty\}}$ we have an analogous theorem.

THEOREM 4.3.1′. *If* $\mu_1^2 < \mu_2^2 < \cdots < \mu_m^2 < 0$ *are the negative eigenvalues of the boundary value problem for* L *with boundary condition*

$$y(0) = 0,$$

then on all functions $F(x)$ *satisfying*

$$\int_0^\infty e^{-|\mu_m|t} |F(t)|\, dt < \infty,$$

the transformation operator $V_{\{LD^2\infty\infty\}}$ *can be written in the form*

$$V_{\{LD^2\infty\infty\}} = \overline{V}_{\{LD^2\infty\infty\}} + \overset{+}{V}_{\{LD^2\infty\infty\}},$$

where

$$\overline{V}_{\{LD^2\infty\infty\}}[F] = \int_0^\infty \sum_{\nu=1}^m \frac{\cosh|\mu_\nu| x \omega_\infty(\mu_\nu^2, t)}{|\mu_\nu| \, \|\omega_\infty(\mu_\nu^2, x)\|_{L^2}^2} F(t)\, dt,$$

$$\overset{+}{V}_{\{LD^2\infty\infty\}}[F] = F(x) + \int_0^\infty \overset{+}{H}_\infty(x, t) F(t)\, dt.$$

The operator $\overset{+}{V}_{\{LD^2\infty\infty\}}$ *is bounded in* M_1:

$$1)\qquad \operatorname*{ess\,sup}_{0\le x<\infty}\int_0^\infty |\overset{+}{H}_\infty(x,t)|\,dt<\infty$$

if and only if $\omega'_\infty(0,\infty)=n_1(0)\ne 0$ (*that is, if and only if* $\omega_\infty(0,x)$ *is not bounded*).

REMARK ON THEOREMS 4.3.1 AND 4.3.1′. From Theorem 3.1.3 we know that the transformation operators $V_{\{LD^2h0\}}$ and $V_{\{LD^2\infty\infty\}}$ map the set of continuous functions $C_{[0,\infty)}$ onto itself. From the form of the operators $\overline{V}_{\{LD^0h0\}}$ and $\overline{V}_{\{LD^2\infty\infty\}}$ it follows that these operators map continuous functions to continuous functions. Therefore the operators

$$\overset{+}{V}_{\{LD2h0\}}=V_{\{LD2h0\}}-\overline{V}_{\{LD2h0\}};\quad \overset{+}{V}_{\{LD2\,\infty\,\infty\}}=V_{\{LD2\,\infty\,\infty\}}-\overline{V}_{\{LD2\,\infty\,\infty\}}$$

map all continuous functions in their domains to continuous functions.

For an arbitrary continuous function $G(x)$ we have

$$\|G\|_{M_1}=\operatorname*{ess\,sup}_{0\le x<\infty}|G(x)|=\sup_{0\le x<\infty}|G(x)|.$$

Consequently for a continuous function $F(x)\in M_1$ the inequality

$$\|\overset{+}{V}[F]\|_{M_1}\le|\overset{+}{V}|_{M_1}\|F\|_{M_1}$$

implies

$$\sup_{0\le x<\infty}|\overset{+}{v}[F(x)]|\le|\overset{+}{V}|_{M_1}\sup_{0\le x<\infty}|F(x)|,$$

where $\overset{+}{V}=\overset{+}{V}_{\{LD^2h0\}}$ or $\overset{+}{V}=\overset{+}{V}_{\{LD^2\infty\infty\}}$. Similarly, for a continuous function $F(x)\in M_x$ we have

$$\sup_{0\le x<\infty}\left|\frac{\overset{+}{V}_x[F(x)]}{1+x}\right|\le|\overset{+}{V}|_{M_{x_0}}\sup_{0\le x<\infty}\left|\frac{F(x)}{1+x}\right|.$$

(Of course these properties of $\overset{+}{V}$ clearly follow from the continuity of the kernel $H(x,t)$ of the transformation operator V about which we remarked at the end of §1 of Chapter 2.)

§4. Some applications

The main applications of the results we have obtained will be made in the theory of generalized almost periodic functions (in the next chapter). But the theory of generalized almost periodic functions is not the only area in which we find questions whose answers depend on the properties of transformation operators examined in this chapter. We will now consider two examples of such questions.

1. **A generalization of Bernstein's inequality**. Let $\underset{x}{L}$ be a differential operator of the form

$$\underset{x}{L}[u]=\frac{d^2}{dx^2}u(x)-q(x)u(x),$$

given on the whole real line, where $q(x)$ is even, real, and continuous and satisfies

$$\int_{-\infty}^{\infty}(1+x^2)\,|q(x)|\,dx<\infty. \tag{4.4.1}$$

THEOREM 4.4.1. *Let an even function $F(x)$ be bounded on the whole real line, and suppose that the operator $\underset{x}{L}$ can be applied to $F(x)$ an unbounded number of times, and that the following conditions are satisfied.*

1) *For arbitrary $a \geqslant 0$*

$$\overline{\lim_{n\to\infty}}\left[\frac{\sup\limits_{|x|<a}|\underset{x}{L^n}[F(x)]|}{2n!}\right]^{\frac{1}{2n}}=0;$$

2)
$$\overline{\lim_{n\to\infty}}\,|\underset{x}{L^n}[F(x)]_{x=0}|^{\frac{1}{2n}}=\sigma<\infty.$$

Then for an arbitrary integer $m \geqslant 0$

$$\sup_{-\infty<x<\infty}|\underset{x}{L^m}[F(x)]|\leqslant C\sigma^{2m}\sup_{-\infty<x<\infty}|F(x)|,$$

where the constant $C<\infty$ does not depend on $F(x)$, σ, or m.

PROOF. Since $q(x)$ and $F(x)$ are even, the function $\underset{x}{L^m}[F(x)]$ is also even. Thus we can restrict our attention to the semi-axis $0\leqslant x<\infty$ and prove

$$\sup_{0\leqslant x<\infty}|\underset{x}{L^m}[F(x)]|\leqslant C\sigma^{2m}\sup_{0\leqslant x<\infty}|F(x)|. \tag{4.4.2}$$

We let $\omega_0(\lambda, x)$ denote the solution of $L[u]+\lambda u=0$ with initial conditions $u(0)=1$ and $u'(0)=0$. Each function $\omega_0(\lambda, x)$ is even.

We consider L on $0\leqslant x<\infty$. By (4.4.1) the boundary value problem for L with boundary condition $y'(0)=0$ has a finite number of negative eigenvalues $\lambda_1<\lambda_2<\cdots<\lambda_n<0$, and the projection operator $P=P_{\{L,0\}}$,

$$P[f]=\int_0^\infty\sum_{\nu=1}^{n}\frac{\omega_0(\lambda_\nu, x)\,\omega_0(\lambda_\nu, t)}{\|\omega_0(\lambda_\nu, x)\|^2_{L^2}}f(t)\,dt,$$

is defined on all functions $f(x)$ such that

$$\int_0^\infty|f(t)|\,e^{-\sqrt{|\lambda_n|}t}\,dt<\infty.$$

Since $F(x)$ is bounded, it satisfies the above condition. Therefore we can write this function in the form

$$F(x) = \overset{-}{F}(x) + \overset{+}{F}(x), \tag{4.4.3}$$

where the functions $\overset{-}{F}(x)$ and $\overset{+}{F}(x)$ are even, and for $x \geqslant 0$

$$\overset{-}{F}(x) = P[F(x)], \quad \overset{+}{F}(x) = F(x) - \overset{-}{F}(x) = (E-P)[F(x)].$$

We consider in detail the function

$$\overset{-}{F}(x) = \int_0^\infty \sum_{\nu=1}^n \frac{\omega_0(\lambda_\nu, x)\,\omega_0(\lambda_\nu, t)}{\| \omega_0(\lambda_\nu, x)\|_{L^2}^2} F(t)\,dt = \sum_{\nu=1}^n a_\nu \omega_0(\lambda_\nu, x)$$

$$\left(a_\nu = \int_0^\infty \frac{\omega_0(\lambda_\nu, t)}{\| \omega_0(\lambda_\nu, x)\|_{L^2}^2} F(t)\,dt \right). \tag{4.4.4}$$

For an arbitrary continuous function $G(x)$ on $0 \leqslant x < \infty$ we introduce the notation

$$\| G \| = \sup_{0 \leqslant x < \infty} |G(x)|,$$

and we see that

$$\| \overset{-}{F} \| \leqslant \sum_{\nu=1}^n |a_\nu| \, \| \omega_0(\lambda_\nu, x) \| \leqslant \| F \| \sum_{\nu=1}^n \| \omega_0(\lambda_\nu, x) \| \int_0^\infty \frac{|\omega_0(\lambda_\nu, t)|}{\| \omega_0(\lambda_\nu, x)\|_{L^2}^2} dt.$$

From Corollary 1 of Lemma 2.1.2 it follows that

$$\sum_{\nu=1}^n \| \omega_0(\lambda_\nu, x) \| \int_0^\infty \frac{|\omega_0(\lambda_\nu, t)|}{\| \omega_0(\lambda_\nu, x)\|_{L^2}^2} dt = C_1 < \infty.$$

Therefore

$$\| \overset{-}{F} \| \leqslant \sum_{\nu=1}^n |a_\nu| \, \| \omega_0(\lambda_\nu, x) \| \leqslant C_1 \| F \|, \tag{4.4.5}$$

$$\| \overset{+}{F} \| = \| F(x) - \overset{-}{F}(x) \| \leqslant \| F \| + \| \overset{-}{F} \| \leqslant \| F \| (1 + C_1), \tag{4.4.5'}$$

where the constant C_1 depends only on $\underset{x}{L}$.

We will show that in (4.4.4) the coefficient $a_\nu = 0$ if $|\lambda_\nu| > \sigma^2$. From conditions 1) and 2) by virtue of the Corollary of Theorem 3.3.2 (see Part I) we know that the function

$$\Phi(x) = W_{\{LD20\}}[F(x)]$$

is an even entire function of exponential type of order σ. Furthermore, for an arbitrary even function $f(x)$ for $x \geqslant 0$ we have

$$W_{\{LD20\}}[f(x)] = V_{\{LD200\}}[f(x)].$$

This equation is clearly true if $f(x) = \omega_0(\lambda, x)$. For an arbitrary even function $f(x)$ it is proved in the usual way, by approximating $f(x)$ by linear combinations of the functions $\omega_0(\lambda, x)$ (where λ is a parameter).

Therefore if $x \geqslant 0$

$$\begin{aligned}\Phi(x) &= W_{\{LD20\}}[F(x)] = V_{\{LD200\}}[F(x)] \\ &= V_{\{LD200\}} P[F(x)] + V_{\{LD200\}}(E-P)[F(x)];\end{aligned}$$

that is, by (4.4.4)

$$\Phi(x) = \sum_{\nu=1}^{n} a_\nu \cosh \sqrt{|\lambda_\nu|}\, x + \overset{+}{V}[F(x)],$$

where, as before, $\overset{+}{V}$ denotes the operator $V_{\{LD^2 00\}}\ (E - P_{\{L,0\}})$.

Since $\Phi(x)$ is a function of exponential type of order σ, it cannot grow on $0 \leqslant x < \infty$ faster than $e^{(\sigma+\epsilon)x}$ for any $\epsilon > 0$. From (4.4.1) and Theorem 4.3.1 it follows that $\overset{+}{V}$ is bounded on M_1. Then by the remark on Theorem 4.3.1 the function $\overset{+}{\Phi}(x) = \overset{+}{V}[F(x)]$ is continuous and bounded:

$$\|\overset{+}{\Phi}(x)\| = \|\overset{+}{V}[F(x)]\| \leqslant |\overset{+}{V}|_{M_1} \|F(x)\|, \tag{4.4.6}$$

which implies that the function $\Sigma_{\nu=1}^{n} a_\nu \cosh \sqrt{|\lambda_\nu|}\, x$ for $x \to \infty$ cannot grow faster than $e^{(\sigma+\epsilon)x}$; that is, $a_\nu = 0$ if $|\lambda_\nu| > \sigma^2$, which we assumed to be true. Furthermore, we see that the function

$$\overset{+}{\Phi}(x) = \overset{+}{V}[F(x)] = V_{\{LD200\}}[F(x) - P[F(x)]] = V_{\{LD200\}}[\overset{+}{F}(x)]$$

is an even, bounded, entire function of exponential type of order $\leqslant \sigma$.

We turn now to the proof of (4.4.2).

We have

$$\underset{x}{L^m}[\bar{F}(x)] = \underset{x}{L^m}\Big[\sum_{\nu=1}^{n} a_\nu \omega_0(\lambda_\nu, x)\Big] = \sum_{\nu=1}^{n} \lambda_\nu^m a_\nu \omega_0(\lambda_\nu, x)$$

and since $a_\nu = 0$ if $|\lambda_\nu| > \sigma^2$, it follows that

$$\|\underset{x}{L^m}[\bar{F}(x)]\| \leqslant \sigma^{2m} \sum_{\nu=1}^{n} |a_\nu| \, \|\omega_0(\lambda_\nu, x)\|;$$

that is, by (4.4.5)

$$\|\underset{x}{L^m}[\bar{F}(x)]\| \leqslant \sigma^{2m} C_1 \|F\|. \tag{4.4.7}$$

For the function $\overset{+}{F}(x)$ we know by Theorem 3.1.2 that

$$\underset{x}{L^m}[\overset{+}{F}(x)] = V_{\{D^2L00\}} \underset{x}{D^{2m}} V_{\{LD^200\}}[\overset{+}{F}(x)] = V_{\{D^2L00\}} \underset{x}{D^{2m}}[\overset{+}{\Phi}(x)],$$

and using the expression for $V_{\{D^2L00\}}$ found in Theorem 4.1.1 we have

$$\underset{x}{L^m}[\overset{+}{F}(x)] = \underset{x}{D^{2m}}[\overset{+}{\Phi}(x)] + \int_0^x K_0(x, t)\{\underset{t}{D^{2m}}[\overset{+}{\Phi}(t)]\}\, dt + c \int_0^x \underset{t}{D^{2m}}[\overset{+}{\Phi}(t)]\, dt,$$

where D denotes differentiation.

Since $\overset{+}{\Phi}(x)$ is even, at zero its derivatives of odd order are zero. Therefore

$$\int_0^x D_t^{2m}[\overset{+}{\Phi}(t)]\,dt = D_t^{2m-1}[\overset{+}{\Phi}(t)]\Big|_0^x = D_x^{2m-1}[\overset{+}{\Phi}(x)] = D_x^{2m}\Big[\int_0^x \overset{+}{\Phi}(t)\,dt\Big]$$

and

$$L_x^m[\overset{+}{F}(x)] = D_x^{2m}\Big[\overset{+}{\Phi}(x) + c\int_0^x \overset{+}{\Phi}(t)\,dt\Big] + \int_0^x K_0(x, t)\, D_t^{2m}[\overset{+}{\Phi}(t)]\,dt. \tag{4.4.8}$$

Furthermore,

$$\overset{+}{F}(x) = V_{\{D^2L00\}}[\overset{+}{\Phi}(x)] = \overset{+}{\Phi}(x) + \int_0^x K_0(x, t)\overset{+}{\Phi}(t)\,dt + c\int_0^x \overset{+}{\Phi}(t)\,dt,$$

which, in the notation of Theorem 4.1.1, shows that

$$\Big\| c\int_0^x \overset{+}{\Phi}(t)\,dt \Big\| \leqslant \|\overset{+}{F}\| + (1 + |K_0|)\,\|\overset{+}{\Phi}\|$$

or by (4.4.5′) and (4.4.6)

$$\Big\| c\int_0^x \overset{+}{\Phi}(t)\,dt \Big\| \leqslant \|F\|\,[1 + C_1 + |\overset{+}{V}|_{M_1}(1 + |K_0|)]. \tag{4.4.9}$$

Since $\overset{+}{\Phi}(x)$ is an even entire function of exponential type of order $\leqslant \sigma$, the function $\int_0^x \overset{+}{\Phi}(t)\,dt$ is an odd entire function of exponential type of order $\leqslant \sigma$. By virtue of the classical Bernstein inequality we have

$$\|D_x^{2m}[\overset{+}{\Phi}(x)]\| \leqslant \sigma^{2m}\|\overset{+}{\Phi}\|,$$

$$\Big\| D_x^{2m}\Big[c\int_0^x \overset{+}{\Phi}(t)\,dt\Big]\Big\| \leqslant \sigma^{2m}\Big\| c\int_0^x \overset{+}{\Phi}(t)\,dt\Big\|;$$

that is, by (4.4.6) and (4.4.9)

$$\|D_x^{2m}[\overset{+}{\Phi}(x)]\| \leqslant \sigma^{2m}|\overset{+}{V}|_{M_1}\|F\|,$$

$$\Big\| D_x^{2m}\Big[c\int_0^x \overset{+}{\Phi}(t)\,dt\Big]\Big\| \leqslant \sigma^{2m}[1 + C_1 + |\overset{+}{V}|_{M_1}(1 + |K_0|)]\,\|F\|.$$

Then from (4.4.8) we obtain

$$\|L_x^m[\overset{+}{F}(x)]\| \leqslant \sigma^{2m}[1 + C_1 + 2|\overset{+}{V}|_{M_1}(1 + |K_0|)]\,\|F\|,$$

or

$$\| \underset{x}{L^m} [\overset{+}{F}(x)] \| \leqslant \sigma^{2m} C_2 \| F \|, \tag{4.4.10}$$

where the constant $C_2 = 1 + C_1 + 2 | \overset{+}{V} |_{M_1} (1 + |K_0|)$ depends only on the operator $\underset{x}{L}$.

From (4.4.3), (4.4.7) and (4.4.10) we have

$$\| \underset{x}{L^m} [F(x)] \| = \| \underset{x}{L^m} [\overline{F}(x)] + \underset{x}{L^m} [\overset{+}{F}(x)] \|$$

$$\leqslant \| \underset{x}{L^m} [\overline{F}(x)] \| + \| \underset{x}{L^m} [\overset{+}{F}(x)] \| \leqslant \sigma^{2m} (C_1 + C_2) \| F \|,$$

and if we set $C = C_1 + C_2$, we obtain

$$\| \underset{x}{L^m} [F(x)] \| \leqslant \sigma^{2m} C \| F \|,$$

where the constant C does not depend on the function $F(x)$ or the numbers σ and m. The proof is complete.

2. **Structure of normed rings generated by generalized shift operators of the second kind.** In Definition 3.2.2 of generalized shift operators of the second kind we assumed that the corresponding differential operators L_1 and L_2 were given on the whole real line. If the differential operators

$$L_1[u] = \frac{d^2}{dx^2} u(x) - q_1(x) u(x), \qquad L_2[u] = \frac{d^2}{dx^2} u(x) - q_2(x) u(x)$$

are given on the half-line $0 \leqslant x < \infty$, the definition should read as follows.

DEFINITION 4.4.1. By the *generalized shift operator of the second kind* associated with the differential operators L_1 and L_2 given on $0 \leqslant x < \infty$ we mean the operator $\overset{\circ}{T}{}_x^y$ which takes each function $F(t) \in \mathfrak{M}^1_{[0,\infty)}$(4) into the function of two variables $\overset{\circ}{T}{}_x^y [F]$ according to the formula

$$\overset{\circ}{T}{}_x^y [F] = \underset{x}{V_1} \underset{y}{V_2} \overset{\circ}{S}{}_x^y [V_1^{-1}[F]] \qquad (x \geqslant 0,\ y \geqslant 0),$$

where

$$V_1 = V_{\{D^2 L_1 0 h_1\}}, \quad V_2 = V_{\{D^2 L_2 0 h_2\}} \qquad (h_1 \neq \infty,\ h_2 \neq \infty)$$

and

$$\overset{\circ}{S}{}_x^y [\Phi] = \frac{1}{2} [\Phi(x+y) + \Phi(|x-y|)] \qquad (x \geqslant 0,\ y \geqslant 0).$$

On twice continuously differentiable functions $F(t)$ satisfying the boundary condition $F'(0) - h_1 F(0) = 0$ the operator $\overset{\circ}{T}{}_x^y$ can be defined as the solution of the partial differential equation

$$\underset{x}{L_1} [v(x, y)] = \underset{y}{L_2} [v(x, y)] \qquad (x \geqslant 0,\ y \geqslant 0),$$

---

(4)The set $\mathfrak{M}^1_{[0,\infty)}$ consists of all functions $F(t)$ defined on $0 \leqslant x < \infty$ and integrable in each finite subinterval $(0, b)$ (see Chapter 1, §2).

with initial conditions

$$\begin{aligned} v(x, 0) &= F(x), & v'_y(x, 0) &= h_1 F(x), & (x \geqslant 0),\\ v(0, y) &= \underset{y}{V}[F(y)], & v'_x(0, y) &= h_2 \underset{y}{V}[F(y)], & (y \geqslant 0), \end{aligned}$$

where $\underset{y}{V} = \underset{y}{V}_{\{L_1 L_2 h_1 h_2\}}$. This can be proved in the same way that we proved (in Chapter 3, §2) the analogous proposition for generalized shift operators of the second kind T_x^y defined on the set $\mathfrak{M}^1_{(-\infty,\infty)}$.

A very interesting special case of Definition 4.4.1 arises when $L_1 = L_2 = L$ and $h_1 = h_2 = h$; that is, when $\overset{\circ}{T}{}_x^y$ is given by the formula

$$\overset{\circ}{T}{}_x^y[F] = \underset{x}{V}\underset{y}{V}\overset{\circ}{S}{}_x^y[V^{-1}[F]], \tag{4.4.11}$$

where

$$V = V_{\{D^2 L 0 h\}}. \tag{4.4.11'}$$

The simplest generalized shift operator is $\overset{\circ}{S}{}_x^y$ which we obtain by setting $L \equiv D^2$ and $h = 0$. The operator $\overset{\circ}{S}{}_x^y$ allows us to introduce in $L^1[0, \infty)$ the operator of multiplication defined as a convolution by the formula

$$f * g(x) = \int_0^\infty \overset{\circ}{S}{}_x^y[f]\, g(y)\, dy.$$

This formula can be written in the form

$$f * g(x) = \frac{1}{2}\int_{-\infty}^{\infty} f(x-y)\, g(y)\, dy,$$

if we extend $f(x)$ and $g(x)$ evenly onto the negative semi-axis.

It is known [21] that relative to this operator the space $L^1[0, \infty)$ (or, equivalently, the subspace of $L^1(-\infty, \infty)$ consisting of the even functions) is a commutative normed ring (without identity), which we will denote by $R_{\overset{\circ}{S}}$.

This ring has been thoroughly investigated, and we will not prove any of its properties here.

We obtain a generalized convolution operation by setting

$$f \circ g(x) = \int_0^\infty \overset{\circ}{T}{}_x^y[f]\, g(y)\, dy, \tag{4.4.12}$$

where $\overset{\circ}{T}{}_x^y$ is a generalized shift operator of the form (4.4.11).

Povzner, who introduced this operation in [10], showed that if the operator L,

$$L[u] = \frac{d^2}{dx^2}u(x) - q(x)\, u(x) \tag{4.4.13}$$

satisfies (4.1.1) and if $\omega_h(0, x)$ is bounded:

$$\sup_{0 \leqslant x < \infty} |\omega_h(0, x)| < \infty,$$

then the space $L^1[0, \infty)$ relative to the operation (4.4.12) is a commutative normed ring (without identity), which we will denote by R_T°.

We will show that the ring R_T° can be written as a direct sum of a finite number of ideals, of which one is isomorphic to R_S° while the others are isomorphic to the field of complex numbers.

We will begin by studying in detail the characteristics of the transformation which led us to the introduction of the operator $\overset{+}{U}_{\{LD^2h0\}}$ in the preceding section (see (4.3.6)).

A boundary value problem for the differential operator (4.4.13) with boundary condition

$$y'(0) - hy(0) = 0$$

has a finite number of negative eigenvalues

$$\lambda_1 < \lambda_2 < \dots < \lambda_n < 0.$$

We denote the set of all linear combinations of the corresponding eigenfunctions $\omega_h(s_\nu^2, x)$ $(s_\nu^2 = \lambda_\nu, \nu = 1, \cdots, n)$ by $\overline{I}_h$. In the preceding section we showed that $\overline{I}_h$ is a finite-dimensional subspace of $L^1[0, \infty)$ and the projection operator onto this subspace is of the form

$$P[f] = \int_0^\infty \sum_{\nu=1}^n \frac{\omega_h(s_\nu^2, x)\,\omega_h(s_\nu^2, t)}{\|\omega_h(s_\nu^2, x)\|_{L^2}^2} f(t)\,dt.$$

Therefore we can decompose $L^1[0, \infty)$ into the direct sum of the subspaces $\overline{I}_h$ and $\overset{+}{I}_h$(5)

$$L^1[0, \infty) = \overline{I}_h \oplus \overset{+}{I}_h,$$

where $\overset{+}{I}_h$ consists of all functions $f(x) \in L^1[0, \infty)$ which the operator $P_{\{Lh\}}$ takes to zero.

From Lemmas 4.3.4 and 4.3.5 it follows that $\overset{+}{U}_{\{LD^2h0\}}$ is a linear operator from $L^1[0, \infty)$ into $\overset{+}{I}_h$.

LEMMA 4.4.1. *If* $\omega_h(0, x)$ *is bounded on* $0 \leqslant x < \infty$, *then* $\overset{+}{U} = \overset{+}{U}_{\{LD^2h0\}}$ *maps* $L^1[0, \infty)$ *bijectively onto* $\overset{+}{I}_h$.

PROOF. Since $\omega_h(0, x)$ is bounded, $\lim_{x\to\infty} \omega_h'(0, x) = 0$ (see Lemma 2.1.2). Then by Theorem 4.1.1 the transformation operator $V = V_{\{D^2L0h\}}$ is of the form

$$V[f] = f(x) + \int_0^\infty K_h(x, t) f(t)\,dt,$$

(5)This decomposition is to be understood in the sense that each function $f(x) \in L^1[0, \infty)$ can be uniquely represented as the sum of two functions $\overline{f}(x) \in \overline{I}_h$ and $\overset{+}{f}(x) \in \overset{+}{I}_h$.

where $K_h(x, t) = 0$ for $x < t$ and

$$\sup_{0 \leq x < \infty} \int_0^\infty |K_h(x, t)|\, dt = |K_h| < \infty. \tag{4.4.14}$$

We introduce the operator $U = U_{\{D^2 L 0 h\}}$:

$$U[f] = f(x) + \int_0^\infty K_h(t, x) f(t)\, dt.$$

From (4.4.14) it follows that this operator is defined on all $L^1[0, \infty)$ and maps this space into itself. Also

$$\begin{aligned}\overset{+}{U} U[f] &= \overset{+}{U}\Big[f(x) + \int_0^\infty K_h(t, x) f(t)\, dt\Big] \\ &= f(x) + \int_0^\infty K_h(t, x) f(t)\, dt + \int_0^\infty \overset{+}{H}_h(t, x) f(t)\, dt \\ &\qquad + \int_0^\infty \int_0^\infty f(t) K_h(t, z) \overset{+}{H}_h(z, x)\, dt\, dz,\end{aligned}$$

or

$$\overset{+}{U} U[f] = f(x) + \int_0^\infty A(t, x) f(t)\, dt,$$

where

$$A(t, x) = \overset{+}{H}_h(t, x) + K_h(t, x) + \int_0^\infty K_h(t, z) \overset{+}{H}_h(z, x)\, dz.$$

By (4.3.5′)

$$\overset{+}{H}_h(t, x) = H_h(t, x) - \sum_{\nu=1}^{n} \frac{\cosh s_\nu t\, \omega_h(s_\nu^2, x)}{\| \omega_h(s_\nu^2, x) \|_{L^2}^2},$$

where $H_h(t, x)$ is the kernel of the operator $V_{\{L D^2 h 0\}}$. Therefore

$$\begin{aligned}A(t, x) &= H_h(t, x) + K_h(t, x) + \int_0^\infty H_h(z, x) K_h(t, z)\, dz \\ &\qquad - \sum_{\nu=1}^{n} \frac{\omega_h(s_\nu^2, x)}{\| \omega_h(s_\nu^2, x) \|_{L^2}^2} \Big(\cosh s_\nu t + \int_0^t K_h(t, z) \cosh s_\nu z\, dz \Big).\end{aligned}$$

Since $H_h(t, x)$ and $K_h(t, x)$ are the kernels of the mutually inverse Volterra operators $V_{\{L D^2 h 0\}}$ and $V_{\{D^2 L 0 h\}}$, we have

$$H_h(t, x) + K_h(t, x) + \int_0^\infty H_h(z, x) K_h(t, z)\, dz = 0.$$

Also,

$$\cosh s_\nu t + \int_0^t K_h(t, z) \cosh s_\nu z\, dz = V_{\{D^2L0h\}}[\cosh s_\nu t] = \omega_h(s_\nu^2, t).$$

Therefore

$$A(t, x) = -\sum_{\nu=1}^{n} \frac{\omega_h(s_\nu^2, x)\, \omega_h(s_\nu^2, t)}{\| \omega_h(s_\nu^2, x) \|_{L^2}^2}$$

and

$$\overset{+}{U} U[f] = f(x) - \int_0^\infty \sum_{\nu=1}^{n} \frac{\omega_h(s_\nu^2, x)\, \omega_h(s_\nu^2, t)}{\| \omega_h(s_\nu^2, x) \|_{L^2}^2} f(t)\, dt = f(x) - P_{\{Lh\}}[f].$$

In particular, if $f(x)$ is in $\overset{+}{I}_h$, then

$$\overset{+}{U} U[f] = f(x),$$

that is, to each $f(x) \in \overset{+}{I}_h$ there corresponds a function $g(x) = U[f] \in L^1[0, \infty)$ such that $\overset{+}{U}[g] = f(x)$.

Consequently $\overset{+}{U}_{\{LD^2h0\}}$ is an injective mapping of $L^1[0, \infty)$ onto $\overset{+}{I}_h$, and the inverse mapping is $U_{\{D^2L0h\}}$.

The lemma is proved.

By Lemma 4.3.4 we have

$$E_{\overset{+}{U}[f]}(s^2) = C_f(s),$$

where we introduce the notation $E_g(s^2)$ to mean the Fourier transform of $g(x)$ relative to $\{\omega_h(s^2, x)\}$, while $C_g(s)$ denotes the Fourier cosine transform of $g(x)$.

From this equation and the fact that the operator $\overset{+}{U}$ is injective (Lemma 4.4.1) we have the important

COROLLARY. *A function $A(s)$ is a generalized Fourier transform $E_g(s^2)$ of some function $g(x) \in \overset{+}{I}_h$ if and only if $A(s)$ is the Fourier cosine transform of some function $f(x) \in L^1[0, \infty)$.*

As in the preceding section, we let $\mathfrak{A}$ be the set of all twice continuously differentiable functions which are equal to zero outside of some finite interval. This set is dense in $L^1[0, \infty)$. From Lemma 4.4.1 we know that the set $\overset{+}{\mathfrak{A}}$, to which $\overset{+}{U}$ maps $\mathfrak{A}$, is dense in $\overset{+}{I}_h$. Hence the set $\overline{I}_h \oplus \overset{+}{\mathfrak{A}}$ of functions $f(x)$ of the form

$$f(x) = \overset{-}{f}(x) + \overset{+}{f}(x), \tag{4.4.15}$$

where

$$\overset{-}{f}(x) = P[f] \in \overline{I}_h, \quad \overset{+}{f}(x) = (E - P)[f] \in \overset{+}{\mathfrak{A}} \tag{4.4.15$'$}$$

is also dense in $L^1[0, \infty)$.

LEMMA 4.4.2. *If $f(x) \in \overline{I}_h \oplus \overset{+}{\mathfrak{A}}$, then*

$$f(x)=\sum_{\nu=1}^{n}\frac{E_f(s_\nu^2)}{\|\omega_h(s_\nu^2,x)\|_{L^2}^2}\,\omega_h(s_\nu^2,x)+\frac{2}{\pi}\int_0^\infty\frac{E_f(s^2)\,s^2}{s^2m^2(s)+n^2(s)}\,\omega_h(s^2,x)\,ds,$$

where

$$E_f(s^2)=\overset{+}{E}_f(s^2)=C_{U[\overset{+}{f}]}(s),$$

and the integral converges absolutely.

PROOF. From the definition of the projection operator $P=P_{\{Lh\}}$ we have

$$\bar{f}(x)=P[f]=\sum_{\nu=1}^{n}\frac{E_f(s_\nu^2)}{\|\omega_h(s_\nu^2,x)\|_{L^2}^2}\,\omega_h(s_\nu^2,x) \tag{4.4.16}$$

and

$$E_f(s^2)=\overset{+}{E}_f(s^2)\quad(\operatorname{Im}s=0).$$

Since $(E-P)[f]=\overset{+}{f}(x)\in\overset{+}{\mathfrak{A}}$, we have $U[f]\in\mathfrak{A}$.

Then by Lemma 4.3.4

$$\overset{+}{f}(x)=(E-P)[f]=\frac{2}{\pi}\int_0^\infty\frac{C_{U[\overset{+}{f}]}(s)\,s^2}{s^2m^2(s)+n^2(s)}\,\omega_h(s^2,x)\,ds, \tag{4.4.17}$$

where

$$C_{U[\overset{+}{f}]}(s)=\overset{+}{E}_f(s^2),$$

and the integral converges absolutely.

If we combine (4.4.15), (4.4.15′), (4.4.16) and (4.4.17), we obtain

$$f(x)=\sum_{\nu=1}^{n}\frac{E_f(s_\nu^2)}{\|\omega_h(s_\nu^2,x)\|_{L^2}^2}\,\omega_h(s_\nu^2,x)+\frac{2}{\pi}\int_0^\infty\frac{E_f(s^2)\,s^2}{s^2m^2(s)+n^2(s)}\,\omega_h(s^2,x)\,ds,$$

where

$$E_f(s^2)=\overset{+}{E}_f(s^2)=C_{U[\overset{+}{f}]}(s),$$

and the integral converges absolutely.

The lemma is proved.

THEOREM 4.4.2. *Let the differential operator* L *of the form* (4.4.13) *satisfy* (4.1.1), *and let the function* $\omega_h(0,x)$ *be bounded on* $0\leqslant x<\infty$. *Then the following assertions are true.*

1) *The operation* (4.4.12) *can be applied to all functions in the space* $L^1[0,\infty)$, *and relative to this operation the space becomes the commutative normed ring* R_T°.

2) *The ring* R_T° *can be written as the direct sum of a finite number of ideals*

$$R_T^\circ=\overset{+}{I}_h\oplus I_h^{(1)}\oplus I_h^{(2)}\oplus\ldots\oplus I_h^{(n)},$$

where the ideal $\overset{+}{I}_h$ *is isomorphic to the ring* R_S°, *and the ideals* $I_h^{(\nu)}$ $(\nu = 1, \cdots, n)$ *are isomorphic to the field of complex numbers.*

PROOF. Let $f_1(x)$ and $f_2(x)$ be arbitrary functions in $\overline{I}_h \oplus \overset{+}{\mathfrak{A}}$. By Lemma 4.4.1

$$f_1(x) = \sum_{\nu=1}^{n} \frac{E_{f_1}(s_\nu^2)}{\| \omega_h(s_\nu^2, x) \|_{L^2}^2} \omega_h(s_\nu^2, x) + \frac{2}{\pi} \int_0^\infty \frac{E_{f_1}(s^2)\, s^2}{s^2 m^2(s) + n^2(s)} \omega_h(s^2, x)\, ds,$$

where the integral converges absolutely. From (4.4.11) which defines $\overset{\circ}{T}{}_x^y$ we have

$$\overset{\circ}{T}{}_x^y [\omega_h(s^2, t)] = \omega_h(s^2, x)\, \omega_h(s^2, y);$$

therefore

$$\overset{\circ}{T}{}_x^y [f_1] = \sum_{\nu=1}^{n} \frac{E_{f_1}(s_\nu^2)}{\| \omega_h(s_\nu^2, x) \|_{L^2}^2} \omega_h(s_\nu^2, x)\, \omega_h(s^2, y)$$
$$+ \frac{2}{\pi} \int_0^\infty \frac{E_{f_1}(s^2)\, s^2}{s^2 m^2(s) + n^2(s)} \omega_h(s^2, x)\, \omega_h(s^2, y)\, ds \cdot$$

and

$$f_1 \circ f_2(x) = \int_0^\infty \overset{\circ}{T}{}_x^y [f_1]\, f_2(y)\, dy$$
$$= \sum_{\nu=1}^{n} \frac{E_{f_1}(s_\nu^2)\, E_{f_2}(s_\nu^2)}{\| \omega_h(s_\nu^2, x) \|_{L^2}^2} \omega_h(s_\nu^2, x) + \frac{2}{\pi} \int_0^\infty \frac{E_{f_1}(s^2)\, E_{f_2}(s^2)\, s^2}{s^2 m^2(s) + n^2(s)} \omega_h(s^2, x)\, ds,$$

where the integral converges absolutely.

Therefore

$$f_1 \circ f_2(x) = P[f_1] \circ P[f_2] + (E - P)[f_1] \circ (E - P)[f_2] \qquad (4.4.18)$$

and

$$\overset{+}{f}_1 \circ \overset{+}{f}_2(x) = (E - P)[f_1] \circ (E - P)[f_2] = \frac{2}{\pi} \int_0^\infty \frac{E_{f_1}(s^2)\, E_{f_2}(s^2)\, s^2}{s^2 m^2(s) + n^2(s)} \omega_h(s^2, x)\, ds,$$

or

$$\overset{+}{f}_1 \circ \overset{+}{f}_2(x) = \frac{2}{\pi} \int_0^\infty \frac{C_{U[\overset{+}{f}_1]}(s)\, C_{U[\overset{+}{f}_2]}(s)\, s^2}{s^2 m^2(s) + n^2(s)} \omega_h(s^2, x)\, ds, \qquad (4.4.18')$$

since by Lemma 4.4.1 we have $E_f(s^2) = C_{U[\overset{+}{f}]}(s)$.

The functions $U[\overset{+}{f}_1]$ and $U[\overset{+}{f}_2]$ belong to $\mathfrak{A}$. Therefore the convolution

$$g(x) = U[\overset{+}{f_1}] * U[\overset{+}{f_2}] \tag{4.4.19}$$

also belongs to $\mathfrak{A}$, and

$$C_g(s) = C_{U[\overset{+}{f_1}]}(s)\, C_{U[\overset{+}{f_2}]}(s),$$

which by Lemma 4.3.4 implies that

$$\overset{+}{U}[g] = \frac{2}{\pi}\int_0^\infty \frac{C_{U[\overset{+}{f_1}]}(s)\, C_{U[\overset{+}{f_2}]}(s)\, s^2}{s^2 m^2(s) + n^2(s)}\, \omega_h(s^2, x)\, ds. \tag{4.4.19'}$$

If we combine (4.4.18), (4.4.18′), (4.4.19) and (4.4.19′), we obtain

$$f_1 \circ f_2(x) = P[f_1] \circ P[f_2] + \overset{+}{U}[U(E-P)[f_1] * U(E-P)[f_2]], \tag{4.4.20}$$

for arbitrary $f_1(x)$ and $f_2(x)$ in $\overline{I}_h \oplus \overset{+}{\mathfrak{A}}$. Since $\overline{I}_h \oplus \overset{+}{\mathfrak{A}}$ is dense in $L^1[0, \infty)$ and the operators P, $\overset{+}{U}$ and U are bounded, the above equation is true for all functions in $L^1[0, \infty)$, which shows that the operation (4.4.12) can be applied to all functions in $L^1[0, \infty)$.

Furthermore, from (4.4.20) we see that the operation is continuous, commutative, and associative.

Thus the space $L^1[0, \infty)$ relative to the operation (4.4.12) is in fact the commutative normed ring R_T° (without identity), and the first assertion of the theorem is proved.

Let $f_1(x) \in \overset{+}{I}_h$ $(f_1(x) \in \overline{I}_h)$; that is, $P[f_1] \equiv 0$ $(P[f_1] \equiv f_1)$. Then from (4.4.20) we have $f_1 \circ f_2(x) \in \overset{+}{I}_h$ $(f_1 \circ f_2(x) \in \overline{I}_h)$ for arbitrary $f_2(x) \in R_T^\circ$. Therefore the subspaces $\overline{I}_h$ and $\overset{+}{I}_h$ are ideals in the ring R_T°, and the decomposition of $L^1[0, \infty)$ into the direct sum of these subspaces is at the same time a decomposition of R_T° into the direct sum of ideals

$$R_T^\circ = \overline{I}_h \oplus \overset{+}{I}_h.$$

We consider the ideal $\overset{+}{I}_h$ as a ring. From (4.4.20) it follows that multiplication in $\overset{+}{I}_h$ is given by the formula $f_1 \circ f_2(x) = \overset{+}{U}[U[f_1] * U[f_2]]$. Consequently

$$U[f_1 \circ f_2] = U\overset{+}{U}[U[f_1] * U[f_2]] = U[f_1] * U[f_2],$$

from which it is clear that U is an isomorphism of $\overset{+}{I}_h$ onto R_S°.

It remains to prove that the ideal $\overline{I}_h$ can be written as the direct sum of a finite number of ideals which are isomorphic to the field of complex numbers. We let $I_h^{(\nu)}$ $(\nu = 1, \cdots, n)$ denote the set of functions $\{C\omega_h(s_\nu^2, x)\}$, where C is an arbitrary constant. From the orthogonality of the eigenfunctions it follows that the sets $I_h^{(\nu)}$ are ideals of the ring R_T°, and their direct sum is the ideal $\overline{I}_h$. It is clear that the operators $P^{(\nu)}$,

$$P^{(\nu)}[f]=\int_0^\infty \omega_h(s_\nu^2,\, t)\, f(t)\, dt \qquad (\nu=1,\, 2,\, \ldots,\, n)$$

are isomorphisms of the ideals $I_h^{(\nu)}$ into the field of complex numbers.

Thus we finally have

$$R_T^\circ=\overset{+}{I}_h\oplus I_h^{(1)}\oplus I_h^{(2)}\oplus\ \ldots\ \oplus I_h^{(n)},$$

where the ideal $\overset{+}{I}_h$ is isomorphic to the ring R_T°, and each of the ideals $I_h^{(\nu)}$ is isomorphic to the field of complex numbers.

The theorem is proved.

CHAPTER 5

GENERALIZED ALMOST PERIODIC FUNCTIONS

§1. Definition and simplest properties of $\{L_1, L_2\}$-almost periodic functions

All differential operators L in this chapter

$$\underset{t}{L}[u(t)]=\frac{d^2}{dt^2}u(t)-q(t)\,u(t) \tag{5.1.1}$$

are assumed to be given on the whole real line $-\infty<t<\infty$, and the corresponding functions $q(t)$ are assumed to be even:

$$q(t)=q(-t).$$

By the generalized shift operator T_x^y associated with the differential operators $\underset{x}{L}_1$ and $\underset{y}{L}_2$ of the form (5.1.1) we will always mean the operator defined by the formula

$$T_x^y[f(t)]=\underset{x}{W}_1\underset{y}{W}_2 S_x^y[\underset{t}{W}_1^{-1}[f(t)]], \tag{5.1.2}$$

where

$$W_1=W_{\{D^2L_1 0\}},\quad W_2=W_{\{D^2L_2 0\}}$$

and

$$S_x^y[\varphi(t)]=\frac{1}{2}[\varphi(x+y)+\varphi(x-y)].$$

According to Definition 3.2.2, for twice-differentiable functions $f(t)$ ($f''(t)\in\mathfrak{M}^1_{(-\infty,\infty)}$) we have

$$T_x^y[f(t)]=v(x,\, y;\, f),$$

where $v(x, y; f)$ is the solution of the partial differential equation

$$\underset{x}{L_1}[v] = \underset{y}{L_2}[v]$$

with initial conditions

$$v(x, y; f)|_{y=0} = f(x), \quad v'_y(x, y; f)|_{y=0} = 0.$$

Let T_x^y be the generalized shift operator of the form (5.1.2) associated with the differential operators $\underset{x}{L_1}$ and $\underset{y}{L_2}$.

DEFINITION 5.1.1. A continuous function $F(t)$ is said to be $\{L_1, L_2\}$-*almost periodic* if the family of functions $T_x^y[F(t)]$ (where y is a parameter) is compact in the sense of uniform convergence on the whole real line $-\infty < x < \infty$; that is, if from any sequence $T_x^{h_1}[F(t)], T_x^{h_2}[F(t)], \cdots$ we can choose a subsequence which converges uniformly on $-\infty < x < \infty$.

The fundamental problem in the theory of $\{L_1, L_2\}$-almost periodic functions is to find a constructive characterization of these functions.

The results of the preceding chapter allow us to solve this problem when the functions $q_1(x)$ and $q_2(y)$ corresponding to the differential operators $\underset{x}{L_1}$ and $\underset{y}{L_2}$ satisfy the inequalities

$$\int_{-\infty}^{\infty} (1 + x^2)\,|q(x)|\,dx < \infty, \qquad \int_{-\infty}^{\infty} (1 + y^2)\,|q(y)|\,dy < \infty. \tag{5.1.3}$$

Therefore we will always assume that (5.1.3) is satisfied.

LEMMA 5.1.1. *For an operator T_x^y of the form* (5.1.2) *we have the following identities*:

1) $$T_x^y[f(t)] = T_x^{-y}[f(t)],$$
2) $$T_x^y[f(-t)] = T_{-x}^y[f(t)].$$

3) *If $f(t)$ is even and $x \geqslant 0$ and $y \geqslant 0$, then*

$$T_x^y[f(t)] = \underset{x}{V_1}\,\underset{y}{V_2}\,S_x^y[V_1^{-1}[f]],$$

where

$$V_1 = V_{\{D^2 L_1 00\}}, \quad V_2 = V_{\{D^2 L_2 00\}}$$

and the function $V_1^{-1}[f]$ is considered to be evenly extended onto the negative semi-axis.

4) *If $f(t)$ is odd and $x \geqslant 0$ and $y \geqslant 0$, then*

$$T_x^y[f(t)] = \underset{x}{V_1}\,\underset{y}{V_2}\,S_x^y[V_1^{-1}[f]],$$

where

$$V_1 = V_{\{D^2 L_1 \infty\infty\}}, \quad V_2 = V_{\{D^2 L_2 00\}}$$

and the function $V_1^{-1}[f]$ *is considered to be oddly extended onto the negative semi-axis.*

PROOF. We let $\omega_0^{(1)}(\lambda, x)$ and $\omega_\infty^{(1)}(\lambda, x)$ $(\omega_0^{(2)}(\lambda, y)$ and $\omega_\infty^{(2)}(\lambda, y))$ denote the solutions of

$$\underset{x}{L_1}[u]+\lambda u=0 \quad (\underset{y}{L_2}[u]+\lambda u=0)$$

with initial conditions

$$\omega_0^{(k)}(\lambda, 0)=1, \quad \frac{d}{dx}\omega_0^{(k)}(\lambda, x)\bigg|_{x=0}=0 \qquad (k=1,\ 2),$$

$$\omega_\infty^{(k)}(\lambda, 0)=0, \quad \frac{d}{dx}\omega_\infty^{(k)}(\lambda, x)\bigg|_{x=0}=1 \qquad (k=1,\ 2).$$

(We will use this notation throughout the rest of this chapter.)

By the definition of T_x^y expressed in (5.1.2) we have

$$T_x^y[\omega_0^{(1)}(\lambda, t)]=\omega_0^{(1)}(\lambda, x)\,\omega_0^{(2)}(\lambda, y), \tag{5.1.4}$$

$$T_x^y[\omega_\infty^{(1)}(\lambda, t)]=\omega_\infty^{(1)}(\lambda, x)\,\omega_0^{(2)}(\lambda, y). \tag{5.1.4'}$$

Since $q_1(x)$ and $q_2(y)$ are even, the functions $\omega_0^{(1)}(\lambda, x)$ and $\omega_0^{(2)}(\lambda, y)$ are even, and $\omega_\infty^{(1)}(\lambda, x)$ is odd. Then from (5.1.4) and (5.1.4′) we find that all four identities which we want to prove are valid for the functions $\omega_0^{(1)}(\lambda, t)$ and $\omega_\infty^{(1)}(\lambda, t)$ for arbitrary values of the parameter λ.

Since the system of functions $\{\omega_0^{(1)}(\lambda, t), \omega_\infty^{(1)}(\lambda, t)\}$ (λ is a parameter) is dense in the set $\mathfrak{M}^1_{(-\infty,\infty)}$, by approximating an arbitrary function $f(t)\in\mathfrak{M}^1_{(-\infty,\infty)}$ by linear combinations of the functions $\omega_0^{(1)}(\lambda, t)$ and $\omega_\infty^{(1)}(\lambda, t)$ we can show that the lemma is true for all $f(t)\in\mathfrak{M}^1_{(-\infty,\infty)}$.

The lemma is proved.

COROLLARY. *If* $F(t)$ *is an* $\{L_1, L_2\}$*-almost periodic function, then its even and odd parts*

$$F_1(t)=\frac{1}{2}[F(t)+F(-t)], \quad F_2(t)=\frac{1}{2}[F(t)-F(-t)]$$

are also $\{L_1, L_2\}$*-almost periodic.*

PROOF. Let $F(t)$ be an arbitrary $\{L_1, L_2\}$-almost periodic function. Then the family $T_x^y[F(t)]$, and therefore also $T_{-x}^y[F(t)]$, is compact in the sense of uniform convergence on the whole real line. Then by identity 2) of Lemma 5.1.1 we know that the family $T_x^y[F(-t)]$ is also compact in the sense of uniform convergence on the whole real line, and the function $F(-t)$ is $\{L_1, L_2\}$-almost periodic. It is clear that the sum or difference of $\{L_1, L_2\}$-almost periodic functions is $\{L_1, L_2\}$-almost periodic. Thus the functions

$$F_1(t)=\frac{1}{2}[F(t)+F(-t)], \quad F_2(t)=\frac{1}{2}[F(t)-F(-t)]$$

are $\{L_1, L_2\}$-almost periodic, which is what we wanted to show.

This property of $\{L_1, L_2\}$-almost periodic functions allows us to treat even and odd functions separately.

In order to solve the fundamental problem we first have to isolate the class of elementary $\{L_1, L_2\}$-almost periodic functions which will play a role analogous to that of the functions $e^{i\lambda x}$ $(-\infty < \lambda < \infty)$ in Bohr's theory of almost periodic functions. The elementary $\{L_1, L_2\}$-almost periodic functions will be the even and odd $\{L_1, L_2\}$-almost periodic solutions of the equations $L_1[u] + \lambda u = 0$.

LEMMA 5.1.2. *The set* Λ_0 (Λ_∞) *of values of* λ *for which an even solution* $\omega_0^{(1)}(\lambda, t)$ *(odd solution* $\omega_\infty^{(1)}(\lambda, t)$*) of* $L_1[u] + \lambda u = 0$ *is an* $\{L_1, L_2\}$-*almost periodic function consists of three parts*:

$$\Lambda_0 = \Lambda_0^+ + \Lambda_0^0 + \Lambda_0^- \quad (\Lambda_\infty = \Lambda_\infty^+ + \Lambda_\infty^0 + \Lambda_\infty^-),$$

where

1) *the set* Λ_0^+ (Λ_∞^+) *consists of all positive values of* λ;

2) *the set* Λ_0^0 (Λ_∞^0) *consists of the single point* 0 *if* $q_2(y) \equiv 0$ *or if* $\omega_0^{(2)}(0, y)$ *and* $\omega_0^{(1)}(0, x)$ $(\omega_\infty^{(1)}(0, x))$ *are uniformly bounded*; *in all other cases this set is empty*;

3) *the set* Λ_0^- (Λ_∞^-) *consists of the common negative eigenvalues of the boundary value problem for* L_1 *with boundary condition* $u'(0) = 0$ $(u(0) = 0)$ *and the boundary value problem for* L_2 *with boundary condition* $u'(0) = 0$ *(this set is finite or empty)*.

PROOF. From (5.1.4) and (5.1.4′) we have

$$T_x^y[\omega_h^{(1)}(\lambda, t)] = \omega_h^{(1)}(\lambda, x)\,\omega_0^{(2)}(\lambda, y) \qquad (h = 0 \ \text{or} \ \infty),$$

where $\omega_0^{(2)}(0, y)$ is an even solution of $\underset{y}{L_2}[u] + \lambda u = 0$.

Then compactness of the family $T_x^y[\omega_h^{(1)}(\lambda, t)]$ implies that the function $\omega_0^{(2)}(\lambda, y)$ is bounded. If $\omega_h^{(1)}(\lambda, x)$ is not bounded and $\omega_0^{(2)}(\lambda, y) \not\equiv$ const, then it is easy to show that the family $T_x^y[\omega_h^{(1)}(\lambda, t)]$ is not compact.[6] Consequently boundedness of $\omega_h^{(1)}(\lambda, x)$ (under the assumption that $\omega_0^{(2)}(\lambda, y) \not\equiv$ const) is necessary for compactness of the family $T_x^y[\omega_h^{(1)}(\lambda, t)]$. On the other hand, if $\omega_h^{(1)}(\lambda, x)$ and $\omega_0^{(2)}(\lambda, y)$ are bounded or $\omega_0^{(2)}(\lambda, y) \equiv$ const, then the family $T_x^y[\omega_h^{(1)}(\lambda, t)]$ is compact.

Thus the function $\omega_h^{(1)}(\lambda, t)$ $(h = 0, \infty)$ is $\{L_1, L_2\}$-almost periodic if and only if both of the functions $\omega_h^{(1)}(\lambda, x)$ and $\omega_0^{(2)}(\lambda, y)$ are bounded or $\omega_0^{(2)}(\lambda, y)$ is constant.

---

(6) We always understand compactness in the sense of uniform convergence on the whole real line.

Now we use condition (5.1.3), which is satisfied by $\underset{x}{L}_1$ and $\underset{y}{L}_2$. By assertion b) of Lemma 2.1.1 condition (5.1.3) implies that $\omega_h^{(1)}(\lambda, x)$ and $\omega_0^{(2)}(\lambda, y)$ are bounded for $\lambda > 0$. These values of λ constitute the set Λ_h^+ ($h = 0$ or ∞). By assertion a) of Lemma 2.1.2 for $\lambda < 0$ the solutions $\omega_h^{(1)}(\lambda, x)$ and $\omega_0^{(2)}(\lambda, y)$ can both be bounded only if λ is a common eigenvalue of the corresponding boundary value problems for $\underset{x}{L}_1$ and $\underset{y}{L}_2$. These common eigenvalues constitute the set Λ_h^- ($h = 0$ or ∞), which by part E of Chapter 2, §1, is finite or empty. Furthermore, the solution $\omega_0^{(2)}(\lambda, y)$ is constant only if $\lambda = 0$ and $q_2(y) \equiv 0$. In this case, and also in the case where $\omega_h^{(1)}(0, x)$ and $\omega_0^{(2)}(0, y)$ are bounded, the set Λ_h^0 ($h = 0$ or ∞) consists of the point 0. In all other cases the set is empty.

The lemma is proved.

We introduce the following examples.

If $q_2(y) \equiv 0$, then $\Lambda_0 = \Lambda_\infty$ and this set consists of all positive values of λ and 0.

If $q_2(y) \geqslant 0$ and $q_2(y) \not\equiv 0$, then $\Lambda_0 = \Lambda_\infty$ and this set consists of all positive values of λ.

If $q_1(t) = q_2(t) \not\equiv 0$ and $\lim_{x\to\infty} |\omega_\infty^{(1)}(0, x)| = \infty$, then Λ_∞ consists of all positive values of λ, and Λ_0 consists of all positive values of λ, the point 0 when $|\omega_0^{(1)}(0, x)| < C$, and possibly a finite number of negative values of λ.

From Definition 5.1.1 it follows that a linear combination of a finite number of $\{L_1, L_2\}$-almost periodic functions is an $\{L_1, L_2\}$-almost periodic function. In particular, functions $\sigma(t)$ of the form

$$\sigma(t) = \sum_{k=1}^{N} a_k \omega_{h_k}^{(1)}(\lambda_k, t), \tag{5.1.5}$$

where $h_k = 0$ or ∞, $\lambda_k \in \Lambda h_k$, and the a_k are arbitrary complex numbers, are $\{L_1, L_2\}$-almost periodic.

If a sequence $\{F_n(t)\}$ ($n = 1, 2, \cdots$) of $\{L_1, L_2\}$-almost periodic functions converges to a function $F(t)$, where

$$\lim_{n\to\infty} \sup_{\substack{-\infty<x<\infty \\ -\infty<y<\infty}} |T_x^y [F(t) - F_n(t)]| = 0,$$

then the limit function $F(t)$ is $\{L_1, L_2\}$-almost periodic.

This simple remark leads to the following important lemma.

LEMMA 5.1.3. *A function $F(t)$ is $\{L_1, L_2\}$-almost periodic if for each $\epsilon > 0$ there exists a function $\sigma_\epsilon(t)$ of the form* (5.1.5) *such that*

$$\sup_{\substack{-\infty<x<\infty \\ -\infty<y<\infty}} |T_x^y [F(t) - \sigma_\epsilon(t)]| < \epsilon.$$

We will see later that this sufficient condition is also necessary. The proof of the fact that the condition is necessary constitutes the principal difficulty in solving the fundamental problem in the theory of $\{L_1, L_2\}$-almost periodic functions.

§2. Reduction of the fundamental problem of the theory of $\{L_1, L_2\}$-almost periodic functions to the simplest case

We have already noted that the Corollary to Lemma 5.1.1 allows us to treat even and odd $\{L_1, L_2\}$-almost periodic functions separately. Since the two cases are analogous, we will investigate even $\{L_1, L_2\}$-almost periodic functions in detail, and for odd ones we will just formulate the final result.

LEMMA 5.2.1. *An arbitrary $\{L_1, L_2\}$-almost periodic function $F(t)$ can be written in the form*

$$F(t) = \sum_{k=1}^{m} a_k \omega_0^{(1)}(\lambda_k, t) + \overset{+}{F}(t),$$

where

1) $\lambda_k \in \Lambda_0^-$, $k = 1, \cdots, m$ *(the set* Λ_0^- *is defined in Lemma* 5.1.2),
2) $\overset{+}{F}(t)$ *is an even* $\{L_1, L_2\}$*-almost periodic function, and*
3) *the function*

$$\Phi(t) = \underset{t}{W}_{\{L_1 D^2 0\}} [\overset{+}{F}(t)] = \underset{t}{W}^{-1}_{\{D^2 L_1 0\}} [\overset{+}{F}(t)]$$

is even and bounded on the whole real line.

PROOF. Since the family $T_x^y[F]$ is compact, the function $T_0^y[F]$ of the parameter y is bounded. Since the projection operator

$$P_x^{(2)}[f(x)] = \sum_{k=1}^{n} \int_0^\infty \frac{\omega_0^{(2)}(\lambda_k, x)\, \omega_0^{(2)}(\lambda_k, t)}{\| \omega_0^{(2)}(\lambda_k, x) \|_{L^2}^2} f(t)\, dt$$

(here $\lambda_1 < \lambda_2 < \lambda_3 < \cdots < \lambda_n < 0$ are the negative eigenvalues of the boundary value problem for L_2 with boundary condition $u'(0) = 0$) is defined on all functions $f(x)$ which are bounded in $[0, \infty)$, then for $y \geqslant 0$ the function $T_0^y[F]$ can be written in the form

$$\begin{aligned} T_0^y[F] &= P_y^{(2)}[T_0^y[F]] + (E - P_y^{(2)})[T_0^y[F]] \\ &= \sum_{k=1}^{n} a_k \omega_0^{(2)}(\lambda_k, y) + (E - P_y^{(2)})[T_0^y[F]], \end{aligned}$$

and then by identity 3) in Lemma 5.1.1 we have

$$T_0^y[F] = \underset{y}{V_2} \underset{y}{V_1^{-1}} [F(y)] = \sum_{k=1}^{n} a_k \omega_0^{(2)}(\lambda_k, y) + (E - P_y^{(2)})[T_0^y[F]],$$

where $V_1 = V_{\{D^2L_100\}}$ and the $a_1, \cdots, a_n$ are constants.

Consequently for $y \geqslant 0$

$$F(y) = \underset{y}{V_1} \underset{y}{V_2^{-1}} \left[\sum_{k=1}^{n} a_k \omega_0^{(2)}(\lambda_k, y) + (E - P_y^{(2)}) [T_0^y [F]] \right]$$

and

$$F(y) = \sum_{k=1}^{n} a_k \omega_0^{(1)}(\lambda_k, y) + \underset{y}{V_1} \underset{y}{V_2^{-1}} (E - P_y^{(2)}) [T_0^y [F]].$$

By Theorem 4.3.1 the operator $\overset{+}{V}_{\{L_2D^200\}} = V_2^{-1}(E - P^{(2)})$ is bounded in the space M_1. Since the function $T_0^y [F]$ is continuous and bounded, the function

$$\Phi(y) = \underset{y}{V_2^{-1}} (E - P_y^{(2)}) [T_0^y [F]] = \overset{+}{V}_{\{L_2D^200\}} [T_0^y [F]]$$

is also continuous and bounded.

Thus for $y \geqslant 0$

$$F(y) = \sum_{k=1}^{n} a_k \omega_0^{(1)}(\lambda_k, y) + \underset{y}{V_1} [\Phi(y)],$$

and if we consider the even extension of this function onto the negative semi-axis we obtain

$$F(y) = \sum_{k=1}^{n} a_k \omega_0^{(1)}(\lambda_k, y) + \underset{y}{W_1} [\Phi(y)], \tag{5.2.1}$$

where $W_1 = W_{\{D^2L_10\}}$, and the function $\Phi(y)$ is continuous, even, and bounded.

We now show that in (5.2.1) the coefficients $a_k = 0$ for all values of $\lambda_k \notin \Lambda_0^-$. We split the first term in (5.2.1) into two parts:

$$\sum_{k=1}^{n} a_k \omega_0^{(1)}(\lambda_k, y) = \sum_{\lambda_j \bar{\in} \Lambda_0^-} a_j \omega_0^{(1)}(\lambda_j, y) + \sum_{\lambda_i \in \Lambda_0^-} a_i \omega_0^{(1)}(\lambda_i, y)$$

and assume, contrary to our assertion, that not all $a_j = 0$. Let λ_{j_0} be the largest λ_j in absolute value for which $a_j \neq 0$. Since $\omega_0^{(2)}(\lambda_{j_0}, y) \not\equiv \text{const}$, there exists a sequence $y_1, y_2, \cdots$ such that

$$\lim_{n \to \infty} y_n = y_0 \neq \infty, \quad y_0 > 0$$

and

$$\omega_0^{(2)}(\lambda_{j_0}, y_0) - \omega_0^{(2)}(\lambda_{j_0}, y_k) \neq 0 \qquad (k = 1, 2, \ldots).$$

From the compact sequence $T_x^{y_k}[F]$ $(k = 1, 2, \cdots)$ we choose a subsequence $T_x^{y_{k'}}[F]$ $(k' \to \infty)$ which converges uniformly on the whole real line. Since the generalized shift operator T_x^y takes continuous functions to continuous functions, this subsequence can only converge to the function $T_x^{y_0}[F]$. Consequently for some $y_{k'} = y' > 0$ we have

$$\sup_{-\infty<x<\infty} |T_x^{y_0}[F] - T_x^{y'}[F]| < 1, \qquad (5.2.2)$$
$$\omega_0^{(2)}(\lambda_{j_0}, y_0) - \omega_0^{(2)}(\lambda_{j_0}, y') \neq 0.$$

On the other hand,

$$\begin{aligned} T_x^{y_0}[F] - T_x^{y'}[F] &= a_{j_0}\omega_0^{(1)}(\lambda_{j_0}, x)[\omega_0^{(2)}(\lambda_{j_0}, y_0) - \omega_0^{(2)}(\lambda_{j_0}, y')] \\ &+ \sum_{\substack{\lambda_j \in \Lambda_0^- \\ j \neq j_0}} a_j \omega_0^{(1)}(\lambda_j, x)[\omega_0^{(2)}(\lambda_j, y_0) - \omega_0^{(2)}(\lambda_j, y')] \\ &+ \sum_{\lambda_i \in \Lambda_0^-} a_i \omega_0^{(1)}(\lambda_i, x)[\omega_0^{(2)}(\lambda_i, y_0) - \omega_0^{(2)}(\lambda_i, y')] \\ &+ \underset{x}{V_1}\underset{y}{V_2} S_x^y[\Phi(t)]|_{y=y_0} - \underset{x}{V_1}\underset{y}{V_2} S_x^y[\Phi(t)]|_{y=y'}. \end{aligned}$$

Since by assumption $\lambda_{j_0} \notin \Lambda_0^-$; that is, λ_{j_0} is not an eigenvalue of the boundary value problem for L_1 with boundary condition $u'(0) = 0$, by Lemma 2.1.2 we have $|\omega_0^{(1)}(\lambda_{j_0}, x)| > C\exp(x\sqrt{|\lambda_{j_0}|})$ for large x. Therefore the first term on the right side of the above equation grows like $\exp(x\sqrt{|\lambda_{j_0}|})$ as $x \to \infty$. Again by Lemma 2.1.2 the second term grows no faster than $\exp(x\sqrt{\alpha})$, where $\alpha = \max_{j\neq j_0} |\lambda_j| < |\lambda_{j_0}|$, and the third term is bounded. The fourth term grows no faster than $|x|$. This follows from Theorem 4.1.1 and the boundedness of $\Phi(t)$, which we established earlier.

Thus the right side of the equation grows like $\exp(x\sqrt{\lambda_{j_0}})$ as $x \to \infty$, and

$$\sup_{-\infty<x<\infty} |T_x^{y_0}[F] - T_x^{y'}[F]| = \infty,$$

which contradicts (5.2.2). This contradiction proves our assertion.

Now (5.2.1) can be written in the form

$$F(t) = \sum_{\lambda_i \in \Lambda_0^-} a_i \omega_0^{(1)}(\lambda_i, t) + \overset{+}{F}(t),$$

where the function

$$\overset{+}{F}(t) = \underset{t}{W_1}[\Phi(t)] = F(t) - \sum_{\lambda_i \in \Lambda_0^-} a_i \omega_0^{(1)}(\lambda_i, t),$$

is the difference of two even $\{L_1, L_2\}$-almost periodic functions and is therefore even and $\{L_1, L_2\}$-almost periodic. The lemma is proved.

The lemma shows that in order to solve the fundamental problem for even $\{L_1, L_2\}$-almost periodic functions $F(t)$ we should concentrate our attention on $\overset{+}{F}(t)$. Our study of $\overset{+}{F}(t)$ is based on the following lemma.

LEMMA 5.2.2. *The function*

$$\Phi(t) = \underset{t}{W^{-1}_{\{D^2L_10\}}}[\overset{+}{F}(t)]$$

introduced in Lemma 5.2.1 *is an even* $\{D^2, D^2\}$*-almost periodic function*; *that is, it is continuous and the family of functions*

$$S_x^y[\Phi] = \frac{1}{2}[\Phi(x+y) + \Phi(x-y)]$$

(y *is a parameter*) *is compact.*

PROOF. We established the continuity and boundedness of $\Phi(t)$ in the preceding lemma. It remains to show that the family $S_x^y[\Phi]$ is compact.

We show first of all that for $x, y \geqslant 0$

$$T_x^y[\overset{+}{F}] = (E - P_x^{(1)})(E - P_y^{(2)})T_x^y[\overset{+}{F}], \tag{5.2.3}$$

where $P^{(1)}$ and $P^{(2)}$ are the projection operators

$$P_x^{(1)}[f] = \sum_{k=1}^{l} \int_0^\infty \frac{\omega_0^{(1)}(\mu_k, x)\,\omega_0^{(1)}(\mu_k, t)}{\|\omega_0^{(1)}(\mu_k, x)\|_{L^2}^2} f(t)\,dt, \tag{5.2.3$'$}$$

$$P_y^{(2)}[f] = \sum_{k=1}^{n} \int_0^\infty \frac{\omega_0^{(2)}(\lambda_k, y)\,\omega_0^{(2)}(\lambda_k, t)}{\|\omega_0^{(2)}(\lambda_k, y)\|_{L^2}^2} f(t)\,dt. \tag{5.2.3$''$}$$

(Here $\mu_1, \cdots, \mu_l$ $(\lambda_1, \cdots, \lambda_n)$ are the negative eigenvalues of the boundary value problem for L_1 (L_2) with boundary condition $u'(0) = 0$.)

By identity 3) in Lemma 5.1.1 for $x, y \geqslant 0$ we have

$$T_x^y[\overset{+}{F}] = \underset{x}{V_1}\underset{y}{V_2}S_x^y[W_1^{-1}[\overset{+}{F}]] = \underset{x}{V_1}\underset{y}{V_2}S_x^y[\Phi],$$

where $V_1 = V_{\{D^2L_100\}}$ and $V_2 = V_{\{D^2L_200\}}$. Since $\Phi(t)$ is bounded, for an arbitrary fixed $y > 0$ the function $V_2S_x^y[\Phi]$ is bounded:

$$\sup_{-\infty<x<\infty} |\underset{y}{V_2}S_x^y[\Phi]| < \infty. \tag{5.2.4}$$

Then by Theorem 4.1.1 we conclude that for an arbitrary fixed $y > 0$ the function

$$T_x^y[\overset{+}{F}] = \underset{x}{V_1}\underset{y}{V_2}S_x^y[\Phi]$$

is either bounded (if $\omega_0^{(1)}(0, x)$ is bounded) or grows no faster than $|x|$ (if $\omega_0^{(1)}(0, x)$ is not bounded). Therefore we can apply to the function $T_x^y[\overset{+}{F}]$ the operator $P_x^{(1)}$, and we obtain

$$\underset{x}{V_1}\underset{y}{V_2}S_x^y[\Phi] = T_x^y[\overset{+}{F}] = P_x^{(1)}T_x^y[\overset{+}{F}] + (E - P_x^{(1)})T_x^y[\overset{+}{F}]$$

and

$$\underset{y}{V_2}S_x^y[\Phi] = \underset{x}{V_1^{-1}}P_x^{(1)}T_x^y[\overset{+}{F}] + \underset{x}{V_1^{-1}}(E - P_x^{(1)})T_x^y[\overset{+}{F}]. \tag{5.2.5}$$

We now observe that in the notation of Theorem 4.3.1 we have $V_1^{-1}(E-P^{(1)})=\overset{+}{V}_{\{L_1D^200\}}$, which by the same theorem implies that the function

$$\underset{x}{V_1^{-1}}(E-P_x^{(1)})\,T_x^y[\overset{+}{F}]=\overset{+}{V}_{\{L_1D^200\}}\,T_x^y[\overset{+}{F}]\,\cdot$$

is either bounded (if $\omega_0^{(1)}(0,x)$ is bounded) or grows no faster than $|x|$ (if $\omega_0^{(1)}(0,x)$ is not bounded). In either case

$$\underset{x}{V_1^{-1}}(E-P_x^{(1)})\,T_x^y[\overset{+}{F}]=O(x). \tag{5.2.6}$$

From (5.2.5) together with the bounds (5.2.4) and (5.2.6) it follows that for an arbitrary fixed $y\geqslant 0$

$$\underset{x}{V_1^{-1}}P_x^{(1)}\,T_x^y[\overset{+}{F}]=O(x). \tag{5.2.7}$$

In the notation of Theorem 4.3.1 we have $V_1^{-1}P^{(1)}=\overline{V}_{\{L_1D^200\}}$. Since the operator $\overline{V}_{\{L_1D^200\}}$ takes an arbitrary function to a finite linear combination of hyperbolic cosines which either equals zero or grows exponentially, the bound (5.2.7) is possible if and only if

$$\underset{x}{V_1^{-1}}P_x^{(1)}\,T_x^y[\overset{+}{F}]\equiv 0.$$

Consequently

$$P_x^{(1)}\,T_x^y[\overset{+}{F}]=\underset{x}{V_1}[\underset{x}{V_1^{-1}}P_x^{(1)}\,T_x^y[\overset{+}{F}]]\equiv 0$$

and

$$T_x^y[\overset{+}{F}]=(E-P_x^{(1)})\,T_x^y[\overset{+}{F}].$$

By a similar argument for $P_y^{(2)}$ we obtain (5.2.3).

We turn to the proof of the fact that the family $S_x^y[\Phi]$ is compact. Since $\Phi(t)$ is even, it suffices to consider only nonnegative values of the parameter y and to show compactness of this family in the sense of uniform convergence on the semi-axis $0\leqslant x<\infty$.

By identity 3) in Lemma 5.1.1 and (5.2.3) we have for $x, y\geqslant 0$

$$\underset{y}{V_2}S_x^y[\Phi]=\underset{x}{V_1^{-1}}T_x^y[\overset{+}{F}]=\underset{x}{V_1^{-1}}(E-P_x^{(1)})\,T_x^y[\overset{+}{F}]. \tag{5.2.8}$$

We have already seen for an arbitrary fixed $y\geqslant 0$ that the function $\underset{y}{V_2}S_x^y[\Phi]$ is bounded and continuous. Consequently the family of functions

$$B(x,y)=\underset{y}{V_2}S_x^y[\Phi]=\underset{x}{V_1^{-1}}(E-P_x^{(1)})\,T_x^y[\overset{+}{F}] \tag{5.2.8$'$}$$

($y \geqslant 0$ is a parameter) consists of bounded, continuous functions. And this family is compact in the sense of uniform convergence on the semi-axis $0 \leqslant x < \infty$. This follows by virtue of (5.2.8′) from the compactness of the family of continuous functions $T_x^y[\overset{+}{F}]$ and the boundedness of the operator $V_1^{-1}(E - P^{(1)}) = \overset{+}{V}_{\{L_1 D^2 00\}}$ in the space $M_{[0,\infty)}$ (Theorem 4.3.1).

By virtue of (5.2.3) we see from (5.2.8) that

$$\underset{y}{V_2} S_x^y[\Phi] = (E - P_y^{(2)}) \underset{x}{V_1^{-1}} (E - P_x^{(1)}) T_x^y[\overset{+}{F}] = (E - P_y^{(2)})[B(x, y)]$$

and

$$S_x^y[\Phi] = \underset{y}{V_2^{-1}} (E - P_y^{(2)})[B(x, y)] = \overset{+}{V}_{\{L_2 D^2 00\}} [B(x, y)].$$

Then by Theorem 4.3.1 we have

$$S_x^y[\Phi] = B(x, y) + \int_0^\infty \overset{+}{H}(y, t) B(x, t)\, dt,$$

where

$$\operatorname*{ess\,sup}_{0 \leqslant y < \infty} \int_0^\infty |\overset{+}{H}(y, t)|\, dt = |\overset{+}{H}| < \infty.$$

Furthermore, since by the remark after Theorem 4.3.1 the operator $\overset{+}{V}_{\{L_2 D^2 00\}}$ takes continuous functions to continuous functions, we have

$$\sup_{0 \leqslant y < \infty} \int_0^\infty |\overset{+}{H}(y, t)|\, dt = \operatorname*{ess\,sup}_{0 \leqslant y < \infty} \int_0^\infty |\overset{+}{H}(y, t)|\, dt = |\overset{+}{H}| < \infty.$$

It is known [5] that since the family $B(x, y)$ is compact, for a given $\epsilon > 0$ the semi-axis $0 \leqslant y < \infty$ can be split into a finite number of mutually disjoint sets $\mathfrak{A}_1, \cdots, \mathfrak{A}_N$ such that

$$\sup_{0 \leqslant x < \infty} |B(x, y') - B(x, y'')| < \varepsilon,$$

if y' and y'' belong to the same set $\mathfrak{A}_i$.

Let $a_1 \in \mathfrak{A}_1, \cdots, a_N \in \mathfrak{A}_N$. Then

$$\int_0^\infty \overset{+}{H}(y, t) B(x, t)\, dt = \sum_{i=1}^N \int_{\mathfrak{A}_i} \overset{+}{H}(y, t) B(x, t)\, dt$$

$$= \sum_{i=1}^N B(x, a_i) \int_{\mathfrak{A}_i} \overset{+}{H}(y, t)\, dt + \delta(x, y),$$

where

$$\delta(x, y) = \sum_{i=1}^{N} \int_{\mathfrak{A}_i} \overset{+}{H}(y, t)[B(x, t) - B(x, a_i)]\, dt.$$

Since

$$\sup_{\substack{0 \leq x < \infty \\ t \in \mathfrak{A}_i}} |B(x, t) - B(x, a_i)| < \varepsilon,$$

it follows that

$$\delta(x, y)| \leqslant \varepsilon \sum_{i=1}^{N} \int_{\mathfrak{A}_i} |\overset{+}{H}(y, t)|\, dt \leqslant \varepsilon |\overset{+}{H}|.$$

Therefore we have

$$S_x^y[\Phi] = B(x, y) + \sum_{i=1}^{N} B(x, a_i) H_i(y) + \delta(x, y),$$

where

$$H_i(y) = \int_{\mathfrak{A}_i} \overset{+}{H}(y, t)\, dt \qquad (i = 1, 2, \ldots, N),$$

$$\sup_{0 \leqslant y < \infty} |H_i(y)| \leqslant |\overset{+}{H}| < \infty \qquad (i = 1, 2, \ldots, N),$$

$$\sup_{\substack{0 \leqslant x < \infty \\ 0 \leqslant y < \infty}} |\delta(x, y)| \leqslant \varepsilon \overset{+}{H}.$$

The family of functions $\{B(x, y) + \Sigma_1^N B(x, a_i)H_i(y)\}$ ($y \geqslant 0$ is a parameter) is compact. In fact, compactness of the family $B(x, y)$ was established earlier, and compactness of the family $\{\Sigma_1^N B(x, a_i)H_i(y)\}$ follows easily from the boundedness of the functions $B(x, a_i)$ and the coefficients $H_i(y)$, $i = 1, \cdots, N$.

Thus for a given $\epsilon > 0$ we can construct a compact family $\{B(x, y) + \Sigma_1^N B(x, a_i)H_i(y)\}$ each function of which differs from the corresponding function in the family $S_x^y[\Phi]$ by less than $\epsilon |\overset{+}{H}|$. Since $\epsilon > 0$ was arbitrary, it follows that the family $S_x^y[\Phi]$ is compact, which is what we wanted to show.

The lemma is proved.

In order to proceed further we must examine in detail the function $\Phi(t) = W_{t\,1}^{-1}[\overset{+}{F}(t)]$; that is, we must solve the fundamental problem for the class of $\{D^2, D^2\}$-almost periodic functions.

The next section is devoted to this problem.

§3. Solution of the fundamental problem for the class of $\{D^2, D^2\}$-almost periodic functions

A constructive characterization of even $\{D^2, D^2\}$-almost periodic functions was first found by Levitan [5], who showed that the class of even $\{D^2, D^2\}$-almost periodic functions coincides with the class of ordinary almost periodic functions of Bohr.

With certain applications in mind, we will consider in this section arbitrary $\{D^2, D^2\}$-almost periodic functions; that is, those continuous functions $F(t)$ for which the family of functions

$$S_x^y[F(t)] = \frac{1}{2}[F(x+y) + F(x-y)]$$

(y is a parameter) is compact in the sense of uniform convergence on the whole real line $-\infty < x < \infty$.

A constructive characterization of these functions is given in the following theorem.

THEOREM 5.3.1. *A function $F(t)$ is $\{D^2, D^2\}$-almost periodic if and only if it is of the form*

$$F(t) = At + \Phi(t),$$

where A is a constant and $\Phi(t)$ is an almost periodic function in the sense of Bohr.

PROOF. SUFFICIENCY. If $F(t)$ is of the indicated form, then it is continuous and

$$S_x^y[F] = AS_x^y[t] + S_x^y[\Phi] = A\,\frac{1}{2}\,[x+y+x-y] + S_x^y[\Phi];$$

that is,

$$S_x^y[F] = Ax + \frac{1}{2}[\Phi(x+y) + \Phi(x-y)].$$

Sufficiency will be proved if we show that the family $[\Phi(x+y) + \Phi(x-y)]$ (y is a parameter) is compact. Since $\Phi(t)$ is an almost periodic function in the sense of Bohr, by Bochner's theorem [22] the families $\{\Phi(x+y)\}$ and $\{\Phi(x-y)\}$ (y is a parameter) are compact, and therefore $[\Phi(x+y) + \Phi(x-y)]$ is also compact. Sufficiency is proved.

NECESSITY. We break the proof into two parts.

In the first part we show that for each $\{D^2, D^2\}$-almost periodic function $F(t)$ we can find a constant A for which the function

$$\Phi(t) = F(t) - At$$

is bounded. Since the function At is clearly $\{D^2, D^2\}$-almost periodic, $\Phi(t)$ is also $\{D^2, D^2\}$-almost periodic.

In the second part of the proof we show that each bounded $\{D^2, D^2\}$-almost periodic function $\Phi(t)$ is almost periodic in the sense of Bohr.

1) If $F(t)$ is $\{D^2, D^2\}$-almost periodic, then, as we have seen earlier, the function

$$F_1(t)=\frac{1}{2}[F(t)+F(-t)],\quad F_2(t)=\frac{1}{2}[F(t)-F(-t)]$$

is also $\{D^2, D^2\}$-almost periodic. Here the even $\{D^2, D^2\}$-almost periodic function $F_1(t)$ is bounded, for if $\lim_{n\to\infty}|F_1(y_n)|=\infty$, then

$$\lim_{n\to\infty}|S_0^{y_n}[F_1]|=\lim_{n\to\infty}\frac{1}{2}|F_1(y_n)+F_1(-y_n)|=\lim_{n\to\infty}|F_1(y_n)|=\infty,$$

which contradicts the compactness of the family $S_x^y[F_1]$. So if we show that for some A the function $F_2(t)-At$ is bounded, we will simultaneously show that $F(t)-At$ is bounded.

We note also that the odd $\{D^2, D^2\}$-almost periodic function $F_2(t)$ is uniformly continuous. For otherwise there would exist a sequence

$$y_1,\ y_2,\ \ldots,\ y_n,\ \ldots;\ x_1,\ x_2,\ \ldots,\ x_n,\ \ldots;\ \lim_{n\to\infty} x_n=0$$

and a positive number $\delta>0$ such that

$$|F_2(y_n+x_n)-F_2(y_n-x_n)|>\delta$$

or, since $F_2(t)$ is odd,

$$|F_2(x_n+y_n)+F_2(x_n-y_n)|>\delta. \tag{5.3.1}$$

Since the family $S_x^y[F]$ is compact, there exist a sequence $\{n_i\}$ and numbers N and y such that for $n_i>N$

$$\sup_{-\infty<x<\infty}|S_x^y[F_2]-S_x^{y_{n_i}}[F_2]|<\frac{1}{4}\delta$$

and therefore

$$|S_x^y[F_2]|>|S_x^{y_{n_i}}[F_2]|-\frac{1}{4}\delta$$

for any x. In particular, for $x=x_{n_i}$ we obtain

$$|F_2(x_{n_i}+y)+F_2(x_{n_i}-y)|>|F_2(x_{n_i}+y_{n_i})+F_2(x_{n_i}-y_{n_i})|-\frac{1}{2}\delta,$$

if $n_i>N$. Then by (5.3.1)

$$\overline{\lim_{n_i\to\infty}}|F_2(x_{n_i}+y)+F_2(x_{n_i}-y)|>\frac{1}{2}\delta.$$

Since $\lim_{n_i\to\infty}x_{n_i}=0$ and $F_2(t)$ is continuous, it follows that

$$|F_2(y)+F_2(-y)|>\frac{1}{2}\delta>0,$$

which contradicts the fact that $F_2(t)$ is odd.

This contradiction shows that $F_2(t)$ must be uniformly continuous.

Now let $F(t)$ be an arbitrary $\{D^2, D^2\}$-almost periodic function. We introduce the family $\{\psi_x(y)\}$:

$$\psi_x(y) = \frac{1}{2}[F(x+y) - F(-x+y)], \tag{5.3.2}$$

which depends on the parameter x $(-\infty < x < \infty)$. Each function in this family is an almost periodic function of y in the sense of Bohr.

For by virtue of Bochner's theorem it suffices to show that for any fixed $x = x_0$ the family $\{\psi_{x_0}(y+h)\}$ (h is a parameter) is compact. We have

$$\psi_{x_0}(y+h) = \frac{1}{2}[F(x_0+y+h) - F(-x_0+y+h)]$$
$$= \frac{1}{2}[F(x_0+y+h) + F(x_0+y-h)] - \frac{1}{2}[F(y+h-x_0) + F(y-h+x_0)];$$

that is,

$$\psi_{x_0}(y+h) = S^h_{x_0+y}[F] - S^{h-x_0}_y[F].$$

Since $F(t)$ is $\{D^2, D^2\}$-almost periodic by assumption, the families $S^h_{x_0+y}[F]$ and $S^{h-x_0}_y[F]$ (h is a parameter) are compact, and therefore the family $\psi_{x_0}(y+h)$ is compact.

To each almost periodic function $\psi_x(y)$ in the sense of Bohr there corresponds a Fourier series

$$\psi_x(y) \sim a_0(x) + a_{\lambda_1}(x)e^{i\lambda_1 y} + \dots + a_{\lambda_k}(x)e^{i\lambda_k y} + \dots,$$

where

$$a_\lambda(x) = \lim_{T\to\infty}\frac{1}{2T}\int_{-T}^{T}\psi_x(y)e^{-i\lambda y}\,dy.$$

For $F(t)$ we choose the function $F_2(t)$. From $F_2(t)$ we construct the family $\{\psi^{(2)}_x(y)\}$ by means of (5.3.2). Since $F_2(t)$ is odd,

$$\psi^{(2)}_x(y) = \frac{1}{2}[F_2(x+y) - F_2(-x+y)] = \frac{1}{2}[F_2(x+y) + F_2(x-y)].$$

The free term in the Fourier series for $\psi^{(2)}_x(y)$ equals

$$a^{(2)}_0(x) = \lim_{T\to\infty}\frac{1}{2T}\int_{-T}^{T}\frac{1}{2}[F_2(x+y) + F_2(x-y)]\,dy,$$

and then, since $F_2(t)$ is uniformly continuous, the coefficient $a^{(2)}_0(x)$ is a uniformly continuous function of the parameter x $(-\infty < x < \infty)$.

Since

$$\psi_x^{(2)}(y) = \frac{1}{2}[F_2(x+y) + F_2(x-y)]$$

$$= -\frac{1}{2}[F_2(-x-y) + F_2(-x+y)] = -\psi_{-x}^{(2)}(y),$$

we have

$$a_0^{(2)}(x) = -a_0^{(2)}(-x).$$

Furthermore,

$$\psi_{x+h}^{(2)}(y) + \psi_{x-h}^{(2)}(y) = \frac{1}{2}[F_2(x+h+y) + F_2(x+h-y)$$

$$+ F_2(x-h+y) + F_2(x-h-y)] = \psi_x^{(2)}(y+h) + \psi_x^{(2)}(y-h),$$

which implies that

$$a_0^{(2)}(x+h) + a_0^{(2)}(x-h) = 2a_0^{(2)}(x).$$

The continuous, odd function $a_0^{(2)}(x)$ which satisfies this functional equation is clearly of the form

$$a_0^{(2)}(x) = Ax,$$

where A is some constant.

From what we showed above, for each fixed x the function $\psi_x^{(2)}(y) - F_2(x)$ is almost periodic in y in the sense of Bohr, and the free term in the Fourier series of this function equals $Ax - F_2(x)$.

By Bessel's inequality,

$$|Ax - F_2(x)|^2 \leqslant \lim_{T\to\infty} \frac{1}{2T}\int_{-T}^{T} |\psi_x^{(2)}(y) - F_2(x)|^2\, dy,$$

which implies that if we can show

$$\sup_{\substack{-\infty<x<\infty\\ -\infty<y<\infty}} |\psi_x^{(2)}(y) - F_2(x)| = M < \infty, \tag{5.3.3}$$

we will simultaneously be showing that $Ax - F_2(t)$ is bounded, and therefore $F(x) - Ax$ is bounded.

Suppose that

$$\sup_{\substack{-\infty<x<\infty\\ -\infty<y<\infty}} |\psi_x^{(2)}(y) - F_2(x)| = \infty. \tag{5.3.3'}$$

Then there exist sequences of numbers $x_1, x_2, \cdots, x_n, \cdots$ and $y_1, y_2, \cdots, y_n, \cdots$ such that

$$\lim_{n\to\infty} |\psi_{x_n}^{(2)}(y_n) - F_2(x_n)| = \infty;$$

that is,

$$\lim_{n\to\infty} \left|\frac{1}{2}[F_2(x_n+y_n) + F_2(x_n-y_n)] - F_2(x_n)\right| = \infty.$$

Since the family $\frac{1}{2}[F_2(x+y_n) + F_2(x-y_n)]$ is compact, there exist a subsequence $\{y_{n_i}\}$ and numbers N and a such that

$$\sup_{-\infty<x<\infty} \frac{1}{2} | F_2(x+y_{n_i}) + F_2(x-y_{n_i}) - F_2(x+a) - F_2(x-a) | < 1,$$

if $n_i > N$. In particular, for $x = x_{n_i}$ we obtain

$$\Big| \Big\{ \frac{1}{2} [F_2(x_{n_i}+y_{n_i}) + F_2(x_{n_i}-y_{n_i})] - F_2(x_{n_i}) \Big\} - \Big\{ \frac{1}{2} [F_2(x_{n_i}+a) + F_2(x_{n_i}-a)] - F_2(x_{n_i}) \Big\} \Big| < 1,$$

which implies that

$$\overline{\lim_{n_i\to\infty}} \left| \frac{1}{2} \{F_2(x_{n_i}+a) + F_2(x_{n_i}-a)\} - F_2(x_{n_i}) \right|$$
$$> \lim_{n_i\to\infty} \left| \frac{1}{2} \{F_2(x_{n_i}+y_{ni}) + F_2(x_{n_i}-y_{n_i})\} - F_2(x_{n_i}) \right| - 1 = \infty;$$

and thus

$$\overline{\lim_{n_i\to\infty}} | F_2(x_{n_i}+a) - F_2(x_{n_i}) | + \overline{\lim_{n_i\to\infty}} | F_2(x_{n_i}-a) - F_2(x_{n_i}) | = \infty,$$

which contradicts the fact that $F_2(x)$ is uniformly continuous.

Therefore (5.3.3′) is false and (5.3.3) is true.

The first part of the proof is complete.

2) In the first part of this proof we showed that for each $\{D^2, D^2\}$-almost periodic function $F(t)$ there exists a constant A such that the function $\Phi(t) = F(t) - At$ is bounded. Since $\Phi(t)$ is $\{D^2, D^2\}$-almost periodic, the family $\{\phi_x(y)\}$ constructed from this function according to (5.3.2)

$$\varphi_x(y) = \frac{1}{2} [\Phi(x+y) - \Phi(-x+y)] \tag{5.3.4}$$

consists of functions of y which are almost periodic in the sense of Bohr. For the Fourier coefficients of $\phi_x(y)$ we have

$$a_\lambda(x) = \lim_{T\to\infty} \frac{1}{2T} \int_{-T}^{T} \varphi_x(y) e^{-i\lambda y}\, dy = \lim_{T\to\infty} \frac{1}{4T} \int_{-T}^{T} [\Phi(x+y) - \Phi(-x+y)] e^{-i\lambda y}\, dy.$$

Furthermore,

$$\int_{-T}^{T} \Phi(x+y) e^{-i\lambda y}\, dy = e^{i\lambda x} \int_{-T+x}^{T+x} \Phi(y) e^{-i\lambda y}\, dy$$
$$= e^{i\lambda x} \left[\int_{T}^{T+x} - \int_{-T}^{-T+x} + \int_{-T}^{T} \Phi(y) e^{-i\lambda y}\, dy \right].$$

Similarly,

$$\int_{-T}^{T} \Phi(-x+y) e^{-i\lambda y}\, dy = e^{-i\lambda x} \left[\int_{-T-x}^{-T} - \int_{T-x}^{T} + \int_{-T}^{T} \Phi(y) e^{-i\lambda y} dy \right].$$

Therefore

$$\lim_{T\to\infty}\frac{1}{2T}\int_{-T}^{T}\varphi_x(y)\,e^{-i\lambda y}\,dy=\lim_{T\to\infty}\left[\frac{e^{i\lambda x}-e^{-i\lambda x}}{2}\cdot\frac{1}{2T}\int_{-T}^{T}\Phi(y)\,e^{-i\lambda y}\,dy+\frac{1}{2T}\mu_x(T)\right],$$

where

$$\mu_x(T)=\frac{e^{i\lambda x}}{2}\left[\int_{T}^{T+x}-\int_{-T}^{-T+x}\Phi(y)\,e^{-i\lambda y}\,dy\right]-\frac{e^{-i\lambda x}}{2}\left[\int_{-T-x}^{-T}-\int_{T-x}^{T}\Phi(y)\,e^{-i\lambda y}\,dy\right].$$

Let $\sup_{-\infty<y<\infty}|\Phi(y)|=m$ (by part 1) of the proof we have $m<\infty$); then

$$|\mu_x(T)|\leqslant 2|x|m,$$

which implies that $\lim_{T\to\infty}\mu_x(T)/2T$ exists and equals zero. Then

$$\lim_{T\to\infty}\frac{1}{2T}\int_{-T}^{T}\varphi_x(y)\,e^{-i\lambda y}\,dy=\frac{e^{i\lambda x}-e^{-i\lambda x}}{2}\lim_{T\to\infty}\frac{1}{2T}\int_{-T}^{T}\Phi(y)\,e^{-i\lambda y}\,dy$$

and the limit on the right side does exist. Using the notation

$$a(\lambda)=\lim_{T\to\infty}\frac{1}{2T}\int_{-T}^{T}\Phi(y)\,e^{-i\lambda y}\,dy, \tag{5.3.5}$$

we obtain

$$a_\lambda(x)=ia(\lambda)\sin\lambda x.$$

Thus to the almost periodic functions $\phi_x(y)$ in the sense of Bohr there correspond the Fourier series

$$\varphi_x(y)\sim i\sum_{k=1}^{\infty}a(\lambda_k)\sin\lambda_k x e^{i\lambda_k y}.$$

We consider the functions

$$S_y^t[\varphi_x]=\frac{1}{2}[\varphi_x(y+t)+\varphi_x(y-t)]$$

(t and x are parameters). These functions are almost periodic in y in the sense of Bohr, and so we have the corresponding Fourier series

$$S_y^t[\varphi_x]\sim\frac{1}{2}i\sum_{k=1}^{\infty}a(\lambda_k)\sin\lambda_k x\,(e^{i\lambda_k t}+e^{-i\lambda_k t})\,e^{i\lambda_k y};$$

that is,

$$S_y^t[\varphi_x]\sim i\sum_{k=1}^{\infty}a(\lambda_k)\sin\lambda_k x\cos\lambda_k t e^{i\lambda_k y}. \tag{5.3.6}$$

Furthermore,

$$\frac{1}{2}|\varphi_x(y+t)+\varphi_x(y-t)|$$
$$=\frac{1}{4}[\Phi(x+y+t)-\Phi(-x+y+t)+\Phi(x+y-t)-\Phi(-x+y-t)]$$
$$=\frac{1}{2}\Big\{\frac{1}{2}[\Phi(x+y+t)+\Phi(x+y-t)]$$
$$-\frac{1}{2}[\Phi(-x+y+t)+\Phi(-x+y-t)]\Big\},$$

or

$$S_y^t[\varphi_x]=\frac{1}{2}S_{x+y}^t[\Phi]-\frac{1}{2}S_{-x+y}^t[\Phi]. \tag{5.3.7}$$

Since the family $S_z^t[\Phi]$ (t is a parameter) is compact, for a given $\epsilon>0$ the real line $-\infty<t<\infty$ can be decomposed into a finite number of mutually disjoint sets $\mathfrak{A}_1, \cdots, \mathfrak{A}_n$ such that

$$\sup_{-\infty<z<\infty}|S_z^t[\Phi]-S_z^{t'}[\Phi]|<\varepsilon, \tag{5.3.8}$$

if t and t' are in the same set $\mathfrak{A}$. Let $\chi_i(t)$ be the characteristic function of the set $\mathfrak{A}_i$. Then $\Sigma_1^n\chi_i(t)=1$ and

$$\lim_{T\to\infty}\frac{1}{2T}\int_{-T}^{T}\sum_{i=1}^{n}\chi_i(t)\,dt=1,$$

which implies that for some $i=i_0$

$$\overline{\lim_{T\to\infty}}\frac{1}{2T}\int_{-T}^{T}\chi_{i_0}(t)\,dt=\delta>0.$$

Thus for the function

$$k(t)=\frac{1}{\delta}\chi_{i_0}(t)$$

there exists a sequence of numbers $T_1, T_2, \cdots$, $\lim_{n\to\infty}T_n=\infty$, such that

$$\lim_{n\to\infty}\frac{1}{2T_n}\int_{-T_n}^{T_n}k(t)\,dt=1. \tag{5.3.9}$$

We consider the sequence of functions

$$\varphi_n(x;\,y)=\frac{1}{2T_n}\int_{-T_n}^{T_n}S_y^t[\varphi_x]\,k(t)\,dt. \tag{5.3.10}$$

The functions $\phi_n(x;y)$ are uniformly bounded. Since the function $\phi_x(y)$ for a fixed x is almost periodic in the sense of Bohr, the set of its ϵ-shifts $E_x(\epsilon)$ is relatively dense. Let $\tau\in E_x(\epsilon)$; that is,

$$\sup_{-\infty<y<\infty} |\varphi_x(y+\tau)-\varphi_x(y)| < \varepsilon.$$

Then

$$\sup_{-\infty<y<\infty} |\varphi_n(x;\ y+\tau)-\varphi_n(x;\ y)|$$

$$= \sup_{-\infty<y<\infty}\left|\frac{1}{2T_n}\int_{-T_n}^{T_n}\{S^t_{y+\tau}[\varphi_x]-S^t_y[\varphi_x]\}\,k(t)\,dt\right.$$

$$\leqslant \sup_{-\infty<y<\infty}\frac{1}{2T_n}\int_{-T_n}^{T_n}\frac{1}{2}\,|\varphi_x(y+\tau+t)+\varphi_x(y+\tau-t)$$

$$-\varphi_x(y+t)-\varphi_x(y-t)|\,k(t)\,dt$$

$$\leqslant \frac{1}{4T_n}\int_{-T_n}^{T_n}2\sup_{-\infty<z<\infty}|\varphi_x(z+\tau)-\varphi_x(z)|\,k(t)\,dt \leqslant \varepsilon\frac{1}{2T_n}\int_{-T_n}^{T_n}k(t)dt.$$

Therefore for a fixed x the functions $\phi_n(x;y)$ form a uniformly bounded, continuous, and almost periodic (in the sense of Bohr) sequence. Thus it is known [22] that for each fixed x the sequence $\phi_n(x;y)$ is compact in the sense of uniform convergence on the whole real line $-\infty<y<\infty$.

This enables us to use the diagonal process to select from $\phi_n(x;y)$ a subsequence $\phi_{n_i}(x;y)$ which converges on a countable dense subset R of $-\infty < x < \infty$ to an almost periodic (in the sense of Bohr) function $\phi(x;y)$ uniformly with respect to y for each fixed $x \in R$.

From (5.3.6) and (5.3.10) it follows that almost periodic (in the sense of Bohr) functions $\phi(x;y)$ $(x\in R)$ have Fourier series

$$\varphi(x;\ y)\sim i\sum_{k=1}^{\infty} a(\lambda_k)\,b(\lambda_k)\sin\lambda_k x e^{i\lambda_k y}, \tag{5.3.11}$$

where

$$b(\lambda)=\lim_{n_i\to\infty}\frac{1}{2T_{n_i}}\int_{-T_{n_i}}^{T_{n_i}}k(t)\cos\lambda t\,dt. \tag{5.3.11'}$$

From (5.3.5) and (5.3.11′) and Bessel's inequality we have

$$\sum_{k=1}^{\infty}|a(\lambda_k)|^2 \leqslant \overline{\lim_{T\to\infty}}\frac{1}{2T}\int_{-T}^{T}|\Phi(u)|^2\,du<\infty,$$

$$\sum_{k=1}^{\infty}|b(\lambda_k)|^2 \leqslant \overline{\lim_{T\to\infty}}\frac{1}{2T}\int_{-T}^{T}|k(u)|^2\,du<\infty.$$

Since $|a(\lambda_k)b(\lambda_k)| \leqslant \tfrac{1}{2}(|a(\lambda_k)|^2+|b(\lambda_k)|^2)$, the above inequalities guarantee absolute convergence of the series on the right side of (5.3.11), and thus we can use the equal sign in that formula:

$$\varphi(x;\, y) = i \sum_{k=1}^{\infty} a(\lambda_k)\, b(\lambda_k) \sin \lambda_k\, x e^{i\lambda_k y}$$

where $x \in R$, and y is arbitrary.

Let t_0 be an arbitrary point in $\mathfrak{A}_{i_0}$. We will find a bound for

$$\eta(x,\, y) = S_y^{t_0}[\varphi_x] - \varphi(x;\, y)$$

for $x \in R$. By (5.3.9) and (5.3.10) we have

$$|\eta(x,\, y)| \leqslant \overline{\lim_{n_i \to \infty}} \frac{1}{2T_{n_i}} \int_{-T_{n_i}}^{T_{n_i}} |S_y^{t_0}[\varphi_x] - S_y^t[\varphi_x]|\, k(t)\, dt.$$

Since $k(t) = \delta^{-1}\chi_{i_0}(t)$, where $\chi_{i_0}(t)$ is the characteristic function of the set $\mathfrak{A}_{i_0}$, we have $k(t) = 0$ for $t \notin \mathfrak{A}_{i_0}$. If $t \in \mathfrak{A}_{i_0}$, then by (5.3.7)

$$|S_y^{t_0}[\varphi_x] - S_y^t[\varphi_x]|$$
$$\leqslant \frac{1}{2} |S_{x+y}^{t_0}[\Phi] - S_{x+y}^t[\Phi]| + \frac{1}{2} |S_{-x+y}^{t_0}[\Phi] - S_{-x+y}^t[\Phi]|,$$

and then by (5.3.8)

$$\sup_{\substack{-\infty < x < \infty \\ -\infty < y < \infty}} |S_y^{t_0}[\varphi_x] - S_y^t[\varphi_x]| < \varepsilon,$$

if $t \in \mathfrak{A}_{i_0}$. Therefore

$$\sup_{\substack{-\infty < y < \infty \\ x \in R}} |\eta(x,\, y)| \leqslant \overline{\lim_{n_i \to \infty}} \frac{1}{2T_{n_i}} \int_{-T_{n_i}}^{T_{n_i}} \varepsilon k(t)\, dt = \varepsilon;$$

that is,

$$\sup_{\substack{-\infty < y < \infty \\ x \in R}} \left| S_y^{t_0}[\varphi_x] - i \sum_{k=1}^{\infty} a(\lambda_k)\, b(\lambda_k) \sin \lambda_k\, x e^{i\lambda_k y} \right| < \varepsilon.$$

The functions in the left side of this inequality are continuous in x (the functions $S_y^{t_0}[\phi_x]$ is continuous by assumption, and the function defined by the series $i\Sigma_1^\infty a(\lambda_k) b(\lambda_k) \sin \lambda_k x e^{i\lambda_k y}$ is continuous because the series converges uniformly) and the set R is dense in $-\infty < x < \infty$. Therefore the inequality holds for all x; that is,

$$\sup_{\substack{-\infty < y < \infty \\ -\infty < x < \infty}} \left| S_y^{t_0}[\varphi_x] - i \sum_{k=1}^{\infty} a(\lambda_k)\, b(\lambda_k) \sin \lambda_k\, x e^{i\lambda_k y} \right| < \varepsilon.$$

For $x = y = u/2$ we obtain

$$\sup_{-\infty<u<\infty}\left|S_{\frac{u}{2}}^{t_0}[\varphi_x]_{x=\frac{u}{2}}-i\sum_{k=1}^{\infty}a(\lambda_k)\,b(\lambda_k)\sin\lambda_k\frac{u}{2}\,e^{i\lambda_k\frac{u}{2}}\right|<\varepsilon. \tag{5.3.12}$$

And by (5.3.4) we have

$$S_y^{t_0}[\varphi_x]=\frac{1}{4}[\Phi(x+y+t_0)+\Phi(x+y-t_0)-\Phi(-x+y+t_0)-\Phi(-x+y-t_0)],$$

which for $x=y=u/2$ implies that

$$S_y^{t_0}[\varphi_x]_{x=y=\frac{u}{2}}=\frac{1}{4}[\Phi(u+t_0)+\Phi(u-t_0)-\Phi(t_0)-\Phi(-t_0)]. \tag{5.3.13}$$

Since $\Sigma_1^\infty|a(\lambda_k)|\,|b(\lambda_k)|<\infty$, we see that

$$G(u)=\frac{1}{2}[\Phi(t_0)+\Phi(-t_0)]+2i\sum_{k=1}^{\infty}a(\lambda_k)\,b(\lambda_k)\sin\lambda_k\frac{u}{2}\,e^{i\frac{\lambda_k u}{2}} \tag{5.3.14}$$

is an almost periodic function in the sense of Bohr. Thus if we let

$$\delta(u)=\frac{1}{2}[\Phi(u+t_0)+\Phi(u-t_0)]-G(u),$$

we obtain

$$\frac{1}{2}[\Phi(u+t_0)+\Phi(u-t_0)]=G(u)+\delta(u), \tag{5.3.15}$$

and by (5.3.12), (5.3.13) and (5.3.14)

$$\sup_{-\infty<u<\infty}|\delta(u)|<2\varepsilon,$$

and $G(u)$ is almost periodic in the sense of Bohr.

On the other hand, we already showed that

$$\psi_{t_0}(u)=\frac{1}{2}[\Phi(t_0+u)-\Phi(-t_0+u)] \tag{5.3.16}$$

is almost periodic in the sense of Bohr. Thus by (5.3.15) and (5.3.16) we have

$$\Phi(u+t_0)=G(u)+\psi_{t_0}(u)+\delta(u)$$

or

$$\Phi(u)=G(u-t_0)+\psi_{t_0}(u-t_0)+\delta(u-t_0),$$

where

$$\sup_{-\infty<u<\infty}|\delta(u-t_0)|<2\varepsilon.$$

Since ϵ is an arbitrarily small positive number and $G(u-t_0)+\psi_{t_0}(u-t_0)$ is almost periodic in the sense of Bohr, the formula above shows that $\Phi(u)$ is the limit of a sequence of almost periodic functions which converges uniformly on $-\infty<u<\infty$. Therefore $\Phi(u)$ is almost periodic in the sense of Bohr, and the second part of the proof is complete.

The theorem is proved.

§4. The fundamental theorem in the theory of $\{L_1, L_2\}$-almost periodic functions

We now have all of the necessary tools for establishing a constructive characterization of even $\{L_1, L_2\}$-almost periodic functions. The solution of this problem is contained in the following theorem.

THEOREM 5.4.1. *A function* $F(t)$ *is an even* $\{L_1, L_2\}$*-almost periodic function if and only if for an arbitrary* $\epsilon > 0$ *there exists a function* $\sigma(t)$ *of the form*

$$\sigma(t) = \sum_{k=1}^{N} a_k \omega_0^{(1)}(\lambda_k, t) \qquad (\lambda_k \in \Lambda_0) \tag{5.4.1}$$

(*a finite linear combination of even* $\{L_1, L_2\}$*-almost periodic solutions of the equations* $L_1[u] + \lambda u = 0$) *such that*

$$\sup_{\substack{-\infty < x < \infty \\ -\infty < y < \infty}} \left| T_x^y [F(t) - \sigma(t)] \right| < \varepsilon.$$

PROOF. Sufficiency was proved in Lemma 5.1.3. We will prove necessity.

Let $F(t)$ be an arbitrary even $\{L_1, L_2\}$-almost periodic function. By Lemma 5.2.1 we have

$$F(t) = \sum_{k=1}^{m} b_k \omega_0^{(1)}(\lambda_k, t) + \overset{+}{F}(t),$$

where $\lambda_k \in \Lambda_0^- \subset \Lambda_0$ and $\overset{+}{F}(t)$ is an even $\{L_1, L_2\}$-almost periodic function such that

$$\Phi(t) = \underset{t}{W_1^{-1}} [\overset{+}{F}(t)] \qquad (W_1^{-1} = W_{\{L_1 D^2 0\}})$$

is even and almost periodic in the sense of Bohr.

If a function $\overset{+}{\sigma}(t)$ of the form (5.4.1) satisfies

$$\sup_{\substack{-\infty < x < \infty \\ -\infty < y < \infty}} \left| T_x^y [\overset{+}{F}(t) - \overset{+}{\sigma}(t)] \right| < \varepsilon,$$

then the function

$$\sigma(t) = \sum_{k=1}^{m} b_k \omega_0^{(1)}(\lambda_k, t) + \overset{+}{\sigma}(t)$$

is also of the form (5.4.1) and

$$\sup_{\substack{-\infty < x < \infty \\ -\infty < y < \infty}} \left| T_x^y [F(t) - \sigma(t)] \right| < \varepsilon.$$

So without loss of generality we can prove necessity just for the function $\overset{+}{F}(t)$.

We also know that $\overset{+}{F}(t)$ and functions $\sigma(t)$ of the form (5.4.1) are even. Then by identities 1) and 2) of Lemma 5.1.1 it follows that the function

$$T_x^y[F(t)-\sigma(t)]$$

is even in x and in y.

So without loss of generality we can restrict our attention to nonnegative values of x and y.

First we outline the general plan of the proof.

By identity 3) of Lemma 5.1.1, for $x, y \geqslant 0$ we have

$$T_x^y[\overset{+}{F}] = \underset{x}{V_1}\underset{y}{V_2}S_x^y[\Phi], \tag{5.4.2}$$

where $V_1 = V_{\{D^2L_100\}}$ and $V_2 = V_{\{D^2L_200\}}$. To the almost periodic (in the sense of Bohr) function $\Phi(t)$ we associate the Fourier series

$$\Phi(t) \sim \sum_{k=1}^{\infty} a(\lambda_k)\cos\lambda_k t,$$

where

$$a(\lambda_k) = 2\lim_{T\to\infty}\frac{1}{2T}\int_{-T}^{T}\Phi(t)\cos\lambda_k t\,dt.$$

Let

$$s_n(t) = \sum_{k=1}^{N_n} b_k^{(n)}\cos\lambda_k t \tag{5.4.3}$$

be the sequence of trigonometric polynomials which we obtain from the Fourier series of $\Phi(t)$ by the method of Fejér and Bochner. Then

$$\lim_{n\to\infty}\ \sup_{-\infty<t<\infty}|\Phi(t)-s_n(t)| = 0. \tag{5.4.3$'$}$$

If we let

$$\sigma_n(t) = \underset{t}{V_1}[s_n(t)] = \sum_{k=1}^{N_n} b_k^{(n)}\,\omega_0^{(1)}(\lambda_k^2, t), \tag{5.4.4}$$

we have

$$T_x^y[\overset{+}{F}(t)-\sigma_n(t)] = \underset{x}{V_1}\underset{y}{V_2}S_x^y[\Phi(t)-s_n(t)] \qquad (x\geqslant 0,\ y\geqslant 0).$$

This identity shows that in order to prove the theorem it suffices to establish two assertions:

A) $\lambda_k^2 \in \Lambda_0^0 + \Lambda_0^+ \subset \Lambda_0$; that is, the functions $\sigma_n(t)$ are linear combinations of even $\{L_1, L_2\}$-almost periodic solutions of the equations $L_1[u] + \lambda u = 0$.

B)
$$\lim_{n\to\infty}\ \sup_{\substack{0\leqslant x<\infty\\ 0\leqslant y<\infty}}|\underset{x}{V_1}\underset{y}{V_2}S_x^y[\Phi(t)-s_n(t)]| = 0.$$

The proof of these assertions is based on Theorem 4.1.1, according to which the transformation operators $V = V_{\{D^2L00\}}$ can be written in the form

$$V = (E+K) + cI, \tag{5.4.5}$$

where

$$(E+K)[f]=f(x)+\int_0^x K(x,t)\,dt, \qquad I[f]=\int_0^x f(t)\,dt,$$

and

$$\sup_{0\leq x\leq\infty}\int_0^x |K(x,t)|\,dt=|K|<\infty$$

and $c=\lim_{x\to\infty}\omega_0'(0,x)$. The operators $E+K$ are bounded in $C[0,\infty)$, and the operators V are bounded in $C[0,\infty)$ if and only if $c=0$; that is, if and only if the functions $\omega_0(0,x)$ are bounded.

We also need two auxiliary lemmas.

LEMMA 5.4.1. *If* $\omega_0^{(2)}(0,y)$ *is not bounded, then*

$$\Phi^{(1)}(y)=\int_0^y \Phi(t)\,dt \qquad (\Phi(t)=\underset{t}{W}_{\{L_1D^2_0\}}[\overset{+}{F}(t)])$$

is almost periodic in the sense of Bohr.

PROOF. From (5.4.2) with $x=0$ we obtain

$$T_0^y[\overset{+}{F}(t)]=\underset{y}{V_2}[\Phi(y)],$$

and then, since the family $T_x^y[\overset{+}{F}]$ (y is a parameter) is compact, the function $\underset{y}{V_2}[\Phi(y)]$ is bounded on $0\leqslant y<\infty$. Using the expression (5.4.5) for V_2, we have

$$\underset{y}{V_2}[\Phi(y)]=\underset{y}{(E+K_2)}[\Phi(y)]+c_2\int_0^y\Phi(t)\,dt,$$

or

$$c_2\int_0^y\Phi(t)\,dt=\underset{y}{V_2}[\Phi(y)]-\underset{y}{(E+K_2)}[\Phi(y)],$$

where $c_2=\lim_{y\to\infty}(d/dy)\omega_0^{(2)}(0,y)\neq 0$. On the right side of this equation stands a bounded function, for we have just shown that $\underset{y}{V_2}[\Phi(y)]$ is bounded, and $\underset{y}{(E+K_2)}[\Phi(y)]$ is bounded because the operator $E+K_2$ is bounded in $C[0,\infty)$ and the continuous, almost periodic (in the sense of Bohr) function $\Phi(y)$ is bounded.

Therefore the left side of the above equation is also a bounded function, and since $c_2\neq 0$, the function

$$\Phi^{(1)}(y) = \int_0^y \Phi(t)\,dt$$

is bounded.

Thus the antiderivative $\Phi^{(1)}(y)$ of the almost periodic (in the sense of Bohr) function $\Phi(y)$ is bounded, and from a well-known theorem of Bohr this antiderivative is also almost periodic (in the sense of Bohr).

The lemma is proved.

LEMMA 5.4.2. *If* $\omega_0^{(1)}(0, x)$ *is not bounded, the family of functions*

$$\int_0^x S_\xi^y [\Phi(t)]\,d\xi \qquad (\Phi(t) = \underset{t}{W}_{\{L_1 D^2 0\}} [\overset{+}{F}(t)])$$

(y is a parameter) is compact.

PROOF. We consider the family of functions

$$A(x, y) = T_x^y [\overset{+}{F}(t)] - \overset{+}{F}(x)$$

(y is a parameter). This family is compact since the family $T_x^y [\overset{+}{F}(t)]$ is compact. Furthermore, for each fixed y the function $A(x, y)$ is bounded. To see this, let $\mathfrak{A}_1$ ($\mathfrak{A}_2$) be the set of values of the parameter y for which $A(x, y)$ is bounded (not bounded). From

$$\lim_{y \to y_0} A(x, y) = A(x, y_0)$$

and the compactness of the family $A(x, y)$ it follows that

$$\lim_{y \to y_0} \sup_{-\infty < x < \infty} |A(x, y) - A(x, y_0)| = 0,$$

for any y_0. Therefore the sets $\mathfrak{A}_1$ and $\mathfrak{A}_2$ are open, and since their union is the whole real line $-\infty < y < \infty$, one of them is empty. The set $\mathfrak{A}_1$ is not empty, since

$$A(x, 0) = T_x^0 [\overset{+}{F}(t)] - \overset{+}{F}(x) \equiv 0;$$

that is, $0 \in \mathfrak{A}_1$. Thus $\mathfrak{A}_2$ is empty and $\mathfrak{A}_1$ is the whole real line $-\infty < y < \infty$.

From (5.2.3) (see Lemma 5.2.2) for $x, y \geqslant 0$ we have

$$T_x^y [\overset{+}{F}(t)] = (E - P_y^{(2)})\, T_x^y [\overset{+}{F}(t)],$$

and then by (5.4.2)

$$\underset{x}{V_1}\underset{y}{V_2} S_x^y [\Phi(t)] = (E - P_y^{(2)}) T_x^y [\overset{+}{F}(t)],$$

where $P_y^{(2)}$ is the projection operator (5.2.3″). Therefore

$$\underset{x}{V_1} S_x^y [\Phi(t)] = \underset{y}{V_2^{-1}} (E - P_y^{(2)}) T_x^y [\overset{+}{F}(t)] = \underset{y}{\overset{+}{V}} T_x^y [\overset{+}{F}(t)],$$

where, as in Theorem 4.3.1, we let $\overset{+}{V}$ denote the operator $V_2^{-1}(E - P^{(2)}) = \overset{+}{V}_{\{L_2 D^2 00\}}$. Thus for $x, y \geqslant 0$ we have

$$\underset{x}{V_1} S_x^y [\Phi(t)] = \underset{y}{\overset{+}{V}} [T_x^y [\overset{+}{F}] - \overset{+}{F}(x)] + \overset{+}{F}(x) \underset{y}{\overset{+}{V}} [1],$$

or

$$\underset{x}{V_1} S_x^y [\Phi(t)] = \underset{y}{\overset{+}{V}} [A(x, y)] + \overset{+}{F}(x) \underset{y}{\overset{+}{V}} [1],$$

from which, by using (5.4.5) for V_1, we obtain

$$(E + \underset{x}{K_1}) S_x^y [\Phi(t)] + c_1 \int_0^x S_\xi^y [\Phi(t)]\, d\xi = \underset{y}{\overset{+}{V}} [A(x, y)] + \overset{+}{F}(x) \underset{y}{\overset{+}{V}} [1]$$

and

$$c_1 \int_0^x S_\xi^y [\Phi(t)]\, d\xi - \overset{+}{F}(x)\, \delta(y) = \underset{y}{\overset{+}{V}} [A(x, y)] - (E + \underset{x}{K_1}) S_x^y [\Phi(t)],$$

where

$$c_1 = \lim_{x \to \infty} \frac{d}{dx} \omega_0^{(1)}(0, x) \neq 0, \quad \delta(y) = \underset{y}{\overset{+}{V}} [1].$$

We consider the right side of this equation. As we showed earlier, the family $A(x, y)$ (y is a parameter) is compact and consists of bounded functions on $-\infty < x < \infty$. By Theorem 4.3.1 the operator $\overset{+}{V}$ is bounded in the space M_1. Therefore (as in the proof of Lemma 5.2.2) the family $\underset{y}{\overset{+}{V}} [A(x,y)]$ is compact.

The family $(E + K_1) S_x^y [\Phi(t)]$ (y is a parameter) is also compact. This follows immediately from the boundedness of $E + K_1$ in $C[0, \infty)$ and the compactness of $S_x^y [\Phi(t)]$.

Consequently the right side of the equation represents a compact family. But then the left side

$$c_1 \int_0^x S_\xi^y [\Phi(t)]\, d\xi - \overset{+}{F}(x)\, \delta(y)$$

(y is a parameter) is also compact. Since $c_1 \neq 0$, in order to prove compactness of the family $\int_0^x S_\xi^y [\Phi(t)]\, d\xi$ $(x \geqslant 0, y \geqslant 0;$ y is a parameter) it suffices to show that the family $\{\overset{+}{F}(x)\delta(y)\}$ is compact.

Since the operator $\overset{+}{V} = \overset{+}{V}_{\{L_2 D^2 00\}}$ is bounded (Theorem 4.3.1), the function $\delta(y) = \underset{y}{\overset{+}{V}}[1]$ is bounded. Thus if $\overset{+}{F}(x)$ is bounded, the family $\{\overset{+}{F}(x)\delta(y)\}$ is compact.

Now suppose $\overset{+}{F}(x)$ is not bounded. We will show that in this case $\delta(y) \equiv$ const, which clearly guarantees that the family $\{\overset{+}{F}(x)\delta(y)\}$ (y is a parameter) is compact.

If $\delta(y)$ is not constant, there exists a sequence $y_1, y_2, \cdots$ of values of the parameter y such that

$$\lim_{n\to\infty} y_n = y_0, \quad \delta(y_n) - \delta(y_0) \neq 0 \qquad (n = 1, 2, \ldots).$$

Since $\delta(y)$ is continuous (see the remark after Theorems 4.3.1 and 4.3.1$'$), we have

$$\lim_{y_n\to y_0} \left\{ c_1 \int_0^x S_\xi^{y_n}[\Phi(t)]\, d\xi - \overset{+}{F}(x)\delta(y_n) \right\} = c_1 \int_0^x S_\xi^{y_0}[\Phi(t)]\, d\xi - \overset{+}{F}(x)\delta(y_0),$$

and by the compactness

$$\lim_{y_n\to y_0} \sup_{0\leqslant x<\infty} \left| c_1 \int_0^x \{S_\xi^{y_n}[\Phi(t)] - S_\xi^{y_0}[\Phi(t)]\}\, d\xi - \overset{+}{F}(x)\{\delta(y_n) - \delta(y_0)\} \right| = 0.$$

Therefore, beginning with some $n = n_0$, we have

$$|\overset{+}{F}(x)\{\delta(y_n) - \delta(y_0)\}| \leqslant 1 + \left| c_1 \int_0^x \{S_\xi^{y_n}[\Phi(t)] - S_\xi^{y_0}[\Phi(t)]\}\, d\xi \right|;$$

that is,

$$|\overset{+}{F}(x)\{\delta(y_n) - \delta(y_0)\}| \leqslant 1 + \frac{|c_1|}{2} \left| \int_{x+y_n}^{x+y_0} \Phi(\xi)\, d\xi + \int_{x-y_n}^{x-y_0} \Phi(\xi)\, d\xi \right|$$

and

$$|\overset{+}{F}(x)|\,|\delta(y_n) - \delta(y_0)| \leqslant 1 + |c_1|\,|y_0 - y_n|\, M \qquad (0 \leqslant x < \infty),$$

where $M = \sup_{-\infty<t<\infty} |\Phi(t)| < \infty$. Since $\delta(y_n) - \delta(y_0) \neq 0$ by assumption, the above inequality shows that $\overset{+}{F}(x)$ is bounded, which contradicts our assumption.

Thus the family $\{\overset{+}{F}(x)\delta(y)\}$ and also the family $\int_0^x S_\xi^y [\Phi(t)]\, d\xi$ $(x \geqslant 0, y \geqslant 0;$ y is a parameter) are compact. Since $\Phi(t)$ is even, the function $\int_0^x S_\xi^y [\Phi(t)]\, d\xi$ is even in y and odd in x. Therefore the family

$\int_0^x S_\xi^y [\Phi(t)]\, d\xi$ $(-\infty < x < \infty, -\infty < y < \infty;$ y is a parameter) is also compact in the sense of uniform convergence on the whole real line.

The lemma is proved.

We now turn to the proof of assertions A) and B). The proof depends on the behavior of the functions $\omega_0^{(1)}(0, x)$ and $\omega_0^{(2)}(0, y)$.

We consider all four possible cases.

1) The functions $\omega_0^{(1)}(0, x)$ and $\omega_0^{(2)}(0, y)$ are bounded. In this case the set $\Lambda_0^0 + \Lambda_0^+$ consists of all nonnegative numbers (Lemma 5.1.2). In (5.4.4) it is clear that $\lambda_k^2 \geqslant 0$, and therefore $\lambda_k^2 \in \Lambda_0^0 + \Lambda_0^+ \subset \Lambda_0$ for all k, which proves assertion A).

Furthermore, since $\omega_0^{(1)}(0, x)$ and $\omega_0^{(2)}(0, y)$ are bounded, by (5.4.5) we know that the operators V_1 and V_2 are bounded in $C[0, \infty)$. If we denote the norms of V_1 and V_2 in $C[0, \infty)$ by $|V_1|$ and $|V_2|$ (we will continue to use this notation for norms of operators acting in the space $C[0, \infty)$ of bounded continuous functions), we have

$$\sup_{\substack{0 \leq x < \infty \\ 0 \leq y < \infty}} |\underset{x}{V_1}\underset{y}{V_2} S_x^y [\Phi(t) - s_n(t)]| \leq |V_1| \cdot |V_2| \sup_{\substack{0 \leq x < \infty \\ 0 \leq y < \infty}} |S_x^y [\Phi(t) - s_n(t)]|,$$

which by (5.4.3′) implies that

$$\lim_{n \to \infty} \sup_{\substack{0 \leq x < \infty \\ 0 \leq y < \infty}} |\underset{x}{V_1}\underset{y}{V_2} S_x^y [\Phi(t) - s_n(t)]| = 0.$$

Thus assertion B) is proved.

2) $\omega_0^{(1)}(0, x)$ is bounded and $\omega_0^{(2)}(0, y)$ is not bounded. By Lemma 5.4.1 we know that since $\omega_0^{(2)}(0, y)$ is not bounded, $\Phi^{(1)}(u) = \int_0^u \Phi(t)\, dt$ is bounded. Therefore the Fourier coefficient $a(0)$ for the function $\Phi(t)$ equals zero and in (5.4.3) all of the λ_k are different from zero. But then in (5.4.4) all of the λ_k^2 are positive, and since by Lemma 5.1.2 the set Λ_0^+ contains all positive numbers, we have $\lambda_k^2 \in \Lambda_0^+ \subset \Lambda_0$ for all k.

Assertion A) is proved.

Since $\omega_0^{(1)}(0, x)$ is bounded, the operator V_1 is bounded in $C[0, \infty)$. Using (5.4.5) for V_2, we have

$$\underset{x}{V_1}\underset{y}{V_2} S_x^y [\Phi(t) - s_n(t)]$$

$$= \underset{x}{V_1} (E + \underset{y}{K_2}) S_x^y [\Phi(t) - s_n(t)] + c_2 \underset{x}{V_1} \int_0^y S_x^\eta [\Phi(t) - s_n(t)]\, d\eta,$$

where V_1 and $E + K_2$ are bounded. Also

$$\sup_{\substack{0\le x<\infty\\0\le y<\infty}} |V_1(E+K_2)\underset{x}{}\,S_x^y\underset{y}{}[\Phi(t)-s_n(t)]|$$

$$\leqslant |V_1|\cdot|E+K_2|\sup_{\substack{0\le x<\infty\\0\le y<\infty}}|S_x^y[\Phi(t)-s_n(t)]|,$$

and since the Fejér-Bochner polynomials $s_n(t)$ converge uniformly on the whole real line to the function $\Phi(t)$, we have

$$\lim_{n\to\infty}\sup_{\substack{0\le x<\infty\\0\le y<\infty}} |\underset{x}{V_1}(E+K_2)\,\underset{y}{S_x^y}[\Phi(t)-s_n(t)]|=0.$$

In order to prove assertion B) it now suffices to show that

$$\lim_{n\to\infty}\sup_{\substack{0\le x<\infty\\0\le y<\infty}}\left|\underset{x}{V_1}\int_0^y S_x^\eta[\Phi(t)-s_n(t)]\,d\eta\right|=0.$$

Since V_1 is bounded, the above equation is true if

$$\lim_{n\to\infty}\sup_{\substack{0\le x<\infty\\0\le y<\infty}}\left|\int_0^y S_x^\eta[\Phi(t)-s_n(t)]\,d\eta\right|=0.$$

If we let

$$\Phi^{(1)}(u)=\int_0^u\Phi(t)\,dt,\qquad s_n^{(1)}(u)=\int_0^u s_n(t)\,dt,$$

then

$$\int_0^y S_x^\eta[\Phi(t)-s_n(t)]\,d\eta$$

$$=\frac{1}{2}\{[\Phi^{(1)}(x+y)-s_n^{(1)}(x+y)]-[\Phi^{(1)}(x-y)-s_n^{(1)}(x-y)]\}$$

and

$$\sup_{\substack{0\le x<\infty\\0\le y<\infty}}\left|\int_0^y S_x^\eta[\Phi(t)-s_n(t)]\,d\eta\right|\leqslant\sup_{-\infty<u<\infty}|\Phi^{(1)}(u)-s_n^{(1)}(u)|.$$

By Lemma 5.4.1 we know that $\Phi^{(1)}(u)=\int_0^u\Phi(t)\,dt$ is almost periodic in the sense of Bohr. Since the Fourier series for $\Phi^{(1)}(u)$ is obtained from the Fourier series for $\Phi(t)$ by termwise integration, the polynomials $s_n^{(1)}(u)=\int_0^u s_n(t)\,dt$ are the Fejér-Bochner polynomials constructed from the Fourier series for $\Phi^{(1)}(u)$. Therefore

$$\lim_{n\to\infty}\sup_{-\infty<u<\infty}|\Phi^{(1)}(u)-s_n^{(1)}(u)|=0$$

and consequently

$$\lim_{n\to\infty} \sup_{\substack{0\leq x<\infty \\ 0\leq y<\infty}} \left| \int_0^y S_x^\eta [\Phi(t) - s_n(t)]\, d\eta \right| = 0,$$

which by what we proved above establishes assertion B).

3) The functions $\omega_0^{(1)}(0, x)$ and $\omega_0^{(2)}(0, y)$ are not bounded. The proof of assertion A) is identical to the proof in the preceding case.

Assertion B) is harder to prove. If we substitute into (5.4.2) the expression (5.4.5) for V_1 and V_2, we obtain

$$T_x^y[\overset{+}{F}] = (E + \underset{x}{K_1})(E + \underset{y}{K_2}) S_x^y [\Phi(t)] + c_1 (E + \underset{y}{K_2}) \int_0^x S_\xi^y [\Phi(t)]\, d\xi$$
$$+ c_2 (E + \underset{x}{K_1}) \int_0^y S_x^\eta [\Phi(t)]\, d\eta + c_1 c_2 \int_0^x \int_0^y S_\xi^\eta [\Phi(t)]\, d\eta\, d\xi,$$

where $E + K_1$ and $E + K_2$ are bounded and $c_1 c_2 \neq 0$.

Since $\Phi(t)$ is even we have

$$\int_0^x S_\xi^y [\Phi(t)]\, d\xi = \frac{1}{2} [\Phi^{(1)}(x + y) + \Phi^{(1)}(x - y)],$$

$$\int_0^y S_x^\eta [\Phi(t)]\, d\eta = \frac{1}{2} [\Phi^{(1)}(x + y) - \Phi^{(1)}(x - y)],$$

$$\int_0^x \int_0^y S_\xi^\eta [\Phi(t)]\, d\eta\, d\xi = \frac{1}{2} [\Phi^{(2)}(x + y) - \Phi^{(2)}(x - y)],$$

where

$$\Phi^{(1)}(u) = \int_0^u \Phi(t)\, dt, \quad \Phi^{(2)}(u) = \int_0^u \Phi^{(1)}(t)\, dt.$$

The above identity now takes the form

$$\begin{aligned} T_x^y[\overset{+}{F}] = {} & (E + \underset{x}{K_1})(E + \underset{x}{K_2}) S_x^y [\Phi(t)] \\ & + \frac{1}{2} c_1 (E + \underset{y}{K_2}) [\Phi^{(1)}(x + y) + \Phi^{(1)}(x - y)] \\ & + \frac{1}{2} c_2 (E + \underset{x}{K_1}) [\Phi^{(1)}(x + y) - \Phi^{(1)}(x - y)] \\ & + \frac{1}{2} c_1 c_2 [\Phi^{(2)}(x + y) - \Phi^{(2)}(x - y)], \end{aligned} \tag{5.4.6}$$

where $x, y \geqslant 0$. Similarly,

$$V_1 V_2 S_x^y [\Phi(t) - s_n(t)] = (E + K_1)(E + K_2) S_x^y [\Phi(t) - s_n(t)]$$
$$+ \frac{1}{2} c_1 (E + K_2) [\Phi^{(1)}(x+y) - s_n^{(1)}(x+y) + \Phi^{(1)}(x-y) - s_n^{(1)}(x-y)]$$
$$+ \frac{1}{2} c_2 (E + K_1) [\Phi^{(1)}(x+y) - s_n^{(1)}(x+y) - \Phi^{(1)}(x-y) + s_n^{(1)}(x-y)]$$
$$+ \frac{1}{2} c_1 c_2 [\Phi^{(2)}(x+y) - s_n^{(2)}(x+y) - \Phi^{(2)}(x-y) + s_n^{(2)}(x-y)], \tag{5.4.7}$$

where $x, y \geqslant 0$ and

$$s_n^{(1)}(u) = \int_0^u s_n(t)\,dt, \qquad s_n^{(2)}(u) = \int_0^u s_n^{(1)}(t)\,dt.$$

Since $\omega_0^{(2)}(0, y)$ is not bounded, $\Phi^{(1)}(u)$ is almost periodic in the sense of Bohr (Lemma 5.4.1) and the $s_n^{(1)}(u)$ are the Fejér-Bochner polynomials constructed from the Fourier series of this function.

Therefore

$$\lim_{n\to\infty} \sup_{-\infty<u<\infty} |\Phi^{(1)}(u) - s_n^{(1)}(u)| = 0,$$

and by virtue of (5.4.7) and the boundedness of $E + K_1$ and $E + K_2$ it follows that

$$\overline{\lim_{n\to\infty}} \sup_{\substack{0\leq x<\infty \\ 0\leq y<\infty}} |V_1 V_2 S_x^y [\Phi(t) - s_n(t)]| \leqslant |c_1 c_2| \overline{\lim_{n\to\infty}} \sup_{-\infty<u<\infty} |\Phi^{(2)}(u) - s_n^{(2)}(u)|.$$

Assertion B) will be proved if we can show that the trigonometric polynomials $s_n^{(2)}(u)$ converge to the function $\Phi^{(2)}(u)$ uniformly on the whole real line.

Since $\Phi^{(2)}(u) = \int_0^u \Phi^{(1)}(t)\,dt$ is the antiderivative of $\Phi^{(1)}(t)$, which is almost periodic in the sense of Bohr, and since the polynomials $s_n^{(2)}(u)$ are the Fejér-Bochner polynomials constructed from the trigonometric series which is obtained by termwise integration of the Fourier series for $\Phi^{(1)}(t)$, it follows from a well-known theorem of Bohr that if we show $\Phi^{(2)}(u)$ to be bounded, we thereby show that it is an almost periodic function in the sense of Bohr and the polynomials $s_n^{(2)}(u)$ converge to this function as $n \to \infty$ uniformly on the whole real line $-\infty < u < \infty$.

Therefore in order to prove assertion B) it suffices to show that $\Phi^{(2)}(u)$ is bounded.

We consider the identity (5.4.6) with y, $0 \leqslant y < \infty$, as a varying parameter. The family of functions $T_x^y[\overset{+}{F}]$ on the left side of (5.4.6) is compact. The first three terms on the right side of this identity also represent a compact family of functions. This is clear from the fact that $\Phi(u)$ and $\Phi^{(1)}(u)$ are almost periodic in the sense of Bohr and $E + K_1$ and $E + K_2$ are bounded. But then the fourth term on the right side of (5.4.6) must also represent a compact family. Since $c_1 c_2 \neq 0$ by assumption, it follows that the family

$$\{\Phi^{(2)}(x+y) - \Phi^{(2)}(x-y)\} \quad (0 \leqslant y < \infty;\ y \text{ is a parameter}) \tag{5.4.8}$$

is compact in the sense of uniform convergence on the semi-axis $0 \leqslant x < \infty$. This family can be compact only if $\Phi^{(2)}(u)$ is bounded. For $\Phi^{(2)}(u)$ is even, and if it were unbounded, we could find a sequence of positive numbers $\{a_n\}$, $n = 1, 2, \cdots$, such that

$$\lim_{n\to\infty} a_n = \infty, \quad \lim_{n\to\infty} |\Phi^{(2)}(a_n)| = \infty. \tag{5.4.9}$$

On the other hand, since the family (5.4.8) is compact, from the sequence of functions

$$\left\{\Phi^{(2)}\left(x + \frac{1}{2}a_n\right) - \Phi^{(2)}\left(x - \frac{1}{2}a_n\right)\right\} \quad (n = 1, 2, \ldots)$$

we can select a subsequence

$$\left\{\Phi^{(2)}\left(x + \frac{1}{2}a_{n_k}\right) - \Phi^{(2)}\left(x - \frac{1}{2}a_{n_k}\right)\right\} \quad (k = 1, 2, \ldots),$$

which is uniformly convergent on $0 \leqslant x < \infty$. Therefore there exist numbers N and c such that

$$\sup_{0\leqslant x<\infty}\left|\Phi^{(2)}\left(x + \frac{1}{2}a_{n_k}\right) - \Phi^{(2)}\left(x - \frac{1}{2}a_{n_k}\right) - \Phi^{(2)}(x+c) + \Phi^{(2)}(x-c)\right| < 1,$$

for arbitrary $k \geqslant N$. In particular, for $x = a_{n_k}/2$ we have

$$\left|\Phi^{(2)}(a_{n_k}) - \Phi^{(2)}(0) - \Phi^{(2)}\left(\frac{1}{2}a_{n_k} + c\right) + \Phi^{(2)}\left(\frac{1}{2}a_{n_k} - c\right)\right| < 1$$

and, since $\Phi^{(2)}(0) = \int_0^0 \Phi^{(1)}(t)\,dt = 0$,

$$|\Phi^{(2)}(a_{n_k})| \leqslant 1 + \left|\Phi^{(2)}\left(\frac{1}{2}a_{n_k} + c\right) - \Phi^{(2)}\left(\frac{1}{2}a_{n_k} - c\right)\right| \quad (k \geqslant N).$$

By the definition of $\Phi^{(2)}(u)$ we have

$$\Phi^{(2)}(x+c) - \Phi^{(2)}(x-c) = \int_{x-c}^{x+c} \Phi^{(1)}(t)\,dt,$$

which implies that

$$|\Phi^{(2)}(x+c) - \Phi^{(2)}(x-c)| \leqslant 2cM \quad (0 \leqslant x < \infty),$$

where $M = \sup_{-\infty<t<\infty} |\Phi^{(1)}(t)| < \infty$.

Consequently for $k \geqslant N$ we have $|\Phi^{(2)}(a_{n_k})| \leqslant 1 + 2cM$ and

$$\overline{\lim_{k\to\infty}} \, |\Phi^{(2)}(a_{n_k})| \leqslant 1 + 2cM < \infty,$$

which contradicts (5.4.9). Therefore $\Phi^{(2)}(u)$ is bounded, and so assertion B) is established.

4) $\omega_0^{(1)}(0, x)$ is not bounded and $\omega_0^{(2)}(0, y)$ is bounded. We have

$$\int_0^x S_\xi^y [\Phi(t)]\, d\xi = \frac{1}{2} [\Phi^{(1)}(x+y) + \Phi^{(1)}(x-y)], \qquad (5.4.10)$$

where

$$\Phi^{(1)}(u) = \int_0^u \Phi(t)\, dt. \qquad (5.4.10')$$

Since $\omega_0^{(1)}(0, x)$ is not bounded, by Lemma 5.4.2 we know that the family of functions (5.4.10) (where y is a parameter) is compact. Then by Theorem 5.3.1 the function $\Phi^{(1)}(u)$ is of the form

$$\Phi^{(1)}(u) = cu + \psi^{(1)}(u), \qquad (5.4.11)$$

where $\psi^{(1)}(u)$ is almost periodic in the sense of Bohr.

If we combine (5.4.11) and (5.4.10′), we obtain

$$c = a(0) = \lim_{T\to\infty} \frac{1}{2T} \int_{-T}^{T} \Phi(t)\, dt; \quad \psi^{(1)}(u) = \int_0^u [\Phi(t) - a(0)]\, dt;$$

that is,

$$\Phi^{(1)}(u) = a(0)\, u + \psi^{(1)}(u),$$

where the functions

$$\psi(u) = \Phi(u) - a(0), \quad \psi^{(1)}(u) = \int_0^u \psi(t)\, dt$$

are almost periodic in the sense of Bohr.

From these equations it follows that if $s_n(t)$ are the Fejér-Bochner polynomials constructed from the Fourier series for $\Phi(t)$, then

$$\hat{s}_n(t) = s_n(t) - a(0) \quad \left(\hat{s}_n^{(1)}(t) = \int_0^t \hat{s}_n(u)\, du \right)$$

are the Fejér-Bochner polynomials constructed from the Fourier series for $\psi(t)$ ($\psi^{(1)}(t)$). Therefore

$$\lim_{n\to\infty} \sup_{-\infty<t<\infty} |\Phi(t) - s_n(t)| = 0, \qquad (5.4.12)$$

$$\lim_{n\to\infty} \sup_{-\infty<t<\infty} |\psi^{(1)}(t) - \hat{s}_n^{(1)}(t)| = 0. \qquad (5.4.12')$$

We will now prove assertions A) and B).

If we substitute into (5.4.2) the expression (5.4.5) for the operator V_1 and note that (by virtue of the boundedness of $\omega_0^{(2)}(0, y)$) the operator V_2 is bounded in $C[0, \infty)$, we obtain

$$\underset{x}{V_1}\underset{y}{V_2} S_x^y [\Phi(t) - s_n(t)] = (E + \underset{x}{K_1}) \underset{y}{V_2} S_x^y [\Phi(t) - s_n(t)] + c_1 \underset{y}{V_2} \int_0^x S_\xi^y [\Phi(t) - s_n(t)]\, d\xi,$$

or

$$\underset{x}{V_1}\underset{y}{V_2} S_x^y [\Phi(t) - s_n(t)] = \frac{1}{2}(E + \underset{x}{K_1}) \underset{y}{V_2} [\Phi(x+y) + \Phi(x-y) - s_n(x+y) - s_n(x-y)] + \frac{1}{2} c_1 \underset{y}{V_2} [\Phi^{(1)}(x+y) + \Phi^{(1)}(x-y) - a(0)(x+y) - a(0)(x-y) - \hat{s}_n^{(1)}(x+y) - \hat{s}_n^{(1)}(x-y)],$$

where $E + K_1$ and V_2 are bounded. Therefore

$$\sup_{\substack{0 \leq x < \infty \\ 0 \leq y < \infty}} |\underset{x}{V_1}\underset{y}{V_2} S_x^y [\Phi(t) - s_n(t)]| \leq |E + K_1||V_2| \sup_{-\infty < t < \infty} |\Phi(t) - s_n(t)| + |c_1||V_2| \sup_{-\infty < t < \infty} |\Phi^{(1)}(t) - a(0)t - \hat{s}_n^{(1)}(t)|,$$

or

$$\sup_{\substack{0 \leq x < \infty \\ 0 \leq y < \infty}} |\underset{x}{V_1}\underset{y}{V_2} S_x^y [\Phi(t) - s_n(t)]| \leq |E + K_1||V_2| \sup_{-\infty < t < \infty} |\Phi(t) - s_n(t)| + |c_1||V_2| \sup_{-\infty < t < \infty} |\psi^{(1)}(t) - \hat{s}_n^{(1)}(t)|,$$

which by (5.4.12) and (5.4.12′) implies that

$$\lim_{n \to \infty} \sup_{\substack{0 \leq x < \infty \\ 0 \leq y < \infty}} |\underset{x}{V_1}\underset{y}{V_2} S_x^y [\Phi(t) - s_n(t)]| = 0.$$

Assertion B) is proved.

For the proof of assertion A) we consider two cases.

If $q_2(y) \equiv 0$ (that is, $\omega_0^{(2)}(0, y) \equiv 1$), then by Lemma 5.1.2 the set $\Lambda_0^0 + \Lambda_0^+$ is the set of all nonnegative numbers, and (since by (5.4.4) it is clear that $\lambda_k^2 \geq 0$) $\lambda_k^2 \in \Lambda_0^0 + \Lambda_0^+$ for arbitrary k, which is what we wanted to show.

If $q_2(y) \not\equiv 0$, then by Lemma 5.1.2 the set $\Lambda_0^0 + \Lambda_0^+$ is the set of all positive numbers. Thus in order to prove assertion A) we have to show that in (5.4.4) we have $\lambda_k^2 > 0$ (that is, in (5.4.3) we have $\lambda_k \neq 0$ for all k). Therefore our problem is reduced to showing that the free term in the Fourier series of $\Phi(t)$

$$a(0) = \lim_{T \to \infty} \frac{1}{2T} \int_{-T}^{T} \Phi(t)\, dt$$

must be zero.

We have

$$\overset{+}{F}(x) = \underset{x}{W}_{\{D^2 L_{10}\}} [\Phi(x)] = a(0)\, \omega_0^{(1)}(0, x) + \underset{x}{W}_{\{D^2 L_{10}\}} [\psi(x)], \tag{5.4.13}$$

where $\psi(x) = \Phi(x) - a(0)$. We will show that $\underset{x}{W}_{\{D^2L_10\}}[\psi(x)]$ is $\{L_1, L_2\}$-almost periodic. By (5.4.2) for $x, y \geqslant 0$ we have

$$T_x^y[W_{\{D^2L_10\}}[\psi]] = \underset{x}{V_1}\underset{y}{V_2} S_x^y[\psi(t)]$$

$$= (E + \underset{x}{K_1}) \underset{y}{V_2} S_x^y[\psi(t)] + c_1 \underset{y}{V_2} \int_0^x S_\xi^y[\psi(t)]\, d\xi,$$

or

$$T_x^y[W_{\{D^2L_10\}}[\psi]] = (E + \underset{x}{K_1}) \underset{y}{V_2} S_x^y[\psi(t)] + \frac{1}{2} c_1 \underset{y}{V_2}[\psi^{(1)}(x+y) + \psi^{(1)}(x-y)],$$

where $\psi^{(1)}(u) = \int_0^u \psi(t)\,dt$. Since $E + K_1$ and V_2 are bounded, and $\psi(t)$ and $\psi^{(1)}(t)$ are almost periodic in the sense of Bohr, the right side of the above equation is a compact family. Consequently the family $T_x^y[W_{\{D^2L_10\}}[\psi]]$ (y is a parameter) is compact, and $\underset{x}{W}_{\{D^2L_10\}}[\psi(x)]$ is an $\{L_1, L_2\}$-almost periodic function.

Then from (5.4.13) it follows that $a(0)\omega_0^{(1)}(0, x)$ is $\{L_1, L_2\}$-almost periodic. By Lemma 5.1.2 this is possible only if $a(0) = 0$ (since $\omega_0^{(1)}(0, x)$ is not bounded and $q_2(y) \not\equiv 0$ by assumption), which proves assertion A).

We have now considered all four possible cases which can occur in Theorem 5.4.1.

Consequently Theorem 5.4.1 has been completely proved, and thus the fundamental problem in the theory of even $\{L_1, L_2\}$-almost periodic functions has been solved.

We will examine several details of the constructive characterization of even $\{L_1, L_2\}$-almost periodic functions which we have just found.

Since $T_x^0[F(t)] = F(x)$, we have

$$\sup_{-\infty<x<\infty} |F(x) - \sigma(x)| \leqslant \sup_{\substack{-\infty<x<\infty \\ -\infty<y<\infty}} |T_x^y[F(t) - \sigma(t)]|.$$

From this inequality and Theorem 5.4.1 we obtain the following important corollary.

COROLLARY 1. *If $F(x)$ is an even $\{L_1, L_2\}$-almost periodic function, it can be uniformly approximated on the whole real line by finite linear combinations of even $\{L_1, L_2\}$-almost periodic solutions of the equations $L_1[u] + \lambda u = 0$.*

It is natural to ask whether this necessary condition for $F(x)$ to be $\{L_1, L_2\}$-almost periodic is also a sufficient condition. In the general case the answer is no. An analysis of the proof of Theorem 5.4.1 leads to the following conclusion.

COROLLARY 2. *If one of the functions* $\omega_0^{(1)}(0, x)$ *or* $\omega_0^{(2)}(0, y)$ *is bounded on the whole real line, then the class of even* $\{L_1, L_2\}$*-almost periodic functions coincides with the class of functions which are uniform limits of sequences of linear combinations of even* $\{L_1, L_2\}$*-almost periodic solutions of the equations* $L_1[u] + \lambda u = 0$.

If neither of the functions $\omega_0^{(1)}(0, x)$ *or* $\omega_0^{(2)}(0, y)$ *is bounded, then there exist functions which are uniform limits of sequences of linear combinations of even* $\{L_1, L_2\}$*-almost periodic solutions of the equations* $L_1[u] + \lambda u = 0$ *but are not* $\{L_1, L_2\}$*-almost periodic.*

This corollary shows that the constructive characterization of even $\{L_1, L_2\}$-almost periodic functions given by Theorem 5.4.1 cannot be simplified in the general case.

Up to this point we have restricted our attention to even $\{L_1, L_2\}$-almost periodic functions, and we have based our study of these functions on the properties of the transformation operators $V_{\{D^2L_100\}}$, $V_{\{D^2L_200\}}$ and $V_{\{L_1D^200\}}$ found in Theorems 4.1.1 and 4.3.1. The study of odd $\{L_1, L_2\}$-almost periodic functions is entirely analogous. The roles of the operators $V_{\{D^2L_100\}}$, $V_{\{D^2L_200\}}$ and $V_{\{L_1D^200\}}$ are played by $V_{\{D^2L_1\infty\infty\}}$, $V_{\{D^2L_200\}}$ and $V_{\{L_1D^2\infty\infty\}}$. Theorems 4.1.1, 4.3.1′ and 5.3.1 allow us to solve the fundamental problem for odd $\{L_1, L_2\}$-almost periodic functions if the solution $\omega_\infty^{(1)}(0, x)$ is not bounded. The corresponding theorem is.

THEOREM 5.4.1′. *Suppose* $\omega_\infty^{(1)}(0, x)$ *is not bounded. Then* $F(x)$ *is an odd* $\{L_1, L_2\}$*-almost periodic function if and only if for each* $\epsilon > 0$ *there exists a function* $\sigma(x)$ *of the form*

$$\sigma(x) = \sum_{k=1}^{N} a_k \omega_\infty^{(1)}(\lambda_k, x) \qquad (\lambda_k \in \Lambda_\infty)$$

(*a finite linear combination of odd* $\{L_1, L_2\}$*-almost periodic solutions of the equations* $L_1[u] + \lambda u = 0$) *such that*

$$\sup_{\substack{-\infty<x<\infty \\ -\infty<y<\infty}} |T_x^y[F(t) - \sigma(t)]| < \varepsilon.$$

By combining Theorems 5.4.1 and 5.4.1′ in the case where $\omega_\infty^{(1)}(0, x)$ is not bounded we obtain the following general result.

A function $F(x)$ *is* $\{L_1, L_2\}$*-almost periodic if and only if for each* $\epsilon > 0$ *there exists a function* $\sigma(x)$ *of the form*

$$\sigma(x) = \sum_{k=1}^{N} a_k \omega_{h_k}^{(1)}(\lambda_k, x) \qquad (\lambda_k \in \Lambda_{h_k},\ h_k = 0 \text{ or } \infty)$$

(a finite linear combination of $\{L_1, L_2\}$*-almost periodic solutions of the equations* $L_1[u] + \lambda u = 0$*) such that*

$$\sup_{\substack{-\infty<x<\infty \\ -\infty<y<\infty}} |T_x^y[F(t) - \sigma(t)]| < \varepsilon.$$

When $\omega_\infty^{(1)}(0, x)$ is bounded we cannot use Theorem 4.3.1′ for studying odd $\{L_1, L_2\}$-almost periodic functions. In this case we have to resort to artificial methods which we will not discuss here.

In conclusion we note that generalized shift operators $\overset{\circ}{T}{}_x^y$ (Definition 4.4.1) allow us to develop the theory of $\{L_1, L_2\}$-almost periodic functions given on $0 \leqslant x < \infty$. In this theory the fundamental theorem is completely analogous to Theorem 5.4.1, and the roles of the functions $\omega_0^{(1)}(\lambda, x)$ are played by the functions $\omega_h^{(1)}(\lambda, x)$, where h is the real parameter occurring in the definition of $\overset{\circ}{T}{}_x^y$.

Received 8 MAY 51

BIBLIOGRAPHY

1. V. A. Steklov, *A cooling problem for a nonhomogeneous rigid rod,* Soobšč. Har'kov. Mat. Obšč. (2) **5** (1896), 136–181. (Russian) Cf. *Problème de refroidissement d'une barre hétérogène,* Ann. Fac. Sci. Univ. Toulouse (2) **3** (1901), 281–313.

2. H. Weyl, *Über gewohnliche Differentialgleichungen mit Singularitäten und die zugehörige Entwicklungen willkürlicher Funktionen,* Math. Ann. **68** (1910), 220–269.

3. E. C. Titchmarsh, *Eigenfunction expansions associated with second-order differential equations,* Clarendon Press, Oxford, 1946. MR **8**, 458.

4. B. M. Levitan, *Expansion in eigenfunctions of differential equations of the second order,* GITTL, Moscow, 1950. (Russian) MR **12**, 183.

5. ———, *The application of generalized displacement operators to second-order linear differential equations.* Uspehi Mat. Nauk **4** (1949), no. 1 (29), 3–112; English transl., Amer. Math. Soc. Transl. (1) **10** (1962), 408–541. MR **11**, 116.

6. ———, *A generalization of almost periodic functions,* Mat. Sb. **24 (66)** (1949), 321–346; English transl., Amer. Math. Soc. Transl. (1) **10** (1962), 378–407. MR **11**, 174.

7. ———, *On a decomposition theorem for characteristic functions of differential equations of the second order,* Dokl. Akad. Nauk SSSR **71** (1950), 605–608. (Russian) MR **11**, 720.

8. J. Delsarte, *Sur une extension de la formule de Taylor,* J. Math. Pures Appl. (9) **17** (1938), 213–231.

9. ———, *Une extension nouvelle de la théorie des fonctions presque-périodiques de Bohr,* Acta Math. **69** (1939), 259–317.

10. A. Ja. Povzner, *On differential equations of Sturm-Liouville type on a half-axis,* Mat. Sb. **23 (65)** (1948), 3-52; English transl., Amer. Math. Soc. Transl. (1) **4** (1962), 24–101. MR **10**, 299.

11. ———, *On equations of the Sturm-Liouville type on a semi-axis,* C. R. (Dokl.) Acad. Sci. URSS **53** (1946), 295–298. MR **8**, 469.

12. N. I. Ahiezer and I. M. Glazman, *Theory of linear operators in Hilbert space*, GITTL, Moscow, 1950; English transl., Ungar, New York, 1961. MR **13**, 358.

13. V. A. Ambarzumian [Ambartsumian], *Über eine Frage der Eigenwerttheorie*, Z. Phys. **53** (1929), 690–695.

14. G. Borg, *Eine Umkehrung der Sturm-Liouvilleschen Eigenwertaufgabe, Bestimmung der Differentialgleichung durch die Eigenwerte*, Acta Math. **78** (1946), 1–96. MR **7**, 382.

15. N. Levinson, *On the uniqueness of the potential in a Schrödinger equation for a given asymptotic phase*, Bull. Amer. Math. Soc. **55** (1949), 517.

16. V. Bargmann, *Remarks on the determination of a central field of force from the elastic scattering phase shifts*, Phys. Rev. (2) **75** (1949), 301–303. MR **11**, 248.

17. M. G. Kreĭn, *On a general method of decomposing Hermite-positive nuclei into elementary products*, C. R. (Dokl.) Acad. Sci. URSS 53 (1946), 3–6. MR **8**, 277.

18. ———, *Solution of the inverse Sturm-Liouville problem*, Dokl. Akad. Nauk SSSR **76** (1951), 21–24. (Russian) MR **12**, 613.

19. N. I. Ahiezer, *Lectures on the theory of approximation*, OGIZ, Moscow, 1947; English transl., Ungar, New York, 1956. MR **10**, 33; **20** #1872.

20. Z. S. Agranovič, *On some questions connected with equations of Sturm-Liouville type on a semi-axis*, Dokl. Akad. Nauk SSSR **66** (1949), 1025–1028. (Russian) MR **11**, 28.

21. I. M. Gel'fand, D. A. Raĭkov and G. E. Šilov, *Commutative normed rings*, Uspehi Mat. Nauk **1** (1946), no. 2 (12), 48–146; English transl., Amer. Math. Soc. Transl. (2) **5** (1957), 115–220. MR **10**, 258; **18**, 714.

22. L. A. Ljusternik, *Basic concepts in functional analysis*, Uspehi Mat. Nauk **1** (1936), 77–140. (Russian)

23. V. V. Nemyckiĭ, *The fixed-point method in analysis*, Uspehi Mat. Nauk **1** (1936), 141–174; English transl., Amer. Math. Soc. Transl. (2) **34** (1963), 1–37.

Translated by:
F. A. Cezus

Amer. Math. Soc. Transl.
(2) Vol. 101, 1973

ON THE ASYMPTOTIC BEHAVIOR OF THE SPECTRAL FUNCTION OF A SELFADJOINT SECOND ORDER DIFFERENTIAL EQUATION*

B. M. LEVITAN

Abstract. In this paper we give an asymptotic formula and error estimate for the spectral function of a selfadjoint second order differential equation. The derivation of the asymptotic formula is essentially based upon Bernstein's classical approximation theorem.

Introduction

Let us consider the differential equation

$$Ly + \lambda y = y'' - q(x)y + \lambda y = 0, \tag{0.1}$$

where $q(x)$ is a real-valued function defined on the interval $(0, b)$, $0 \leqslant b \leqslant \infty$, and integrable on each subinterval $[0, b']$ where $0 \leqslant b' < b$.

If h is an arbitrary real number, then $\phi(x, \lambda)$ denotes the solution to (0.1) satisfying the initial conditions

$$\phi(0, \lambda) = 1, \quad \phi'(0, \lambda) = h. \tag{0.2}$$

If $h = \infty$, then $\phi(x, \lambda)$ denotes the solution to (0.1) which satisfies the initial conditions

$$\varphi(0, \lambda) = 0, \quad \varphi'(0, \lambda) = \sqrt{\lambda}. \tag{0.2'}$$

One can show (cf. [1], [2], [3]) that given h there exists at least one monotone function $\rho(\lambda)$ which is bounded in each finite interval and has the property that Parseval's equation

$$\int_0^b f^2(x)\,dx = \int_{-\infty}^{\infty} E^2(\lambda)\,d\rho(\lambda), \qquad E(\lambda) = \underset{b' \to b}{\text{l.i.m.}} \int_0^{b'} f(x)\,\varphi(x, \lambda)\,dx$$

holds for each function $f(x) \in L^2(0, b)$.

$\rho(\lambda)$ is called the spectral function for equation (0.1).

It is of interest, especially in connection with the inverse Sturm-Liouville problem, to determine which monotone functions $\rho(\lambda)$ can be spectral functions

---

*Translation of Izv. Akad. Nauk SSSR Ser. Mat. 16 (1952), 325–352.

AMS (MOS) subject classifications (1970). Primary 34A10, 34A25, 34B25; Secondary 41A50.

for (0.1). An answer to this problem is given in [4] where in fact both necessary and sufficient conditions are given for a monotone function $\rho(\lambda)$ to be a spectral function for equation (0.1). These conditions are as follows:

1) $\int_{-\infty}^{0} e^{\sqrt{\lambda}x}\, d\rho(\lambda) < \infty$ for all $x < 2b$.

2) Let

$$\sigma(\lambda) = \rho(\lambda) - \frac{2}{\pi}\sqrt{\lambda}$$

for $\lambda > 0$, and define

$$a(x) = \int_{1}^{\infty} \frac{\cos\sqrt{\lambda}x}{\lambda}\, d\sigma(\lambda).$$

Then in order for $\rho(\lambda)$ to be a spectral function for (0.1), where $q(x)$ is continuous, it is necessary that $a(x)$ have a continuous third derivative on $[0, 2b)$, and it suffices that $a(x)$ have a continuous fourth derivative on $[0, 2b)$.

The necessity of the first condition had already been obtained by V. A. Marčenko ([5], [6]).

Since the differentiability of the function $a(x)$ may be very difficult to verify, it would be valuable to find necessary conditions on the asymptotic behavior of the spectral function $\rho(\lambda)$ as $\lambda \to \infty$. We obtained the first such condition in [7]. We showed that as $\lambda \to \infty$

$$\rho(\lambda) < C\sqrt{\lambda},$$

where C is some absolute constant. Then Marčenko obtained the interesting asymptotic formula (cf. [5], [6])

$$\rho(\lambda) = \frac{2}{\pi}\sqrt{\lambda} + o(\sqrt{\lambda}). \tag{0.3}$$

In this paper we give a new derivation of (0.3). Our methods also yield a more precise error estimate than in (0.3).

For $\lambda > 0$, $\lambda = \mu^2$, set $\rho(\lambda) = \sigma(\mu)$ and then define $\sigma(\mu)$ on the negative semi-axis so as to be odd. We will show that

$$\sigma(\mu + a) - \sigma(a) = \frac{2}{\pi}\mu + O(\ln\mu), \tag{0.4}$$

$$\frac{1}{\mu}\int_{0}^{\mu}\left\{\frac{1}{2}\left[\sigma(\nu + a) + \sigma(\nu - a)\right] - \frac{2}{\pi}\nu\right\} d\nu = O(1) \tag{0.5}$$

uniformly in a as $\mu \to \infty$.

If there is no negative spectrum and either $h = 0$ or $h = \infty$, and if

$$\int_0^x |q(s)|\,ds = O(x^\alpha)$$

as $x \to 0, \alpha > 0$, then

$$\sigma(\mu + a) - \sigma(a) = \frac{2}{\pi}\mu + O(1) \tag{0.6}$$

uniformly in a as $\mu \to \infty$.

We will primarily use Bernstein's classical approximation theorem in proving these asymptotic formulas.

§1. Some auxiliary results from the theory of Fourier integrals

Suppose $\sigma(\mu)$ is a complex-valued function defined for all real μ and satisfies(1)

$$\sup_{-\infty<\mu<\infty} \overset{\mu+1}{\underset{\mu}{V}} \{\sigma(\mu)\} = M < \infty. \tag{1.1}$$

Set

$$E_2(\alpha) = \frac{1}{\sqrt{2\pi}}\left\{\int_{-\infty}^{-1} \frac{e^{-i\alpha\mu}}{-\mu^2}\,d\sigma(\mu) + \int_{-1}^{1} \frac{e^{-i\alpha\mu} - 1 + i\alpha\mu}{-\mu^2}\,d\sigma(\mu) + \int_{1}^{\infty} \frac{e^{-i\alpha\mu}}{-\mu^2}\,d\sigma(\mu)\right\}. \tag{1.2}$$

It is easy to see that in each finite interval (1.1) guarantees the absolute and uniform convergence of the integral in (1.2). Hence $E_2(\alpha)$ must be continuous. $E_2(\alpha)$ is called the Bochner-Stieltjes transform of $\sigma(\mu)$. If $\sigma(\mu)$ is differentiable, then (1.2) gives the Bochner transform of $\sigma'(\mu)$ (see [8], German p. 110). For the remainder of this paper we will distinguish between the Bochner and Bochner-Stieltjes transforms. We shall see that the basic properties of the Bochner transform carry over to the Bochner-Stieltjes transform.

In this section we prove two basic lemmas.

LEMMA 1.1. *Let*

$$\rho_n(\mu) = \frac{3}{2\pi}\int_{-\infty}^{\infty} \frac{\sin^4 na}{n^3a^4}\,d\sigma(\mu + a), \tag{1.3}$$

(1) $V_\mu^{\mu+1}$ denotes the variation of σ on $(\mu, \mu + 1)$.

$$E_{2,n}(\alpha) = \frac{1}{\sqrt{2\pi}} \left\{ \int_{-\infty}^{-1} \frac{e^{-i\alpha\mu}}{-\mu^2} \rho_n(\mu)\, d\mu \right.$$

$$\left. + \int_{-1}^{1} \frac{e^{-i\alpha\mu} - 1 + i\alpha\mu}{-\mu^2} \rho_n(\mu)\, d\mu + \int_{1}^{\infty} \frac{e^{-i\alpha\mu}}{-\mu^2} \rho_n(\mu)\, d\mu \right\}. \tag{1.4}$$

Then the following assertions are true.

1) *At each point of continuity of* $\sigma(\mu)$

$$\lim_{n\to\infty} \int_0^{\mu} \rho_n(\nu)\, d\nu = \sigma(\mu) - \frac{1}{2}\{\sigma(+0) + \sigma(-0)\}. \tag{1.5}$$

2)

$$\sup_{-\infty<\mu<\infty} \int_{\mu}^{\mu+1} |\rho_n(\nu)|\, d\nu \leqslant M, \tag{1.6}$$

where M *is the same constant appearing in condition* (1.1).

3)

$$\lim_{n\to\infty} E_{2,n}(\alpha) = E_2(\alpha) \tag{1.7}$$

uniformly in each finite interval.

PROOF. 1) It follows from (1.1) that

$$|\sigma(a)| = O(a)$$

for large a. Thus

$$\lim_{a\to\infty} \frac{\sigma(\mu + a)}{a^4} = 0.$$

for each fixed μ. Integrating (1.3) by parts, we obtain

$$\rho_n(\mu) = -\frac{3}{2\pi} \int_{-\infty}^{\infty} \sigma(\mu + a)\, d\left(\frac{\sin^4 na}{n^3 a^4}\right).$$

If we integrate this equation between 0 and μ and then integrate by parts, we get

$$\int_0^{\mu} \rho_n(\nu)\, d\nu = -\frac{3}{2\pi} \int_{-\infty}^{\infty} \left\{\int_0^{\mu} \sigma(\nu + a)\, d\nu\right\} d\left(\frac{\sin^4 na}{n^3 a^4}\right)$$

$$= -\frac{3}{2\pi} \int_{-\infty}^{\infty} \left\{ \int_a^{a+\mu} \sigma(\nu)\, d\nu\right\} d\left(\frac{\sin^4 na}{n^3 a^4}\right)$$

$$= \frac{3}{2\pi} \int_{-\infty}^{\infty} \frac{\sin^4 na}{n^3 a^4} \{\sigma(a + \mu) - \sigma(a)\}\, da.$$

Suppose now that ν is continuous at μ. Then as $n \to \infty$ we obtain (1.5) (cf. [8], German p. 20).

2) Define the monotone function $\sigma_1(\mu)$ as follows:

$$\sigma_1(\mu) = \begin{cases} \overset{\mu}{\underset{+0}{V}} \{\sigma(\mu)\}, & \mu > 0, \\ -\overset{\mu}{\underset{-0}{V}} \{\sigma(\mu)\}, & \mu < 0. \end{cases}$$

Then from the definition of the Stieltjes integral we get

$$|\rho_n(\nu)| \leqslant \frac{3}{2\pi} \int_{-\infty}^{\infty} \frac{\sin^4 na}{n^3 a^4} \, d\sigma_1(\nu + a).$$

Proceeding as in the proof of 1), this inequality gives

$$\int_{\mu}^{\mu+1} |\rho_n(\nu)| \, d\nu \leqslant \frac{3}{2\pi} \int_{-\infty}^{\infty} \frac{\sin^4 na}{n^3 a^4} \{\sigma_1(\mu + a + 1) - \sigma_1(\mu + a)\} \, da \leqslant M$$

and since μ is arbitrary, we have proved (1.6).

3) (1.7) follows from 1) and 2) and from Helly's theorem on passing to the limit under the Stieltjes integral sign.

LEMMA 1.2. *Suppose* $K(a)$ *satisfies*

$$K(a) = O\left(\frac{1}{a^4}\right)$$

for large a. *If*

$$\gamma(\alpha) = \frac{1}{\sqrt{2\pi}} \int_{-\infty}^{\infty} K(a) e^{-i\alpha a} \, da,$$

$$\tau(\mu) = \frac{1}{\sqrt{2\pi}} \int_{-\infty}^{\infty} K(a) \, d\sigma(\mu + a),$$

$$\Phi_2(\alpha) = \frac{1}{\sqrt{2\pi}} \left\{ \int_{\infty}^{-1} \frac{e^{-i\alpha\mu}}{-\mu^2} \tau(\mu) \, d\mu \right.$$

$$\left. + \int_{-1}^{1} \frac{e^{-i\alpha\mu} - 1 + i\alpha\mu}{-\mu^2} \tau(\mu) \, d\mu + \int_{1}^{\infty} \frac{e^{-i\alpha\mu}}{-\mu^2} \tau(\mu) \, d\mu \right\},$$

then

$$\Phi_2(\alpha) = \gamma(\alpha) E_2(\alpha) - 2\int_0^\alpha E_2(\alpha)\gamma'(\alpha)\,d\alpha + \int_0^\alpha d\alpha \int_0^\alpha E_2(\alpha)\gamma''(\alpha)\,d\alpha + c_0 + c_1\alpha, \tag{1.8}$$

where c_0 *and* c_1 *are constants.*

PROOF. Let $\rho_n(\mu)$ be the function in Lemma 1.1. Set

$$\tau_n(\mu) = \frac{1}{\sqrt{2\pi}} \int_{-\infty}^{\infty} K(a)\rho_n(\mu + a)\,da,$$

and let $\Phi_{2,n}(\alpha)$ and $E_{2,n}(\alpha)$ be the Bochner transforms of $\tau_n(\mu)$ and $\rho_n(\mu)$ respectively, i. e.

$$\Phi_{2,n}(\alpha) = \frac{1}{\sqrt{2\pi}}\left\{\int_{-\infty}^{-1} \frac{e^{-i\alpha\mu}}{-\mu^2}\tau_n(\mu)\,d\mu \right.$$

$$\left. + \int_{-1}^{1} \frac{e^{-i\alpha\mu} - 1 + i\alpha\mu}{-\mu^2}\tau_n(\mu)\,d\mu + \int_1^\infty \frac{e^{-i\alpha\mu}}{-\mu^2}\tau_n(\mu)\,d\mu\right\},$$

$$E_{2,n}(\alpha) = \frac{1}{\sqrt{2\pi}}\left\{\int_{-\infty}^{-1} \frac{e^{-i\alpha\mu}}{-\mu^2}\rho_n(\mu)\,d\mu \right. \tag{1.9}$$

$$\left. + \int_{-1}^{1} \frac{e^{-i\alpha\mu} - 1 + i\alpha\mu}{-\mu^2}\rho_n(\mu)\,d\mu + \int_1^\infty \frac{e^{-i\alpha\mu}}{-\mu^2}\rho_n(\mu)\,d\mu\right\}.$$

It is known that

$$\Phi_{2,n}(\alpha) = \gamma(\alpha) F_{2,n}(\alpha) - 2\int_0^\alpha E_{2,n}(\alpha)\gamma'(\alpha)\,d\alpha$$

$$+ \int_0^\alpha d\alpha \int_0^\alpha E_{2,n}(\alpha)\gamma''(\alpha)\,d\alpha + c_0^{(n)} + c_1^{(n)}\alpha. \tag{1.10}$$

According to Lemma 1.1, $\tau_n(\mu)$ and $E_{2,n}(\alpha)$ converge uniformly in each finite interval to $\tau(\mu)$ and $E_2(\alpha)$ respectively. We will show that $\Phi_{2,n}(\alpha)$ converges to $\Phi_2(\alpha)$, and thus (1.8) will follow from (1.10). Now

$$\sup_{-\infty<\mu<\infty} \int_\mu^{\mu+1} |\tau_n(\nu)|\,d\nu \leqslant \frac{1}{\sqrt{2\pi}} \int_{-\infty}^{\infty} |K(a)|\,da \cdot \sup_{-\infty<\mu<\infty} \int_\mu^{\mu+1} |\rho_n(\nu)|\,d\nu.$$

Passing to the limit under the integral sign in (1.9), we get

$$\lim_{n\to\infty} \Phi_{2,n}(\alpha) = \Phi_2(\alpha).$$

There are three results of equation (1.8) which will play an important role.

1. *If* $E_2(\alpha)$ *is linear in some interval* (a, b), *then* $\Phi_2(\alpha)$ *is also linear in* (a, b).

This assertion follows from differentiating (1.8) twice.

2. *If* $\gamma(\alpha) = 0$ *in some interval* (a, b), *then* $\Phi_2(\alpha)$ *is linear in* (a, b) *for any* $E_2(\alpha)$ *whatsoever.*

3. *If* $\gamma(\alpha) = 1$ *in some interval* (a, b), *then*

$$\Phi_2(\alpha) = E_2(\alpha) + c_0 + c_1\alpha$$

in (a, b).

§2. Proof of an inequality

To prove a theorem analogous to Bernstein's approximation theorem, we will need an inequality analogous to Bohr's classical inequality [9]. This latter inequality is:

BOHR'S INEQUALITY. *Let* $f(\mu)$ *be a measurable function bounded on each real ray. Suppose the Bochner transform of* $f(\mu)$ *is linear in the interval* $(-\Lambda, \Lambda)$, $\Lambda > 0$. *Let* $f^{-[1]}(\mu)$ *denote an indefinite integral of* $f(\mu)$ *whose Bochner transform is linear in* $(-\Lambda, \Lambda)$. *Then*

$$|f^{-[1]}(\mu)| \leqslant \frac{\pi}{2} \cdot \frac{1}{\Lambda} \cdot \sup_{-\infty<\mu<\infty} |f(\mu)|. \tag{2.1}$$

Bohr proved (2.1) for periodic functions. The generalization of Bohr's inequality to arbitrary measurable bounded functions is given in [10]. In the inequality which we need, the estimate on the integral is not given by

$$\sup_{-\infty<\mu<\infty} f\,|(\mu)|,$$

but rather by

$$\sup_{-\infty<\mu<\infty} \overset{\mu+1}{\underset{\mu}{V}} \{\sigma(\mu)\}.$$

THEOREM 2.1. *Suppose* $\sigma(\mu)$ *satisfies*

$$\sup_{-\infty<\mu<\infty} \overset{\mu+1}{\underset{\mu}{V}} \{\sigma(\mu)\} = M < \infty \tag{2.2}$$

and has a Bochner-Stieltjes transform which is linear in $(-\Lambda, \Lambda)$, $0 < \Lambda \leqslant 1$. *Choose* K *to be a constant such that the Bochner transform of* $\sigma(\mu) + K$ *is linear in* $(-\Lambda, \Lambda)$.(2) *Then the inequality*

---

(2) We will see this is possible.

$$|K+\sigma(\mu)| \leqslant C\frac{M}{\Lambda} \tag{2.3}$$

holds, where C *is some absolute constant.*

PROOF. We first prove (2.3) for trigonometric polynomials. Then the general case easily follows (see [10]). Hence, suppose

$$\sigma(\mu)=\int_0^\mu\left(\sum_{k=1}^n a_k e^{i\lambda_k\nu}\right)d\nu=\int_0^\mu S_n(\nu)\,d\nu$$

with real λ_k. An elementary calculation shows (see [8], German p. 115) that the condition $|\lambda_k| \geqslant \Lambda$ is equivalent to the linearity of the Bochner-Stieltjes transform of $\sigma(\mu)$ in $(-\Lambda, \Lambda)$.

So we need to show that if $|\lambda_k| \geqslant \Lambda$, then

$$\left|K+\int_0^\mu S_n(\nu)\,d\nu\right| \leqslant \frac{C}{\Lambda}\sup_{-\infty<\mu<\infty}\int_\mu^{\mu+1}|S_n(\nu)|\,d\nu, \tag{2.4}$$

where the constant K is chosen so that the trigonometric polynomial

$$K+\int_0^\mu S_n(\nu)\,d\nu$$

has no constant term. We now show that without loss of generality we may suppose $\Lambda = 1$. Indeed, the trigonometric polynomial

$$t_n(\mu)=S_n\left(\frac{\mu}{\Lambda}\right)=\sum_{k=1}^n a_k e^{i\frac{\lambda_k}{\Lambda}\mu}=\sum_{k=1}^n a_k e^{i\nu_k\mu}$$

satisfies the condition $|\nu_k| \geqslant 1$. Let

$$N=\sup_{-\infty<\mu<\infty}\int_\mu^{\mu+1}|t_n(\nu)|\,d\nu$$

and suppose it has already been shown that

$$\left|A+\int_0^\mu t_n(\nu)\,d\nu\right| \leqslant CN, \tag{2.5}$$

where the constant A has been chosen so that the trigonometric polynomial

$$A+\int_0^\mu t_n(\nu)\,d\nu$$

has no constant term. Then (2.5) implies

$$\left|A+\int_0^{\mu} S_n\left(\frac{\nu}{\Lambda}\right)d\nu\right| = \left|A+\Lambda\int_0^{\frac{\mu}{\Lambda}} S_n(\nu)\,d\nu\right| \leqslant C\cdot N;$$

dividing by Λ, we get

$$\left|\frac{A}{\Lambda}+\int_0^{\frac{\mu}{\Lambda}} S_n(\nu)\,d\nu\right| \leqslant C\cdot\frac{N}{\Lambda},$$

which is equivalent to the inequality

$$\left|K+\int_0^{\mu} S_n(\nu)\,d\nu\right| \leqslant C\cdot\frac{N}{\Lambda}, \tag{2.6}$$

where $K = A/\Lambda$, since μ is arbitrary.

Moreover,

$$N = \sup_{-\infty<\mu<\infty}\int_{\mu}^{\mu+1} |t_n(\nu)|\,d\nu = \sup_{-\infty<\mu<\infty}\int_{\mu}^{\mu+1}\left|S_n\left(\frac{\nu}{\Lambda}\right)\right|d\nu$$

$$= \Lambda\sup_{-\infty<\mu<\infty}\int_{\frac{\mu}{\Lambda}}^{\frac{\mu}{\Lambda}+\frac{1}{\Lambda}} |S_n(\nu)|\,d\nu = \Lambda\sup_{-\infty<\mu<\infty}\int_{\mu}^{\mu+\frac{1}{\Lambda}} |S_n(\nu)|\,d\nu.$$

Set

$$\frac{1}{\Lambda} = n+\theta \quad (\Lambda\leqslant 1),$$

where n is a positive integer and $0\leqslant\theta<1$. Then

$$\Lambda\sup_{-\infty<\mu<\infty}\int_{\mu}^{\mu+\frac{1}{\Lambda}} |S_n(\nu)|\,d\nu = \frac{1}{n+\theta}\sup_{-\infty<\mu<\infty}\int_{\mu}^{\mu+n+\theta} |S_n(\nu)|\,d\nu$$

$$\leqslant \frac{1}{n}\sup_{-\infty<\mu<\infty}\int_{\mu}^{\mu+n+1} |S_n(\nu)|\,d\nu \leqslant \left(\frac{n+1}{n}\right)\sup_{-\infty<\mu<\infty}\int_{\mu}^{\mu+1} |S_n(\nu)|\,d\nu < 2M.$$

Thus (2.4) follows from (2.6) by replacing $2C$ by C.

We now prove inequality (2.5). Set

$$\psi(\lambda) = \begin{cases} \dfrac{\lambda}{i} & \text{for } 0\leqslant\lambda\leqslant 1, \\[2mm] \dfrac{1}{i\lambda} & \text{for } \lambda\geqslant 1 \end{cases} \qquad \psi(-\lambda) = -\psi(\lambda).$$

We now calculate the Fourier transform of $\psi(\lambda)$. We get

$$g(a)=\frac{1}{\sqrt{2\pi}}\int_{-\infty}^{\infty}\psi(\lambda)e^{-i\lambda a}\,d\lambda=\sqrt{\frac{2}{\pi}}\int_0^1\lambda\sin\lambda a\,d\lambda+\sqrt{\frac{2}{\pi}}\int_1^{\infty}\frac{\sin\lambda a}{\lambda}\,d\lambda.$$

Since

$$\int_1^{\infty}\frac{\sin\lambda a}{\lambda}\,d\lambda=\int_a^{\infty}\frac{\sin\lambda}{\lambda}\,d\lambda,$$

the function $g(a)$ is bounded in a neighborhood of the point $a=0$ and at $a=0$, g has a discontinuity of the first kind with a jump equal to $\sqrt{2\pi}$. Integrating by parts twice, we get

$$g(a)=\sqrt{\frac{2}{\pi}}\,\frac{\sin a}{a^2}-\sqrt{\frac{2}{\pi}}\,\frac{1}{a^2}\int_1^{\infty}\frac{\sin\lambda a}{\lambda^3}\,d\lambda.$$

Thus $g(a)\in L(-\infty,\infty)$ and

$$g(a)=O\left(\frac{1}{a^2}\right) \tag{2.7}$$

as $a\to\infty$.

Consider the function

$$F_n(\mu)=\frac{1}{\sqrt{2\pi}}\int_{-\infty}^{\infty}t_n(\mu+a)\,g(a)\,da.$$

We will show that

$$F_n(\mu)=A+\int_0^{\mu}t_n(\nu)\,d\nu,$$

where the constant A has been chosen so that the constant term of the polynomial $F_n(\mu)$ is zero.

Indeed, by the inverse Fourier transform,

$$F_n(\mu)=\sum_{k=1}^{n}a_ke^{i\nu_k\mu}\frac{1}{\sqrt{2\pi}}\int_{-\infty}^{\infty}e^{i\nu_k a}\,g(a)\,da=\sum_{k=1}^{n}\frac{a_k}{i\nu_k}e^{i\nu_k\mu}=A+\int_0^{\mu}t_n(\nu)\,d\nu.$$

Since

$$\frac{1}{\sqrt{2\pi}}\int_{-\infty}^{\infty}t_n(\mu+a)\,g(a)\,da=\sum_{k=-\infty}^{\infty}\frac{1}{\sqrt{2\pi}}\int_k^{k+1}t_n(\mu+a)\,g(a)\,da,$$

the estimate (2.7) gives

$$\left|A+\int_0^\mu t_n(\nu)\,d\nu\right| \leqslant C_1\int_{-1}^{1}|t_n(\mu+a)|\,da$$

$$+C_2\left\{\sum_{k=-\infty}^{-1}+\sum_{k=1}^{\infty}\frac{1}{k^2}\int_k^{k+1}|t_n(\mu+a)|\,da\right\}$$

$$< C_3\sup_{-\infty<\mu<\infty}\int_\mu^{\mu+1}|t_n(\nu)|\,d\nu = C_3N.$$

Using a rough calculation, we can show that $C_3 < \pi + 3$.

Consider now the general case. Suppose $\sigma(\mu)$ satisfies (2.2) and suppose its Bochner-Stieltjes transform is linear in $(-\Lambda, \Lambda)$, $\Lambda \leqslant 1$. For arbitrary positive n, set

$$\rho_n(\mu) = \frac{3}{2\pi}\int_{-\infty}^{\infty}\frac{\sin^4 na}{n^3a^4}\,d\sigma(\mu+a).$$

Since the Fourier transform of $(\sin^4 na)/n^3a^4$ vanishes outside the interval $(-4n, 4n)$, Lemma 1.2 implies the Bochner transform of $\rho_n(\mu)$ is linear in the intervals $(-\infty, 4n)$, $(-\Lambda, \Lambda)$, and $(4n, \infty)$. As was shown in [10], there exists a sequence of finite trigonometric polynomials $S_{n,N}(\mu)$ with coefficients outside of the interval $(-\Lambda, \Lambda)$ which converges to $\rho_n(\mu)$ uniformly in each finite interval as $N \to \infty$. Using the representation of the sum $S_{n,N}(\mu)$ given in [10], we get

$$\sup_{-\infty<\mu<\infty}\int_\mu^{\mu+1}|S_{n,N}(\nu)|\,d\nu \leqslant \sup_{-\infty<\mu<\infty}\int_\mu^{\mu+1}|\rho_n(\nu)|\,d\nu$$

and hence by part 2) of Lemma 1.1

$$\sup_{-\infty<\mu<\infty}\int_\mu^{\mu+1}|S_{n,N}(\nu)|\,d\nu \leqslant M.$$

Inequality (2.4) gives

$$\left|K_{n,N}+\int_0^\mu S_{n,N}(\nu)\,d\nu\right| \leqslant \frac{C}{\Lambda}\cdot M. \tag{2.8}$$

Suppose now that μ and n are fixed. We first show that the limit

$$K_n = \lim_{N\to\infty} K_{n,N}$$

exists. Now

$$K_{n,N}+\int_0^{\mu} S_{n,N}(\nu)\,d\nu=\frac{2}{\sqrt{2\pi}}\int_{-8}^{\infty} S_{n,N}(\mu+a)\,g(a)\,da$$

and letting $N\to\infty$ in this equation, we get

$$\lim_{N\to\infty} K_{n,N}+\int_0^{\mu}\rho_n(\nu)\,d\nu=\frac{1}{\sqrt{2\pi}}\int_{-\infty}^{\infty}\rho_n(\mu+a)\,g(a)\,da. \tag{2.9}$$

Hence

$$K_n=\lim_{N\to\infty} K_{n,N}$$

exists.

Passing to the limit in (2.8) as $N\to\infty$, we get

$$\left|K_n+\int_0^{\mu}\rho_n(\nu)\,d\nu\right|\leqslant\frac{C}{\Lambda}M. \tag{2.10}$$

Clearly the Bochner transform of $K_n+\int_0^{\mu}\rho_n(\nu)d\nu$ is linear in $(-\Lambda,\Lambda)$. Suppose that σ is continuous at μ. Passing to the limit in (2.9) as $n\to\infty$, we obtain from Lemma 1.1

$$K+\sigma(\mu)=\int_{-\infty}^{\infty} g(a)\,d\sigma(\mu+a),$$

where

$$K=\lim_{n\to\infty} K_n-\frac{1}{2}\{\sigma(+0)+\sigma(-0)\}.$$

Passing to the limit in (2.10), we get

$$|K+\sigma(\mu)|\leqslant\frac{C}{\Lambda}\cdot M.$$

Clearly the Bochner transform of $K+\sigma(\mu)$ is linear in $(-\Lambda,\Lambda)$. Thus the theorem is proved.

REMARK 1. Inequality (2.3) can be iterated. However, starting with $r=1$, Bohr's inequality can be used since the function $K+\sigma(\mu)$ is bounded.

REMARK 2. If the constant C_1 can be chosen so that $C_1+\sigma(\mu)$ is an odd function, then the Bochner transform (but not the Bochner-Stieltjes transform!) of $C_1+\sigma(\mu)$ is linear in $(-\Lambda,\Lambda)$. Indeed, suppose first that

$$\sigma'(\mu)=\sum_{k=1}^{n} C_k\cos\lambda_k\mu;\qquad |\lambda_k|\geqslant\Lambda.$$

In this case

$$\int_0^\mu \sigma'(\mu)\,d\mu = \sum_{k=1}^n C_k \frac{\sin \lambda_k \mu}{\lambda_k} = \sigma(\mu) - \sigma(0).$$

Since the trigonometric polynomial $\Sigma_1^n C_k (\sin \lambda_k \mu / \lambda_k)$ has no constant term, the Bochner transform of $\sigma(\mu) - \sigma(0)$ is linear in $(-\Lambda, \Lambda)$. Clearly $\sigma(\mu) - \sigma(0)$ is odd. Consider now the general case. Let

$$\rho_n(\mu) = \frac{3}{2\pi} \int_{-\infty}^{\infty} \frac{\sin^4 na}{n^3 a^4}\, d\sigma(\mu + a).$$

If $\sigma(\mu)$ is odd, then clearly $\rho_n(\mu)$ is even. Thus (see the proof of the preceding theorem) the trigonometric polynomials $S_{n,N}(\mu)$ are also even, and since $\lim_{N\to\infty} S_{n,N}(\mu) = \rho_n(\mu)$, the Bochner transform for $\int_0^\mu \rho_n(\nu)\,d\nu$ is linear in $(-\Lambda, \Lambda)$. In view of Lemma 1.1, if σ is continuous at μ, then

$$\lim_{n\to\infty} \int_0^\mu \rho_n(\nu)\,d\nu = \sigma(\mu) - \frac{1}{2}\{\sigma(+0) + \sigma(-0)\} = \sigma(\mu) + C_1.$$

The function $\sigma(\mu) + C_1$ is odd and its Bochner transform is linear in $(-\Lambda, \Lambda)$.

§3. Proof of a theorem analogous to Bernstein's theorem

In this section we will consider a theorem which lays the foundation for the asymptotic formulas (0.4), (0.5), and (0.6).

Suppose $\sigma(\mu)$ satisfies (2.2). Let us introduce the following notation: $\sigma^{-[0]}(\mu)$ denotes the function $\sigma(\mu) + C_0$, where C_0 is an arbitrary constant; $\sigma^{-[1]}(\mu)$ denotes an indefinite integral of $\sigma^{-[0]}(\mu)$, i.e. the function

$$\int_0^\mu \sigma(\nu)\,d\nu + C_0\mu + C_1,$$

where C_1 is an arbitrary constant; and in general, $\sigma^{-[r]}(\mu)$ denotes an r-fold indefinite integral of $\sigma^{-[0]}(\mu)$. Thus $\sigma^{-[r]}(\mu)$ is defined uniquely up to an rth degree polynomial. By a lemma due to Esclangon ([13], p. 213 = 375) it follows that if $\sigma(\mu)$ is bounded and if some one of the functions $\sigma^{-[r]}(\mu)$ $(r > 1)$ is bounded (denote it by $\sigma_0^{-[r]}(\mu)$), then the functions

$$[\sigma_0^{-[r]}(\mu)]' = \sigma_0^{-[r-1]}(\mu), \ldots, [\sigma_0^{-[2]}(\mu)]' = \sigma_0^{-[1]}(\mu)$$

are also bounded.

THEOREM 3.1. *Suppose* $\sigma(\mu)$ *satisfies* (2.2). *Suppose that for each* $\epsilon > 0$ *there is a function* $\sigma_\epsilon(\mu)$ *whose Bochner-Stieltjes transform is linear in the interval* $(-\epsilon, \epsilon)$ *and which satisfies the inequality*

$$\sup_{-\infty<\mu<\infty} \overset{\mu+1}{\underset{\mu}{V}} \{\sigma(\mu) - \sigma_\varepsilon(\mu)\} < K \cdot \varepsilon^{r+1-\alpha}, \tag{3.1}$$

where the constants $K > 0$, $0 < \alpha < 1$, *and* $r \geqslant 0$ (r *integral*) *are independent of* ϵ. *Then among the functions* $\sigma^{-[r-1]}(\mu)$ *is a bounded one, denoted by* $\sigma_0^{-[r-1]}(\mu)$, *satisfying*

$$\left| \int_0^\mu \sigma_0^{-[r-1]}(\nu + a)\, d\nu \right| \leqslant A + B\,|\mu|^\alpha, \tag{3.2}$$

where the constants A *and* B *are independent of* μ *and* a. *If* $\alpha = 0$, *then*(3)

$$\left| \int_0^\mu \sigma_0^{-[r-1]}(\nu + a)\, d\nu \right| \leqslant A + B \ln|\mu|. \tag{3.3}$$

PROOF. Define

$$\varepsilon_k = \frac{1}{2^k} \quad (k = 0, 1, 2, \ldots), \qquad \sigma_k(\mu) = \sigma_{\varepsilon_k}(\mu),$$

$$\tau_k(\mu) = \sigma_k(\mu) - \sigma_{k-1}(\mu), \qquad \tau_0(\mu) = \sigma_0(\mu)$$

and consider the infinite series

$$\sigma_0(\mu) + \{\sigma_1(\mu) - \sigma_0(\mu)\} + \cdots = \tau_0(\mu) + \tau_1(\mu) + \cdots = \lim_{k\to\infty} \sigma_k(\mu). \tag{3.4}$$

The series (3.4) clearly converges to $\sigma(\mu) + C$ at each point of continuity of $\sigma(\mu)$. Let

$$\tau_k(\mu) = \sigma_k(\mu) - \sigma_{k-1}(\mu).$$

The Bochner-Stieltjes transform of $\tau_k(\mu)$ is linear in the interval $(-2^{-k}, 2^{-k})$. Since

$$\sup_{-\infty<\mu<\infty} \overset{\mu+1}{\underset{\mu}{V}} \{\tau_k(\mu)\} \leqslant \sup_{-\infty<\mu<\infty} \overset{\mu+1}{\underset{\mu}{V}} \{\sigma(\mu) - \sigma_k(\mu)\}$$

$$+ \sup_{-\infty<\mu<\infty} \overset{\mu+1}{\underset{\mu}{V}} \{\sigma(\mu) - \sigma_{k-1}(\mu)\} \leqslant \frac{K}{2^{k(r+1-\alpha)}} + \frac{K}{2^{(k-1)(r+1-\alpha)}} = \frac{K_1}{2^{k(r+1-\alpha)}},$$

(2.3) gives

---

(3) For $r = 0$, we suppose by definition

$$\int_0^\mu \sigma_0^{-[0]}(\nu)\, d\nu = \sigma(\mu) + C,$$

and consequently we have

$$|\sigma(\mu + a) - \sigma(a)| \leqslant A + B(\mu)^\alpha,$$

$$|\sigma(\mu + a) - \sigma(a)| \leqslant A + B \ln|\mu|.$$

$$|\tau_k^{-[r-1]}(\mu)| \leqslant \frac{K_2}{2^{k(1-\alpha)}}. \tag{3.5}$$

Thus the series

$$\tau_0^{-[r-1]}(\mu) + \sum_{k=1}^{\infty} \tau_k^{-[r-1]}(\mu) = g(\mu) \tag{3.6}$$

converges absolutely and uniformly. We now show that

$$g(\mu) = \sigma_0^{-[r-1]}(\mu).$$

Indeed, differentiating the series for $g(\mu)$ $r-1$ times, we obtain the uniformly convergent series (3.4) whose sum equals $\sigma(\mu) + C$ almost everywhere. Let us estimate

$$\int_0^{\mu} \sigma_0^{-[r-1]}(\nu + a)\, d\nu.$$

Let μ_0 be an arbitrary real number and let the positive integer m be determined by the inequalities

$$2^{m-1} \leqslant \mu_0 < 2^m.$$

Moreover, set

$$\int_0^{\mu_0} g(\nu + a)\, d\nu = \int_0^{\mu_0} \tau_0^{-[r-1]}(\nu + a)\, d\nu + \sum_{k=1}^{m-1} \int_0^{\mu_0} \tau_k^{-[r-1]}(\nu + a)\, d\nu$$

$$+ \sum_{k=m}^{\infty} \int_0^{\mu_0} \tau_k^{-[r-1]}(\nu + a)\, d\nu = P(\mu_0, a) + Q(\mu_0, a) + R(\mu_0, a).$$

We will estimate each summand separately. (3.5) gives us

$$R(\mu_0, a) \leqslant K_2 \sum_{k=m}^{\infty} \frac{1}{2^{k(\alpha-1)}} \left| \int_0^{\mu_0} d\nu \right| = K_2 |\mu_0| \sum_{k=m}^{\infty} \frac{1}{2^{k(\alpha-1)}}$$

$$\leqslant K_3 |\mu_0| \frac{1}{2^{m(1-\alpha)}} \leqslant K_4 |\mu_0| \cdot |\mu_0|^{\alpha-1} = K_4 |\mu_0|^{\alpha}. \tag{3.7}$$

Again we use (2.3) to estimate $Q(\mu_0, a)$. We obtain

$$\left| C_k^{(r)} + \int_0^{\mu} \tau_k^{-[r-1]}(\nu)\, d\nu \right| \leqslant CK_2 2^{k\alpha} = K_5 2^{k\alpha}.$$

Consequently

$$\left| \int_0^{\mu_0} \tau_k^{-[r-1]}(\nu + a)\, d\nu \right| \leqslant \left| C_k^{(r)} + \int_0^{\mu_0 + a} \tau_k^{-[r-1]}(\nu)\, d\nu \right| +$$

$$+\left| C_k^{(r)} + \int_0^a \tau_k^{-[r-1]}(\nu)\, d\nu \right| \leqslant 2K_5 2^{k\alpha}.$$

Thus

$$|Q(\mu_0, a)| \leqslant 2K_5 \sum_{k=1}^{m-1} 2^{k\alpha} \leqslant K_6 2^{(m-1)\alpha} \leqslant K_6 |\mu_0|^\alpha. \tag{3.8}$$

Finally we estimate $P(\mu_0, a)$. Since

$$\tau_0(\mu) = \sigma_0(\mu) = \sigma(\mu) - \{\sigma(\mu) - \sigma_0(\mu)\},$$

we get

$$\sup_{-\infty<\mu<\infty} \overset{\mu+1}{\underset{\mu}{V}} \{\tau_0(\mu)\} \leqslant \sup_{-\infty<\mu<\infty} \overset{\mu+1}{\underset{\mu}{V}} \{\sigma(\mu)\}$$
$$+ \sup_{-\infty<\mu<\infty} \overset{\mu+1}{\underset{\mu}{V}} \{\sigma(\mu) - \sigma_0(\mu)\} \leqslant M + K = M_1. \tag{3.9}$$

Since the Bochner transform of $\sigma_0(\mu)$ is linear in the interval $(-1, 1)$, (2.3) and (3.9) yield

$$|P(\mu_0, a)| \leqslant \left| \int_0^{\mu_0} \tau_0^{-[r-1]}(\nu + a)\, d\nu \right| \leqslant A, \tag{3.10}$$

where the constant A is independent of μ_0 and a. Combining inequalities (3.7), (3.8) and (3.10), we obtain (3.2).

If $\alpha = 0$, the estimates for $P(\mu_0, a)$ and $R(\mu_0, a)$ still hold. We get a different estimate for $Q(\mu_0, a)$, namely

$$|Q(\mu_0, a)| \leqslant 2K_5 \sum_{k=1}^{m-1} 2^0 = 2K_5(m-1) \leqslant K_7 \ln |\mu_0|.$$

Thus the estimate (3.3) holds for the case $\alpha = 0$.

REMARK 1. If $\alpha = 0$, then $g(\mu)$ is the sum of the series (3.4).

REMARK 2. Theorem 3.1 is analogous to the converse theorem in approximation theory (Bernstein's theorem). It is easy in our case to obtain a theorem analogous to Jackson's theorem. However, we will not dwell on this point since the result is not needed here.

Using a method suggested by A. Zygmund [11] for making Bernstein's theorem sharper, we can get still another estimate on $\sigma_0^{-[r]}(\mu)$ for the case $\alpha = 0$.

THEOREM 3.2. *Suppose the conditions of Theorem* 3.1 *are satisfied for* $\alpha = 0$. *Then among the functions* $\sigma^{-[r]}(\mu)$ *there is a function* $\sigma_0^{-[r]}(\mu)$

satisfying(4)

$$\left| \frac{1}{\mu} \int_0^\mu \frac{1}{2} \{\sigma_0^{-[r]}(a+\nu) - \sigma_0^{-[r]}(a-\nu)\} d\nu \right| \leqslant K, \tag{3.11}$$

where the constant K *is independent of* μ *and* a.

PROOF. First consider the case $r > 0$. Let a and μ_0 be the same as in the preceding proof. Then (3.6) gives

$$\frac{1}{2}\{g(a+\nu) + g(a-\nu)\} = \frac{1}{2}\{\tau_0^{-[r-1]}(a+\nu) + \tau_0^{-[r-1]}(a-\nu)\}$$

$$+ \sum_{k=1}^{\infty} \frac{1}{2}\{\tau_k^{-[r-1]}(a+\nu) + \tau_k^{-[r-1]}(a-\nu)\} = \omega_0(a, \nu) + \sum_{k=1}^{\infty} \omega_k(a, \nu).$$

For a fixed a the functions $\omega_k(a, \nu)$, $k = 0, 1, 2, \cdots$, are even. Thus (see Remark 2 at the end of the preceding section) the Bochner transforms of the functions

$$\omega_k^{-[1]}(a, \mu) = \int_0^\mu \omega_k(a, \nu)\, d\nu$$

are linear in a neighborhood of zero.(5) Now choose constants $C_k(a)$ such that the Bochner transforms of

$$C_k(a) + \int_0^\mu \omega_k^{-[1]}(a, \nu)\, d\nu = \omega_k^{-[2]}(a, \nu)$$

are linear in a neighborhood of zero. Applying inequality (2.3) and the estimate (3.5), we get

$$|\omega_k^{-[2]}(a, \mu)| \leqslant C \cdot K_2 2^k = K_7 . 2^k.$$

Reasoning as before, we obtain

$$\left| \int_0^\mu d\mu \int_0^\mu \omega_k(a, \nu)\, d\nu \right| \leqslant 2K_7 . 2^k. \tag{3.11'}$$

Now let

---

(4) If $r = 0$, then $\sigma_0^{-[0]}(\mu) = \sigma(\mu) + C$, and the estimate (3.11) takes the form

$$\left| \frac{1}{\mu} \int_0^\mu \frac{1}{2} \{\sigma(a+\nu) - \sigma(a-\nu)\}\, d\nu \right| \leqslant K.$$

(5) If the Bochner transform of $f(\mu)$ is linear in a neighborhood of zero, then it is also linear for $f(\mu + a)$ for any constant a. This is obvious for a trigonometric polynomial; the general case is obtained by approximating the function by trigonometric polynomials.

$$\int_0^{\mu_0} d\mu \int_0^{\mu} \frac{1}{2}\{g(a+\nu)+g(a-\nu)\}\,d\nu = \int_0^{\mu_0} d\mu \int_0^{\mu} \omega_0(a,\nu)\,d\nu$$

$$+\sum_{k=1}^{m-1}\int_0^{\mu_0} d\mu \int_0^{\mu} \omega_k(a,\nu)\,d\nu + \sum_{k=m}^{\infty}\int_0^{\mu_0} d\mu \int_0^{\mu} \omega_k(a,\nu)\,d\nu = P(a,\mu_0)+Q(a,\mu_0)+R(a,\mu_0).$$

In view of (3.11′),

$$|P(a,\mu_0)| \leqslant 2K_7 = A_1,$$

$$|Q(a,\mu_0)| \leqslant 2K_7 \sum_{k=1}^{m-1} \cdot 2^k \leqslant K_8 2^m \leqslant K_8 |\mu_0|. \tag{3.12}$$

Finally, (3.5) implies

$$|R(a,\mu_0)| \leqslant K_2 \sum_{k=m}^{\infty} \frac{1}{2^k} \int_0^{|\mu_0|} d\mu \int^{|\mu|} d\nu$$

$$\leqslant K_9 |\mu_0|^2 \sum_{k=m}^{\infty} \frac{1}{2^k} \leqslant K_{10} |\mu_0|. \tag{3.13}$$

Now (3.12) and (3.13) give the estimate

$$\left| \frac{1}{\mu} \int_0^{\mu} d\nu \int_0^{\nu} \frac{1}{2}\{g(a+\nu_1)+g(a-\nu_1)\}\,d\nu_1 \right| \leqslant K, \tag{3.14}$$

where the constant K is independent of a and μ.

Let

$$g(\mu) = \sigma_0^{-[r-1]}(\mu) \quad \text{and} \quad \sigma_0^{-[r]} = \int_0^{\mu} \sigma_0^{-[r-1]}(\nu)\,d\nu.$$

Then

$$\int_0^{\nu} \frac{1}{2}\{g(a+\nu)+g(a-\nu)\}\,d\nu = \frac{1}{2}\left\{ \int_a^{a+\nu} g(\nu)\,d\nu - \int_a^{a-\nu} g(\nu)\,d\nu \right\}$$

$$= \frac{1}{2}\{\sigma_0^{-[r]}(a+\nu) - \sigma_0^{-[r]}(a-\nu)\}.$$

Thus the inequality (3.14) is equivalent to (3.11).

Now let $r=0$ and $\alpha=0$. Then (3.1) takes the form

$$\sup_{-\infty<\mu<\infty} \overset{\mu+1}{\underset{\mu}{V}} \{\sigma(\mu)-\sigma_\varepsilon(\mu)\} < K.\varepsilon. \tag{3.15}$$

Let n be an arbitrary positive number and define

$$\rho_n(\mu) = \frac{3}{2\pi} \int_{-\infty}^{\infty} \frac{\sin^4 na}{n^3 a^4} \, d\sigma(\mu + a),$$

$$\rho_{n,\varepsilon}(\mu) = \frac{3}{2\pi} \int_{-\infty}^{\infty} \frac{\sin^4 na}{n^3 a^4} \, d\sigma_\varepsilon(\mu + a).$$

Then Lemma 1.2 implies that the Bochner transform of $\rho_{n,\epsilon}(\mu)$ is linear in the interval $(-\epsilon, \epsilon)$, and Lemma 1.1 and inequality (3.15) imply

$$\sup_{-\infty<\mu<\infty} \int_{\mu}^{\mu+1} |\rho_n(\nu) - \rho_{n,\varepsilon}(\nu)| \, d\nu < K.\varepsilon.$$

Set

$$\rho_n(\mu) = \rho_{n,0}(\mu) + \{\rho_{n,1}(\mu) - \rho_{n,0}(\mu)\} + \cdots = \lim_{k\to\infty} \rho_{n,k}(\mu),$$

where

$$\rho_{n,k}(\mu) = \rho_{n,\varepsilon_k}(\mu), \quad \varepsilon_k = \frac{1}{2^k}.$$

Then we can show

$$\left| \frac{1}{\mu} \int_0^{\mu} d\nu \int_0^{\nu} \frac{1}{2} \{\rho_n(a + \nu_1) + \rho_n(a - \nu_1)\} \, d\nu_1 \right| \leqslant K, \tag{3.16}$$

where the constant K is independent of μ and a. Now Lemma 1.1 gives

$$\lim_{n\to\infty} \int_0^{\nu} \frac{1}{2} \{\rho_n(a + \nu_1) + \rho_n(a - \nu_1)\} \, d\nu_1$$

$$= \lim_{n\to\infty} \frac{1}{2} \left\{ \int_0^{a+\nu} \rho_n(\nu_1) \, d\nu_1 - \int_0^{a-\nu} \rho_n(\nu_1) \, d\nu_1 \right\} = \frac{1}{2} \{\sigma(a + \nu) - \sigma(a - \nu)\}.$$

Thus, passing to the limit in (3.16) as $n \to \infty$, we conclude

$$\left| \frac{1}{\mu} \int_0^{\mu} \frac{1}{2} \{\sigma(a + \nu) - \sigma(a - \nu)\} \, d\nu \right| \leqslant K,$$

as was to be shown.

§4. Some auxiliary estimates for the spectral function of a selfadjoint second order differential equation

1. Consider the differential equation

$$y'' + \{\lambda - q(x)\} y = 0, \tag{4.1}$$

in the interval $(0, b)$, $0 \leqslant b \leqslant \infty$, where $q(x)$ is a real-valued function integrable on each finite subinterval $(0, b')$, $b' < b$. Let h be an arbitrary real number and let $\phi(x, \lambda)$ be the solution to (4.1) satisfying the initial conditions

$$\varphi(0, \lambda) = 1, \quad \varphi'(0, \lambda) = h. \tag{4.2}$$

h may equal infinity, in which case we take $\phi(x, \lambda)$ to be the solution to (4.1) satisfying the initial conditions

$$\varphi(0, \lambda) = 0, \quad \varphi'(0, \lambda) = \sqrt{\lambda}. \tag{4.2'}$$

We will investigate the case $h \neq \infty$ in detail. Changes needed for the case $h = \infty$ will be given at the end of the paper.

It is well known [1] that given h there exists at least one monotone function $\rho(\lambda)$ which is bounded in each finite interval and satisfies Parseval's equation

$$\int_0^b f^2(x)\,dx = \int_{-\infty}^{\infty} E^2(\lambda)\,d\rho(\lambda), \quad E(\lambda) = \underset{b' \to b}{\text{l.i.m.}} \int_0^{b'} f(x)\,\varphi(x, \lambda)\,dx$$

for each $f(x) \in L^2(0, b)$.

$\rho(\lambda)$ is called the spectral function for equation (4.1). In order to develop the asymptotic formulas indicated in the Introduction, we need some preliminary estimates for the spectral function $\rho(\lambda)$. These estimates will be derived in this section.

LEMMA 4.1 (MARČENKO [5], [6]). *For each fixed positive number* $x_0 < 2b$

$$\int_{-\infty}^{0} \cos\sqrt{\lambda}\,x_0\,d\rho(\lambda) < \infty.$$

PROOF.(6) As was shown in [12], there exists a continuous function $K(t, s)$, $s \leqslant t < b$, such that

$$\cos\sqrt{\lambda}\,t = \varphi(t, \lambda) + \int_0^t K(t, s)\,\varphi(s, \lambda)\,ds. \tag{4.3}$$

(6) We give here a different proof than Marčenko's original one.

Integrating both sides of equation (4.3) from 0 to x_0 with respect to t and interchanging the order of integration, we get

$$\frac{\sin\sqrt{\lambda}\,x_0}{\sqrt{\lambda}} = \int_0^{x_0} \varphi(t, \lambda)\,dt + \int_0^{x_0} dt \int_0^t K(t, s)\,\varphi(s, \lambda)\,ds$$

$$= \int_0^{x_0} \left\{1 + \int_t^{x_0} K(s, t)\,ds\right\} \varphi(t, \lambda)\,dt = \int_0^{x_0} H(x_0, t)\,\varphi(t, \lambda)\,dt.$$

From this formula we conclude that the function $\sin\sqrt{\lambda}\,x_0/\sqrt{\lambda}$ is the Fourier transform (with respect to the eigenfunctions $\phi(t, \lambda)$) of the function equal to $H(x_0, t)$ for $t < x_0$ and equal to zero for $t > x_0$. Parseval's equation now gives

$$\int_{-\infty}^{\infty} \left\{\frac{\sin\sqrt{\lambda}\,x_0}{\sqrt{\lambda}}\right\}^2 d\rho(\lambda) = \int_0^{x_0} H^2(x_0, t)\,dt.$$

Marčenko's lemma follows immediately from this equation.

2. Let $\epsilon < b/2$ be an arbitrary positive number and let a be an arbitrary real number. Consider the function

$$g_\varepsilon(t; a) = \begin{cases} \dfrac{1}{\varepsilon^2}(2\varepsilon - t)\cos at, & 0 \leqslant t \leqslant 2\varepsilon, \\ 0, & 2\varepsilon \leqslant t \leqslant b. \end{cases}$$

Multiply both sides of equation (4.3) by $g_\epsilon(t)$ and integrate from 0 to b with respect to t. Changing the order of integration, we get

$$\psi_\varepsilon(\lambda, a) = \frac{1}{\varepsilon^2}\int_0^{2\varepsilon} (2\varepsilon - t)\cos\sqrt{\lambda}\,t \cdot \cos at\,dt$$

$$= \int_0^{2\varepsilon} g_\varepsilon(t; a)\,\varphi(t, \lambda)\,dt + \int_0^{2\varepsilon} g_\varepsilon(t, a)\,dt \int_0^t K(t, s)\,\varphi(s, \lambda)\,ds$$

$$= \int_0^{2\varepsilon} \left\{g_\varepsilon(s, a) + \int_s^{2\varepsilon} K(t, s)\,g_\varepsilon(t, a)\,dt\right\} \varphi(s, \lambda)\,ds. \tag{4.4}$$

Integration by parts gives

$$\psi_\varepsilon(\lambda, a) = \frac{1}{2\varepsilon^2}\int_0^{2\varepsilon} (2\varepsilon - t)\cos(\sqrt{\lambda} + a)\,t\,dt + \frac{1}{2\varepsilon^2}\int_0^{2\varepsilon} (2\varepsilon - t)\cos(\sqrt{\lambda} - a)\,t\,dt$$

$$= \left\{\frac{\sin\varepsilon(\sqrt{\lambda} + a)}{\varepsilon(\sqrt{\lambda} + a)}\right\}^2 + \left\{\frac{\sin\varepsilon(\sqrt{\lambda} - a)}{\varepsilon(\sqrt{\lambda} - a)}\right\}^2.$$

Equation (4.4) shows that $\psi_\epsilon(\lambda, a)$ is the Fourier transform of the function

$$g_\varepsilon(s, a) + \int_s^{2\varepsilon} K(t, s)\, g_\varepsilon(t, a)\, dt.$$

Thus Parseval's equation gives

$$\int_{-\infty}^{\infty} \psi_\varepsilon^2(\lambda, a)\, d\rho(\lambda) = \int_0^{2\varepsilon} \left\{ g_\varepsilon(s, a) + \int_0^{2\varepsilon} K(t, s)\, g_\varepsilon(t, a)\, dt \right\}^2 ds. \tag{4.5}$$

From the representation of $\psi_\epsilon(\lambda, a)$ as a definite integral it is easy to see that, given $\lambda < 0$, we have

$$|\psi_\varepsilon(\lambda, a)| \leqslant \cosh \sqrt{|\lambda|} \cdot 2\varepsilon$$

uniformly in a (a is a real number). Thus Lemma 4.1 applies to give

$$\int_{-\infty}^{0} \psi_\varepsilon^2(\lambda, a)\, d\rho(\lambda) = O(1)$$

uniformly in a as $\epsilon \to 0$. Thus equation (4.5) implies

$$\int_0^{\infty} \psi_\varepsilon^2(\lambda, a)\, d\rho(\lambda) = \int_0^{2\varepsilon} g_\varepsilon^2(s, a)\, ds + 2 \int_0^{2\varepsilon} g_\varepsilon(s, a)\, ds \int_s^{2\varepsilon} K(t, s)\, g_\varepsilon(t, a)\, dt$$
$$+ \int_0^{2\varepsilon} ds \left\{ \int_s^{2\varepsilon} K(t, s)\, g_\varepsilon(t, a)\, dt \right\}^2 + O(1). \tag{4.6}$$

We now estimate the second and third terms on the right-hand side of (4.6). Consider the second term. Interchanging the order of integration, we get

$$\int_0^{2\varepsilon} g_\varepsilon(s, a)\, ds \int_t^{2\varepsilon} K(t, s)\, g_\varepsilon(t, a)\, dt = \int_0^{2\varepsilon} g_\varepsilon(t, a)\, dt \int_0^{t} K(t, s)\, g_\varepsilon(s, a)\, ds. \tag{4.7}$$

Let us estimate the integral $\int_0^t K(t, s)\, ds$ as $t \to 0$.

Along with formula (4.3) we have

$$\varphi(t, \lambda) = \cos \sqrt{\lambda}\, t - \int_0^t K_1(t, s) \cos \sqrt{\lambda}\, s\, ds. \tag{4.3'}$$

Inverting (4.3′), we get (4.3). Thus

$$K(t, s) = K_1(t, s) + \int_0^s K_1(t, r)\, K_1(r, s)\, dr + \cdots. \tag{4.8}$$

The representation of $K_1(t, s)$ obtained by Marčenko [5] immediately implies that

$$|K_1(t, s)| = |h| + O\left(\int_0^t |q(s)|\,ds\right) \tag{4.9}$$

as $t \to 0$.(7) Thus (4.8) implies that

$$\int_0^t |K(t, s)|\,ds = |h|\cdot t + O\left(t\int_0^t |q(s)|\,ds\right). \tag{4.10}$$

as $t \to 0$. Since $|g_\epsilon(t, a)| = O(1/\epsilon)$ uniformly in a, we conclude from (4.10) and (4.7) that

$$\left|\int_0^{2\varepsilon} g_\varepsilon(s, a)\,ds \int_t^{2\varepsilon} K(t, s)\, g_\varepsilon(t, a)\,dt\right| = \left|\int_0^{2\varepsilon} g_\varepsilon(t, a)\,dt \int_0^t K(t, s)\, g_\varepsilon(s, a)\,ds\right|$$

$$= O\left(\frac{1}{\varepsilon^2}\int_0^{2\varepsilon} dt \int_0^t |K(t, s)|\,ds\right) = O\left\{\frac{1}{\varepsilon^2}\int_0^{2\varepsilon} dt\left(|h|t + O\left(t\int_0^t |q(s)|\,ds\right)\right)\right\}$$

$$= O(|h|) + O\left(\int_0^{2\varepsilon} |q(s)|\,ds\right).$$

We now estimate the third term in (4.6). Since

$$\int_s^{2\varepsilon} K(t, s)\, g_\varepsilon(t, a)\,dt = O(1),$$

we conclude that as $\epsilon \to 0$

$$\int_0^{2\varepsilon} ds \left\{\int_s^{2\varepsilon} K(t, s)\, g_\varepsilon(t, a)\,dt\right\}^2 = O\left(\int_0^{2\varepsilon} ds \int_s^{2\varepsilon} |K(t, s)|\,|g_\varepsilon(t, a)|\,dt\right)$$

$$= O\left(\int_0^{2\varepsilon} |g_\varepsilon(t, a)|\,dt \int_0^t |K(t, s)|\,ds\right) = O\left\{\frac{1}{\varepsilon}\int_0^{2\varepsilon} dt\left(|h|\cdot t + O\left(t\int_0^t q(s)\,ds\right)\right)\right\}$$

$$= O(\varepsilon|h|) + O\left(\varepsilon\int_0^{2\varepsilon} |q(s)|\,ds\right) = o(h) + o\left(\int_0^{2\varepsilon} |q(s)|\,ds\right).$$

Thus (4.6) implies

$$\int_0^\infty \varphi_\varepsilon^2(\lambda, a)\,d\rho(\lambda) = \int_0^{2\varepsilon} g_\varepsilon^2(s, a)\,ds + O(|h|) + O\left(\int_0^{2\varepsilon} |q(s)|\,ds\right) + O(1). \tag{4.11}$$

---

(7) One can also obtain (4.9) by analyzing the partial differential equation for $K_1(t, s)$ given in [12] or [4].

If there is no negative spectrum, the term $O(1)$ is not needed in (4.11).

For $\lambda > 0$, let $\lambda = \mu^2$ and $\rho(\lambda) = \rho(\mu^2) = \sigma(\mu)$, and extend $\sigma(\mu)$ to negative values of μ to be odd. Then (4.11) can be rewritten in the form

$$\frac{1}{2}\int_{-\infty}^{\infty} \psi_\varepsilon^2(\mu^2, a)\, d\sigma(\mu) = \int_0^{2\varepsilon} g_\varepsilon^2(s, a)\, ds + O(|h|) + O\left(\int_0^{2\varepsilon} |q(s)|\, ds\right) + O(1). \tag{4.11'}$$

Marčenko's important lemma [5] follows immediately from (4.11′).

LEMMA 4.2. *There exists a constant* C *independent of* a *such that*

$$|\sigma(a+1) - \sigma(a)| < C. \tag{!}$$

PROOF. Substitute $\epsilon = \epsilon_0$ into (4.11′), where $\epsilon_0 = \min(1, b/2)$. The monotonicity of $\sigma(\mu)$ implies that

$$\int_a^{a+1} \left\{\frac{\sin \varepsilon_0 (\mu - a)}{\varepsilon_0 (\mu - a)}\right\}^4 d\sigma(\mu) = \int_0^1 \left(\frac{\sin \varepsilon_0 \mu}{\varepsilon_0 \mu}\right)^4 < C_1.$$

Since $\sin \nu/\nu > 2/\pi$ for $0 \leqslant \nu \leqslant 1$, the preceding estimate gives

$$\frac{16}{\pi^4}\int_0^1 d\sigma(\mu + a) < C_1,$$

whereupon (!) holds.

3. We now consider the function *

$$\psi_\varepsilon^*(\mu, a) = \frac{\sin \varepsilon(\mu + a)}{\varepsilon(\mu + a)} \cdot \frac{\sin \varepsilon(\mu - a)}{\varepsilon(\mu - a)},$$

where ϵ and a are fixed.

Observe that $\psi_\epsilon^*(\mu, a)$ is even. ** Now consider the Fourier transform of this function: ***

$$g_\epsilon^*(t, a) = \frac{2}{\pi}\int_0^\infty \psi_\epsilon^*(\mu, a) \cos \mu t\, d\mu.$$

Although an explicit formula for $g_\epsilon^*(t, a)$ can be calculated, we only need to know that $g_\epsilon^*(t, a)$ has the following two properties:

---

* *Translator's note.* Perhaps the author intends $\psi_\epsilon^*(\mu^2, a) = \dfrac{\sin \epsilon(\mu + a)}{\epsilon(\mu + a)} \cdot \dfrac{\sin \epsilon(\mu - a)}{\epsilon(\mu - a)}$.

** *Translator's note.* Perhaps he intends to say $\psi_\epsilon^*(\mu^2, a)$ is even.

*** *Translator's note.* Probably he intends $g_\epsilon^*(t, a) = \frac{2}{\pi}\int_0^\infty \psi_\epsilon^*(\mu^2, a) \cos \mu t\, d\mu$.

1) $g_\epsilon^*(t, a)$ vanishes outside $(0, 2\epsilon)$ for $t > 0$.

2) $g_\epsilon^*(t, a) = O(1/\epsilon)$ uniformly in a.

The first property holds since, for fixed ϵ and a, $\psi_\epsilon^*(t, a)$ is an even entire function of exponential type with power 2ϵ, and thus by the Paley-Wiener Theorem its Fourier cosine transform vanishes outside $(0, 2\epsilon)$.

The second property follows from the estimates

$$|g_\varepsilon^*(t, a)| \leqslant \frac{2}{\pi}\int_0^\infty |\psi_\varepsilon^*(\mu^2, a)|\,d\mu = \frac{1}{\pi}\int_{-\infty}^{\infty} |\psi_\varepsilon^*(\mu^2, a)|\,d\mu$$

$$\leqslant \frac{1}{\pi}\left\{\int_{-\infty}^{\infty}\left[\frac{\sin\varepsilon(\mu+a)}{\varepsilon(\mu+a)}\right]^2 d\mu\right\}^{1/2}\left\{\int_{-\infty}^{\infty}\left[\frac{\sin\varepsilon(\mu-a)}{\varepsilon(\mu-a)}\right]^2 d\mu\right\}^{1/2}$$

$$= \frac{1}{\pi\varepsilon}\int_{-\infty}^{\infty}\frac{\sin^2\varepsilon\mu}{\varepsilon\mu^2}\,d\mu = O\left(\frac{1}{\varepsilon}\right).$$

Using these two properties of $g_\epsilon^*(t, a)$, we can get an estimate analogous to (4.11′).

$$\frac{1}{2}\int_{-\infty}^{\infty}\psi_\varepsilon^{*2}(\mu^2, a)\,d\sigma(\mu) = \int_0^{2\varepsilon} g_\epsilon^{*2}(s, a)\,ds + O(|h|) + O\left(\int_0^{2\varepsilon}|q(s)|\,ds\right) + O(1). \tag{4.11″}$$

Subtracting twice (4.11″) from (4.11′), we get

$$\frac{1}{2}\int_{-\infty}^{\infty}\left\{\left[\frac{\sin\varepsilon(\mu+a)}{\varepsilon(\mu+a)}\right]^4 + \left[\frac{\sin\varepsilon(\mu-a)}{\varepsilon(\mu-a)}\right]^4\right\}d\sigma(\mu) = \int_0^{2\varepsilon} g_\varepsilon^2(s, a)\,ds$$
$$-2\int_0^{2\varepsilon} g_\varepsilon^{*2}(s, a)\,ds + O(h) + O\left(\int_0^{2\varepsilon}|q(s)|\,ds\right) + O(1). \tag{4.11‴}$$

By virtue of Parseval's equation for ordinary Fourier integrals,

$$\int_0^{2\varepsilon} g^2(s, a)\,ds = \frac{2}{\pi}\int_0^\infty\left\{\left[\frac{\sin\varepsilon(\mu+a)}{\varepsilon(\mu+a)}\right]^2 + \left[\frac{\sin\varepsilon(\mu-a)}{\varepsilon(\mu-a)}\right]^2\right\}^2 d\mu,$$

$$\int_0^{2\varepsilon} g_\varepsilon^{*2}(s, a)\,ds = \frac{2}{\pi}\int_0^\infty\left\{\frac{\sin\varepsilon(\mu+a)}{\varepsilon(\mu+a)}\,\frac{\sin\varepsilon(\mu-a)}{\varepsilon(\mu-a)}\right\}^2 d\mu.$$

Thus

$$\int_0^{2\varepsilon} g_\varepsilon^2(s, a)\,ds - 2\int_0^{2\varepsilon} g_\varepsilon^*(s, a)\,ds = \frac{2}{\pi}\int_0^\infty\left\{\left[\frac{\sin\varepsilon(\mu+a)}{\varepsilon(\mu+a)}\right]^4 + \left[\frac{\sin\varepsilon(\mu-a)}{\varepsilon(\mu-a)}\right]^4\right\}d\mu =$$

$$=\frac{2}{\pi}\int_0^\infty\left[\frac{\sin\varepsilon(\mu+a)}{\varepsilon(\mu+a)}\right]^4 d\mu+\frac{2}{\pi}\int_{-\infty}^0\left[\frac{\sin\varepsilon(\mu+a)}{\varepsilon(\mu+a)}\right]^4 d\mu=\frac{2}{\pi}\int_{-\infty}^\infty\left[\frac{\sin\varepsilon(\mu+a)}{\varepsilon(\mu+a)}\right]^4 d\mu$$

$$=\frac{2}{\pi}\int_{-\infty}^\infty\left(\frac{\sin\varepsilon\mu}{\varepsilon\mu}\right)^4 d\mu=\frac{2}{\pi\varepsilon}\int_{-\infty}^\infty\frac{\sin^4\mu}{\mu^4}\,d\mu=\frac{2}{\pi.\varepsilon}\cdot\frac{2\pi}{3}\,.$$

Since $\sigma(\mu)$ is odd, (4.11‴) gives

$$\int_{-\infty}^\infty\left[\frac{\sin\varepsilon(\mu-a)}{\varepsilon(\mu-a)}\right]^4 d\sigma(\mu)=\frac{2}{\pi\cdot\varepsilon}\frac{2\pi}{3}+O(h)+O\left(\int_0^{2\varepsilon}|q(s)|\,ds\right)+O(1).$$

Multiplying both sides by $3\epsilon/2\pi$, we get

$$\frac{3}{2\pi}\int_{-\infty}^\infty\frac{\sin^4\varepsilon\mu}{\varepsilon^3\mu^4}\,d\sigma(\mu+a)=\frac{2}{\pi}+O(\varepsilon|h|)+O\left(\varepsilon\int_0^{2\varepsilon}|q(s)|\,ds\right)+O(\varepsilon),$$

or

$$\frac{3}{2\pi}\int_{-\infty}^\infty\frac{\sin^4\varepsilon\mu}{\varepsilon^3\mu^4}\,d\mu=1,$$

since

$$\frac{3}{2\pi}\int_{-\infty}^\infty\frac{\sin^4\varepsilon\mu}{\varepsilon^3\mu^4}\,d\left\{\sigma(\mu+a)-\frac{2}{\pi}(\mu+a)\right\}$$

$$=O(\varepsilon h)+O\left(\varepsilon\int_0^{2\varepsilon}|q(s)|\,ds\right)+O(\varepsilon). \tag{4.12}$$

If there is no negative spectrum and $h=0$, then (4.12) implies

$$\frac{3}{2\pi}\int_{-\infty}^\infty\frac{\sin^4\varepsilon\mu}{\varepsilon^3\mu^4}\,d\left\{\sigma(\mu+a)-\frac{2}{\pi}(\mu+a)\right\}=O\left(\varepsilon\int_0^{2\varepsilon}|q(s)|\,ds\right). \tag{4.13}$$

The estimates (4.12) and (4.13) will play an essential part in deriving the asymptotic formulas for the spectral function $\rho(\lambda)$. The preceding analysis shows that these estimates hold uniformly in a.

§5. Asymptotic behavior of the spectral function $\rho(\lambda)$

1. In this section we derive the asymptotic formulas claimed in the Introduction for the spectral function of a selfadjoint second order differential equation.

Let ϵ be an arbitrary positive number, and form the Jackson kernel

$$D(\alpha, \varepsilon) = \begin{cases} 1 - \frac{3}{8}\frac{\alpha^2}{\varepsilon^2} + \frac{3}{32}\frac{|\alpha|^3}{\varepsilon^3}, & \text{if} \quad 0 \leqslant |\alpha| \leqslant 2\varepsilon, \\ \frac{1}{32}\left(4 - \frac{|\alpha|}{\varepsilon}\right)^3, & \text{if} \quad 2\varepsilon \leqslant |\alpha| \leqslant 4\varepsilon, \\ 0, & \text{if} \quad |\alpha| \geqslant 4\varepsilon. \end{cases}$$

Taking the Fourier transform of $D(\alpha, \epsilon)$, we get

$$d(\mu, \varepsilon) = \frac{1}{\sqrt{2\pi}} \int_{-\infty}^{\infty} D(\alpha, \varepsilon) e^{-i\alpha\mu} d\alpha = \frac{3}{\sqrt{2\pi}} \cdot \frac{\sin^4 \varepsilon\mu}{\varepsilon^3\mu^4}.$$

Construct the function

$$H(\alpha, \varepsilon) = \frac{4}{9}\left\{D(\alpha, \varepsilon) + 8D(\alpha, 2\varepsilon) - \frac{27}{4} D\left(\alpha, \frac{3}{2}\varepsilon\right)\right\}.$$

Using the explicit representation for $D(\alpha, \epsilon)$, it is easy to show that $H(\alpha, \epsilon) = 1$ for $0 \leqslant |\alpha| \leqslant 2\epsilon$. Let

$$h(\mu, \varepsilon) = \frac{4}{9}\left\{d(\mu, \varepsilon) + 8d(\mu, 2\varepsilon) - \frac{27}{4} d\left(\mu, \frac{3}{2}\varepsilon\right)\right\}.$$

The inverse Fourier transform gives

$$H(\alpha, \varepsilon) = \frac{1}{\sqrt{2\pi}} \int_{-\infty}^{\infty} h(\mu, \varepsilon) e^{i\alpha\mu} d\mu.$$

Now suppose that $\tau(\mu)$ satisfies

$$\sup_{-\infty<\mu<\infty} \overset{\mu+1}{\underset{\mu}{V}} \{\tau(\mu)\} < \infty. \tag{5.1}$$

Define

$$\rho_\varepsilon(\mu) = \frac{1}{\sqrt{2\pi}} \int_{-\infty}^{\infty} h(a, \varepsilon) d\tau(\mu + a),$$

$$\tau_\varepsilon(\mu) = \tau(\mu) - \int_0^\mu \rho_\varepsilon(\nu) d\nu.$$

Lemma 1.2 implies that the Bochner-Stieltjes transform of $\tau_\epsilon(\mu)$ is linear in the interval $(-2\epsilon, 2\epsilon)$.

Lemma 4.2 implies that the function $\tau(\mu) = \sigma(\mu) - 2\mu/\pi$ satisfies (5.1). Moreover, (4.12) gives

$$\sup_{-\infty<\mu<\infty} \overset{\mu+1}{\underset{\mu}{V}} \{\tau(\mu) - \tau_\varepsilon(\mu)\} = \sup_{-\infty<\mu<\infty} \int_\mu^{\mu+1} |\rho_\varepsilon(\nu)| d\nu \leqslant$$

$$\leqslant \sup_{-\infty<\mu<\infty} |\rho_\varepsilon(\mu)| = O(\varepsilon|h|) + O\left(\varepsilon\int_0^{4\varepsilon}|q(s)|\,ds\right) + O(\varepsilon).$$

Thus, applying Theorems 3.1 and 3.2 ($r = 0$ and $\alpha = 0$) to $\tau(\mu) = \sigma(\mu) - 2\mu/\pi$, we obtain

THEOREM 5.1.

$$\sigma(\mu + a) - \sigma(a) = \frac{2}{\pi}\mu + O(\ln|\mu|),$$

$$\frac{1}{\mu}\int_0^\mu \left\{\frac{1}{2}[\sigma(a+\nu) - \sigma(a-\nu)] - \frac{2}{\pi}\nu\right\} d\nu = O(1)$$

uniformly in a *as* $\mu \to \infty$.

Since $\sigma(\nu)$ is odd, the estimate in (0.5) holds.

2. If $h = 0$ and there is no negative spectrum, and if

$$\int_0^t |q(s)|\,ds = O(t^\alpha),$$

as $t \to 0$, where $\alpha < 0$, then (4.13) gives

$$\sup_{-\infty<\mu<\infty} \overset{\mu+1}{\underset{\mu}{V}} \{\tau(\mu) - \tau_\varepsilon(\mu)\} = O(\varepsilon^{1+\alpha}),$$

where $\tau(\mu) = \sigma(\mu) - 2\mu/\pi$. Thus, again applying Theorem 4.1, we get the following theorem.

THEOREM 5.2. *If there is no negative spectrum, if* $h = 0$, *and if*

$$\int_0^t |q(s)|\,ds = O(t^\alpha),$$

as $t \to 0$ *where* $\alpha < 0$, *then*

$$\sigma(\mu + a) - \sigma(a) = \frac{2}{\pi}\mu + O(1)$$

uniformly in a *as* $\mu \to \infty$.

3. Consider briefly the case $h = \infty$. Then $\phi(x, \lambda)$ denotes the solution to (4.1) satisfying the initial conditions

$$\varphi(0, \lambda) = 0, \quad \varphi'(0, \lambda) = \sqrt{\lambda}.$$

In this case we use the formulas

$$\varphi(t, \lambda) = \sin\sqrt{\lambda}\, t + \int_0^t K(t, s) \sin\sqrt{\lambda}\, s\, ds,$$

$$\sin\sqrt{\lambda}\, t = \varphi(t, \lambda) - \int_0^t K_1(t, s)\, \varphi(s, \lambda)\, ds.$$

All the preceding estimates hold, except instead of the function

$$g_\varepsilon(t, a) = \frac{1}{\varepsilon^2}(2\varepsilon - t) \cos at$$

we use

$$\frac{1}{\varepsilon^2}(2\varepsilon - t) \sin at.$$

The final estimate then takes the form

$$\frac{3}{2\pi} \int_{-\infty}^{\infty} \frac{\sin^4 \varepsilon\mu}{\varepsilon^3 \mu^4}\, d\left\{\sigma(\mu + a) - \frac{2}{\pi}(\mu + a)\right\} = O(\varepsilon) + O\left(\varepsilon \int_0^{4\varepsilon} |q(s)|\, ds\right). \tag{5.3}$$

If there is no negative spectrum, the term $O(\epsilon)$ is not needed. Note also that if $h = \infty$, Lemma 4.2 follows from (4.11‴). The estimate (5.3) proves Theorems 5.1 and 5.2 in this case.

Received 18/DEC/51

BIBLIOGRAPHY

1. B. M. Levitan, *Expansion in characteristic functions of differential equations of the second order*, GITTL, Moscow, 1950. (Russian) MR 12, 183.

2. E. C. Titchmarsh, *Eigenfunction expansions associated with second order differential equations*, Clarendon Press, Oxford, 1946. MR 8, 458.

3. H. Weyl, *Über gewöhnliche Differentialgleichungen mit Singularitäten und die zugehörigen Entwicklungen willkürlicher Funktionen*, Math. Ann. **68** (1910), 220–269.

4. I. M. Gel'fand and B. M. Levitan, *On the determination of a differential equation from its spectral function*, Izv. Akad. Nauk SSSR Ser. Mat. **15** (1951), 309–360; English transl., Amer. Math. Soc. Transl. (2) **1** (1955), 253–304. MR **13**, 558; **17**, 489.

5. V. A. Marčenko, *Some questions in the theory of one-dimensional second order linear differential operators*. I, Trudy Moskov. Mat. Obšč. **1** (1952), 327–420; English transl., Amer. Math. Soc. Transl. (2) **101** (1973), 1–104. MR **15**, 315.

6. ———, *On the transformation formulas generated by a second order linear differential operator*, Dokl. Akad. Nauk SSSR **74** (1950), 657–660. (Russian) MR **12**, 707.

7. B. M. Levitan, *On a decomposition theorem for characteristic functions of differential equations of the second order*, Dokl. Akad. Nauk SSSR **71** (1950), 605–608. (Russian) MR **11**, 720.

8. S. Bochner, *Vorlesungen über Fouriersche Integrale*, Akademische Verlagsgesellschaft, Leipzig, 1932; English transl., Ann. of Math. Studies, no. 42, Princeton Univ. Press, Princeton, N. J., 1959. MR **21** #5851.

9. H. Bohr, *Un théorème général sur l'intégration d'un polynôme trigonométrique,* C. R. Acad. Sci. Paris **200** (1935), 1276–1277.

10. B. M. Levitan, *Über eine Verallgemeinerung der Ungleichung von S. Bernstein und H. Bohr,* C. R. (Dokl.) Acad. Sci. URSS **15** (1937), 169–172.

11. A. Zygmund, *Smooth functions,* Duke Math. J. **12** (1945), 47–76. MR **7**, 60.

12. B. M. Levitan, *An application of generalized displacement operators to linear second order differential equations,* Uspehi Mat. Nauk **4** (1949), no. 1 (29), 3–112; English transl., Amer. Math. Soc. Transl. (1) **10** (1962), 408–541. MR **11**, 116.

13. ———, *Some questions in the theory of almost periodic functions.* II, Uspehi Mat. Nauk **2** (1947), no. 6 (22), 174–214; English transl., Amer. Math. Soc. Transl. (1) **10** (1962), 326–377. MR **10**, 293.

Translated by:
W. W. Walker

Amer. Math. Soc. Transl.
(2) Vol. 101, 1973

ON A SPECIAL TAUBERIAN THEOREM*

B. M. LEVITAN

Abstract. In the paper a special Tauberian theorem is proved (with an estimate of the remainder), on the basis of which it is possible to obtain the asymptotic behavior of the spectral function of a selfadjoint partial differential equation.

Introduction

The present work is closely connected with the first three sections of [1], in which we obtained an asymptotic formula for the spectral function of an ordinary selfadjoint differential equation of second order, having made more precise, as a preliminary, a special Tauberian theorem of Wiener [2].

In order to obtain the asymptotic behavior of the spectral function for a partial differential equation, one must extend the Tauberian theorem mentioned to the case of functions which grow at infinity like a polynomial. This is done in the present paper. The application of the results to the study of the asymptotic behavior of the spectral function for a partial differential equation will be given in later papers.

In order to avoid misunderstanding, we note that the letter C will denote a constant which is not always the same and varies with context.

§1. Some information on Bochner-Stieltjes integrals

1. Let $\sigma(\mu)$ denote a complex function defined for all real values μ, and which satisfies the condition[1]

$$\sup_{-\infty<a<\infty} \frac{1}{(1+|a|)^k} \overset{a+1}{\underset{a}{V}} \{\sigma(\mu)\} \leqslant M, \tag{1.1}$$

where M is some constant and k is a fixed natural number.

The case $k=0$ was examined by us in [1].

We assume

---

*Translation of Izv. Akad. Nauk SSSR Ser. Mat. 17 (1953), 269–284.

AMS (MOS) subject classifications (1970). Primary 40E05, 42A76; Secondary 34B25, 35P20.

(1) By $V_a^{a+1}\{\sigma(\mu)\}$ we denote the variation of the function $\sigma(\mu)$ on the interval $(a, a+1)$.

$$E_{k+2}(\alpha)=\frac{1}{\sqrt{2\pi}}\left\{\int_{-\infty}^{-1}+\int_{1}^{\infty}\frac{e^{-i\alpha\mu}}{(-i\mu)^{k+2}}d\sigma(\mu)+\int_{-1}^{1}\frac{e^{i\alpha\mu}-L_{k+1}(\alpha\mu)}{(i\mu)^{k+2}}d\sigma(\mu)\right\},\tag{1.2}$$

where

$$L_{k+1}(\beta)=1-i\beta-\frac{\beta^2}{2!}+i\frac{\beta^3}{3!}+\cdots+\frac{(-i\beta)^{k+2}}{(k+1)!}.$$

It is easily seen that condition (1.1) ensures absolute and uniform convergence of the integral (1.2) on each finite interval. In what follows we shall call the function $E_{k+2}(\alpha)$ the Bochner-Stieltjes transform of order $k+2$ of the function $\sigma(\mu)$. If $\sigma(\mu)$ is a differentiable function, then formula (1.2) defines the Bochner transform of the function $\sigma'(\mu)$ of order $k+2$ (see [3], §28, formula (4)). In the present section, two lemmas are proved which show that the basic properties of the Bochner transform carry over to the Bochner-Stieltjes transform.

LEMMA 1.1. *Assume*

$$\rho_n(\mu)=\frac{n}{r_k}\int_{-\infty}^{\infty}\left(\frac{\sin na}{na}\right)^{2k+4}d\sigma(\mu+a),\tag{1.3}$$

where

$$r_k=n\int_{-\infty}^{\infty}\left(\frac{\sin na}{na}\right)^{2k+4}da=\int_{-\infty}^{\infty}\left(\frac{\sin a}{a}\right)^{2k+4}da,$$

and denote the Bochner transform of the function $\rho_n(\mu)$ *by* $E^{(n)}_{k+2}(\alpha)$; *that is, we assume*

$$E^{(n)}_{k+2}(\alpha)=\frac{1}{\sqrt{2\pi}}\left\{\int_{-\infty}^{-1}+\int_{1}^{\infty}\frac{e^{-i\alpha\mu}}{(-i\mu)^{k+2}}\rho_n(\mu)\,d\mu+\int_{-1}^{1}\frac{e^{-i\alpha\mu}-L_{k+1}(\alpha\mu)}{(-i\mu)^{k+2}}\rho_n(\mu)\,d\mu\right\}.$$

Then the following assertions are valid:

1) *At each point of continuity of the function* $\sigma(\mu)$

$$\lim_{n\to\infty}\int_0^{\mu}\rho_n(\nu)\,d\nu=\sigma(\mu)-\frac{1}{2}\{\sigma(+0)+\sigma(-0)\}.\tag{1.4}$$

2)

$$\sup_{-\infty<\mu<\infty}\frac{1}{(1+|\mu|)^k}\int_{\mu}^{\mu+1}|\rho_n(\nu)|\,d\nu\leqslant C(k)M,\tag{1.5}$$

where the number M *is the same as in condition* (1.1), *and the constant* $C(k)$ *depends only on* k.

3) *Uniformly on each finite interval*

$$\lim_{n\to\infty} E_{k+2}^{(n)}(\alpha) = E_{k+2}(\alpha). \tag{1.6}$$

PROOF. 1) From the estimate (1.1) it follows that for large a

$$\sigma(a+1) - \sigma(a) = O(|a|^k)$$

and hence

$$\sigma(a) = O(|a|^{k+1}).$$

Therefore for each fixed μ

$$\lim_{a\to\infty} \frac{\sigma(\mu + a)}{a^{2k+4}} = 0.$$

Integrating by parts, we find from formula (1.3) that

$$\rho_n(\mu) = -\frac{n}{r_k} \int_{-\infty}^{\infty} \sigma(a+\mu)\, d\left(\frac{\sin na}{na}\right)^{2k+4}.$$

Integrating the last equality with respect to μ from 0 to μ, we obtain

$$\int_0^\mu \rho_n(\nu)\, d\nu = -\frac{n}{r_k} \int_{-\infty}^{\infty} \left\{\int_0^\mu \sigma(\nu + a)\, d\nu\right\} d\left(\frac{\sin na}{na}\right)^{2k+4}.$$

Integrating by parts in inverse order, we find

$$\int_0^\mu \rho_n(\nu)\, d\nu = -\frac{n}{r_k} \int_{-\infty}^{\infty} \left\{\int_a^{a+\mu} \sigma(\nu)\, d\nu\right\} d\left(\frac{\sin na}{na}\right)^{2k+4}$$

$$= \frac{n}{r_n} \int_{-\infty}^{\infty} \left(\frac{\sin na}{na}\right)^{2k+4} \{\sigma(a+\mu) - \sigma(a)\}\, da.$$

By virtue of well-known theorems formula (1.4) follows from the last formula when $n \to \infty$.

2) We define a monotonically increasing function $\sigma_1(\mu)$, letting

$$\sigma_1(\mu) = \begin{cases} \overset{\mu}{\underset{+0}{V}} \{\sigma(\mu)\}, & \mu > 0, \\ \overset{\mu}{\underset{-0}{V}} \{\sigma(\mu)\}, & \mu < 0. \end{cases}$$

From the definition of the Stieltjes integral it follows easily that

$$|\rho_n(\nu)| \leqslant \frac{n}{r_k} \int_{-\infty}^{\infty} \left(\frac{\sin na}{na}\right)^{2k+4} d\sigma_1(a+\nu).$$

From these inequalities with the help of integration by parts it follows (in just the same way as in the proof of subsection 1) that

$$\int_{\mu}^{\mu+1} |\rho_n(\nu)|\, d\nu \leqslant \frac{n}{r_k} \int_{-\infty}^{\infty} \left(\frac{\sin na}{na}\right)^{2k+4} \{\sigma_1(\mu+a+1) - \sigma_1(\mu+a)\}\, da$$

$$= \frac{n}{r_k} \int_{-\infty}^{\infty} \left(\frac{\sin na}{na}\right)^{2k+4} \mathop{V}_{a+\mu}^{a+\mu+1} \{\sigma(\mu)\}\, da.$$

By virtue of estimate (1.1),

$$\mathop{V}_{a+\mu}^{a+\mu+1} \{\sigma(\mu)\} \leqslant M(1+|a+\mu|)^k \leqslant M(1+|a|+|\mu|).$$

Therefore

$$\int_{\mu}^{\mu+1} |\rho_n(\nu)|\, d\nu \leqslant \frac{nM}{r_k} \int_{-\infty}^{\infty} \left(\frac{\sin na}{na}\right)^{2k+4} (1+|a|+|\mu|)^k\, da$$

$$= \frac{nM}{r_k} \left\{ \int_{-\infty}^{-1} + \int_{-1}^{1} + \int_{1}^{\infty} \left(\frac{\sin na}{na}\right)^{2k+4} (1+|a|+|\mu|)^k\, da \right\} = I_1 + I_2 + I_3.$$

Further,

$$I_2 \leqslant \frac{nM}{r_k} \int_{-1}^{1} \left(\frac{\sin na}{na}\right)^{2k+4} (2+|\mu|)^k\, da$$

$$\leqslant C'M(1+|\mu|)^k \frac{n}{r_k} \int_{\mp\infty}^{\infty} \left(\frac{\sin na}{na}\right)^{2k+4} da = C'M(1+|\mu|)^k;$$

$$I_3 \leqslant \frac{nM}{r_k} \int_{1}^{\infty} \frac{1}{(na)^{2k+4}} (1+a+|\mu|)^k\, da.$$

Because in the last integral $a \geqslant 1$ and $1+|\mu| \geqslant 1$, we get

$$(1+a+|\mu|) = a(1+|\mu|)\left(\frac{1}{a} + \frac{1}{1+|\mu|}\right) \leqslant 2a(1+|\mu|).$$

Therefore

$$(1+a+|\mu|)^k \leqslant 2^k a^k (1+|\mu|)^k.$$

Consequently

$$I_3 \leqslant \frac{2^k M (1+|\mu|)^k}{n^{2k+3}} \int_1^\infty \frac{da}{a^{k+4}} = C'' M (1+|\mu|)^k.$$

One can estimate I_1 in exactly the same way. These inequalities prove (1.5).

3) Equality (1.6) follows from 1) and 2) and from Helly's theorem on passing to the limit under the sign of a Stieltjes integral.

2. LEMMA 1.2. *As* $|a| \to \infty$ *let the function* $K(a)$ *satisfy the condition*

$$K(a) = O\left(\frac{1}{|a|^{k+4}}\right).$$

Let

$$\gamma(\alpha) = \frac{1}{\sqrt{2\pi}} \int_{-\infty}^{\infty} K(a) e^{-i\alpha a} da,$$

$$\tau(\mu) = \frac{1}{\sqrt{2\pi}} \int_{-\infty}^{\infty} K(a) d\sigma(\mu + a).$$

Further, denote the Bochner transform of order $k+2$ *of the function* $\tau(\mu)$ *by* $\Phi_{k+2}(\alpha)$; *that is, assume*

$$\Phi_{k+2}(\alpha) = \frac{1}{\sqrt{2\pi}} \left\{ \int_{-\infty}^{-1} + \int_1^{\infty} \frac{e^{-i\alpha\mu}}{(-i\mu)^{k+2}} \tau(\mu) d\mu + \int_{-1}^{1} \frac{e^{-i\alpha\mu} - L_{k+1}(\alpha\mu)}{(-i\mu)^{k+2}} \tau(\mu) d\mu \right\};$$

and denote the Bochner-Stieltjes transform (of order $k+2$*) of the function* $\sigma(\mu)$ *by* $E_{k+2}(\alpha)$. *Then the functions* $E_{k+2}(\alpha)$ *and* $\Phi_{k+2}(\alpha)$ *are connected by the following equations*:

$$\Phi_{k+2}(\alpha) = \gamma(\alpha) E_{k+2}(\alpha)$$
$$+ \sum_{l=1}^{k+1} (-1)^l C_{k+1}^l \int_0^{\alpha} d\alpha_1 \int_0^{\alpha_1} d\alpha_2 \ldots \int_0^{\alpha_l} \gamma^{(l)}(\alpha_l) E_{k+2}(\alpha_l) d\alpha_l + P_{k+1}(\alpha), \tag{1.7}$$

where C_{k+1}^l *are the binomial coefficients, and* $P_{k+1}(\alpha)$ *are polynomials in* α *of order* $k+1$.

PROOF. Define the function $\rho_n(\mu)$ in exactly the same way as in the preceding lemma. We assume that

$$\tau_n(\mu)=\frac{1}{\sqrt{2\pi}}\int_{-\infty}^{\infty}K(a)\rho_n(\mu+a)\,da,$$

and denote the Bochner transform of order $k+2$ of the function $\tau_n(\mu)$ by $\Phi_{k+2}^{(n)}(\alpha)$. We estimate $\int_\mu^{\mu+1}|\tau_n(\nu)|\,d\nu$. On the basis of part 2 of Lemma 1.1, for large μ the following estimate is valid:

$$\int_\mu^{\mu+1}|\tau_n(\nu)|\,d\nu\leqslant\int_{-\infty}^{\infty}|K(a)|\,da\int_\mu^{\mu+1}|\rho_n(\nu+a)|\,d\nu$$

$$=\int_{-\infty}^{\infty}|K(a)|\,da\int_{\mu+a}^{\mu+a+1}|\rho_n(\nu)|\,d\nu\leqslant C(k)M\int_{-\infty}^{\infty}|K(a)|(1+|\mu+a|)^k da=O(|\mu|^k). \tag{1.8}$$

From this estimate it follows that the function $\Phi_{k+2}^{(n)}(\alpha)$ exists. Denote the Bochner transform of the function $\rho_n(\mu)$ by $E_{k+2}^{(n)}(\alpha)$. As is known (see [3], §30, formula (19)), $\Phi_{k+2}^{(n)}(\alpha)$ and $E_{k+2}^{(n)}(\alpha)$ are connected by the equality

$$\Phi_{k+2}^{(n)}(\alpha)=\gamma(\alpha)E_{k+2}^{(n)}(\alpha)$$

$$+\sum_{l=1}^{k+1}(-1)^l C_{k+1}^l\int_0^\alpha d\alpha_1\ldots\int_0^{\alpha_l}\gamma^{(l)}(\alpha_l)E_{k+2}^{n}(\alpha_l)\,d\alpha_l+P_{k+1}^{(n)}(\alpha), \tag{1.9}$$

where $P_{k+1}^{(n)}(\alpha)$ are polynomials of order $k+1$.

We will suppose that $n\to\infty$. By Lemma 1.1 and Helly's theorem, $\tau_n(\mu)$ tends uniformly to $\tau(\mu)$ on each finite interval, and $E_{k+2}^{(n)}(\alpha)$ tends to $E_{k+2}(\alpha)$. If we show that $\Phi_{k+2}^{(n)}(\alpha)$ tends to $\Phi_{k+2}(\alpha)$, then formula (1.7) will follow from formula (1.9).(2) But this follows directly from estimate (1.8), because by virtue of this estimate it is possible to pass to the limit under the integral sign in the formula for. $\Phi_{k+2}^{(n)}(\alpha)$, and in the result we obtain the function $\Phi_{k+2}(\alpha)$.

---

(1) From the existence of the limits $\lim_{n\to\infty}E_{k+2}^{(n)}(\alpha)$ and $\lim_{n\to\infty}\Phi_{k+2}^{(n)}(\alpha)$ follows the existence of the limit $\lim_{n\to\infty}P_{k+1}^{(n)}(\alpha)=P_{k+1}(\alpha)$. Representing $P_{k+1}^{(n)}(\alpha)$ by the Lagrange interpolation formula with arbitrary interpolation points, it is easily shown that $P_{k+1}(\alpha)$ is also a polynomial of order $k+1$.

3. Following Bochner, we will say that the functions $\phi(\alpha)$ and $\psi(\alpha)$ are $(k+2)$-equivalent on the interval (a, b), and write

$$\varphi(\alpha) \overset{k+2}{\asymp} \psi(\alpha),$$

if their difference $\phi(\alpha) - \psi(\alpha)$ on this interval is a polynomial of order $k+1$.

We indicate three consequences of equality (1.7) which play an essential role in what follows:

1°. *If on some interval (a, b) the function $E_{k+2}(\alpha)$ is $(k+2)$-equivalent to zero (that is, it is a polynomial of order $k+1$), then the function $\Phi_{k+2}(\alpha)$ is also $(k+2)$-equivalent to zero in the interval (a, b).*

Indeed, in order to verify this assertion, it is sufficient to differentiate both parts of equality (1.7) $k+2$ times. We obtain (see [3], §30, formula (4)) $(a \leqslant \alpha \leqslant b)$:

$$\Phi_{k+2}^{(k+2)}(\alpha) = \gamma(a) E_{k+2}^{(k+2)}(\alpha),$$

which was required.

2°. *If $\gamma(\alpha) = 1$ on some interval (a, b), then, whatever the function $E_{k+2}(\alpha)$ might be on this interval,*

$$\Phi_{k+2}(\alpha) \overset{k+2}{\asymp} 0.$$

3°. *If $\gamma(\alpha) = 1$ on some interval (a, b), then on this interval*

$$\Phi_{k+2}(\alpha) \overset{k+2}{\asymp} E_{k+2}(\alpha).$$

Assertions 2° and 3° follow immediately from (1.7).

§2. Proof of an auxiliary inequality

In the present section an inequality is proved which should be considered as a certain extension of a classical inequality of H. Bohr.

THEOREM 2.1. *Let the function $\sigma(\mu)$ satisfy condition (1.1). Assume that the Bochner-Stieltjes transform of the function $\sigma(\mu)$ on the interval $(-\Lambda, \Lambda)$ $(0 < \Lambda \leqslant 1)$ is $(k+2)$-equivalent to zero (that is, it is a polynomial of order $k+1$). Select a constant A such that the Bochner transform of order $k+2$ of the function $\sigma(\mu) + A$ will be $(k+2)$-equivalent to zero on the interval*(3) *$(-\Lambda, \Lambda)$. Under these conditions,*

$$\sup_{\mu} \frac{1}{(1+|\mu|)^k} |A + \sigma(\mu)| \leqslant \frac{CM}{\Lambda^{k+1}}, \tag{2.1}$$

(3) We shall see from the proof of the theorem that this is possible.

where M *has the same meaning as in condition* (1.1), *and* C *is some constant depending only on* k (*and not on* M *or* Λ).

PROOF. 1. Let $\rho_n(\mu)$ have the same meaning as in Lemma 1.1. By virtue of Lemma 1.2, the Bochner transform of the function $\rho_n(\mu)$ on the interval $(-\Lambda, \Lambda)$ is $(k+2)$-equivalent to zero. We first prove inequality (2.1) for the function

$$\sigma_n(\mu) = \int_0^\mu \rho_n(\nu)\, d\nu,$$

assuming that $\Lambda = 1$. Here the constant C also turns out to be independent of n. As we show next, the general case can be reduced to this special one.

Consider an auxiliary function $\psi(\lambda)$ satisfying the following conditions:

1) $\psi(-\lambda) = -\psi(\lambda)$.

2) $\psi(\lambda) = 1/i\lambda$ for $\lambda > 1$.

3) $\psi(\lambda)$ is extended on the interval $(-1, 1)$ so that a continuous $(k+2)$th derivative of $\psi(\lambda)$ exists.

Consider the Fourier transform of the function $\psi(\lambda)$:

$$g(a) = \frac{1}{\sqrt{2\pi}} \int_{-\infty}^{\infty} \psi(\lambda) e^{-i\lambda a}\, d\lambda = \sqrt{\frac{2}{\pi}}\, i \int_0^\infty \psi(\lambda) \sin \lambda a\, d\lambda$$
$$= i\sqrt{\frac{2}{\pi}} \int_0^1 \psi(\lambda) \sin \lambda a\, d\lambda + \sqrt{\frac{2}{\pi}} \int_0^\infty \frac{\sin \lambda \alpha}{\lambda}\, d\lambda.$$

Since

$$\int_1^\infty \frac{\sin \lambda a}{\lambda}\, d\lambda = \int_a^{a\cdot\infty} \frac{\sin \lambda}{\lambda}\, d\lambda,$$

it follows that in a neighborhood about the point $a = 0$ the function $g(a)$ is bounded and has a discontinuity of the first kind with jump equal to $\sqrt{2\pi}$.

Integrating $k+2$ times by parts, we obtain that when $|a| \to \infty$ the following estimate holds:

$$g(a) = O\left(\frac{1}{|a|^{k+2}}\right). \tag{2.2}$$

2. We will show that

$$\frac{1}{\sqrt{2\pi}} \int_{-\infty}^{\infty} \rho_n(\mu + a)\, g(a)\, da = A_n + \int_0^\mu \rho_n(\nu)\, d\nu, \tag{2.3}$$

where the constant A_n possesses the property that the Bochner transform of the function

$$A_n + \int_0^\mu \rho_n(\nu)\, d\nu$$

on the interval $(-\Lambda, \Lambda)$ is $(k+2)$-equivalent to zero.

If $\rho_n(\mu)$ is a finite trigonometrical polynomial whose exponents lie outside of the interval $(-1, 1)$, then formula (2.3) follows immediately from Fourier inversion formulas. We will examine the general case. For each fixed n, $\rho_n(\mu)$ is an entire analytic function of finite order. As is known (see [5]), it is possible to find a sequence of finite trigonometrical polynomials $S_N^{(n)}(\mu)$ which converge as $N \to \infty$ to $\rho_n(\mu)$ uniformly on each finite interval and whose Fourier exponents lie outside the interval $(-1, 1)$. Applying formula (2.3) to $S_N^{(n)}(\mu)$, we obtain

$$\frac{1}{\sqrt{2\pi}} \int_{-\infty}^{\infty} S_N^{(n)}(\mu + a)\, g(a)\, da = A_N^{(n)} + \int_0^\mu S_N^{(n)}(\nu)\, d\nu. \tag{2.4}$$

Letting $N \to \infty$, we obtain (2.3)(4) from (2.4). From formula (2.3) follows the estimate(5)

$$\left| A_n + \int_0^\mu \rho_n(\nu)\, d\nu \right| \leqslant \frac{1}{\sqrt{2\pi}} \int_{-\infty}^{\infty} |\rho_n(\mu + a)|\, |g(a)|\, da$$

$$= \frac{1}{\sqrt{2\pi}} \sum_{r=-\infty}^{\infty} \int_r^{r+1} |\rho_n(\mu + a)|\, |g(a)|\, da = \frac{1}{\sqrt{2\pi}} \int_0^1 |\rho_n(\mu + a)|\, |g(a)|\, da$$

$$+ \frac{1}{\sqrt{2\pi}} \left\{ \sum_{r=-\infty}^{-1} + \sum_{r=1}^{\infty} \int_r^{r+1} |\rho_n(\mu + a)|\, |g(a)|\, da \right\} \leqslant C \int_\mu^{\mu+1} |\rho_n(a)|\, da$$

$$+ C \left\{ \sum_{r=-\infty}^{-1} \frac{1}{|r|^{k+2}} \int_{r+\mu}^{r+\mu+1} |\rho_n(a)|\, da + \sum_{r=1}^{\infty} \frac{1}{r^{k+2}} \int_{r+\mu}^{r+\mu+1} |\rho_n(a)|\, da \right\} \leqslant$$

---

(4) From formula (1) of [5] and properties of the polynomials $S_N^{(n)}(\mu)$ indicated in the same place it follows easily that

$$\int_\mu^{\mu+1} \left| S_N^{(n)}(\nu) \right| d\nu \leqslant C\,(1 + |\mu|)^k,$$

where C does not depend on N. From this estimate and the estimate (2.2) it follows that we can pass to the limit under the integral sign in (2.4).

(5) C denotes a constant which is not always the same one.

$$\leqslant CM\left\{(1+|\mu|)^k+\sum_{r=1}^{\infty}\frac{1}{r^{k+2}}(1+r+|\mu|)^k\right\}$$

$$\leqslant CM(1+|\mu|)^k\left\{1+\sum_{r+1}^{\infty}\frac{1}{r^2}\left[\frac{r+1+|\mu|}{r(1+|\mu|)}\right]^k\right\}$$

$$\leqslant CM(1+|\mu|)^k\left\{1+2^k\sum_{r=1}^{\infty}\frac{1}{r^2}\right\}=C\cdot M(1+|\mu|)^k. \tag{2.5}$$

The estimate for the case we are considering is equivalent to estimate (2.1).

3. We examine the case $\Lambda < 1$. As a preliminary we estimate

$$\sup_{-\infty<\mu<\infty}\frac{1}{(1+|\mu|)^k}\int_{\mu}^{\mu+1}\left|\rho_n\left(\frac{\nu}{\Lambda}\right)\right|d\nu.$$

We assume that

$$\tau_n(\mu)=\frac{1}{(1+|\mu|)^k}\rho_n(\mu).$$

We have

$$\sup_{-\infty<\mu<\infty}\int_{\mu}^{\mu+1}|\tau_n(\nu)|\,d\nu=\sup_{-\infty<\mu<\infty}\int_{\mu}^{\mu+1}\frac{1}{(1+|\nu|)^k}|\rho_n(\nu)|\,d\nu$$

$$\leqslant C\sup_{-\infty<\mu<\infty}\frac{1}{(1+|\mu|)^k}\int_{\mu}^{\mu+1}|\rho_n(\nu)|\,d\nu\leqslant C\cdot M, \tag{2.6}$$

where C is a constant. As we proved in [1], from the estimate (2.6) we can infer the following estimate:(6)

$$\sup_{-\infty<\mu<\infty}\int_{\mu}^{\mu+1}\left|\tau_n\left(\frac{\nu}{\Lambda}\right)\right|d\nu<C\cdot M.$$

Therefore

$$\sup_{-\infty<\mu<\infty}\frac{1}{(1+|\mu|)^k}\int_{\mu}^{\mu+1}\left|\rho_n\left(\frac{\nu}{\Lambda}\right)\right|d\nu$$

$$=\sup_{-\infty<\mu<\infty}\frac{1}{(1+|\mu|)^k}\int_{\mu}^{\mu+1}\left(1+\frac{|\nu|}{\Lambda}\right)^k\left|\tau_n\left(\frac{\nu}{\Lambda}\right)\right|d\nu\leqslant$$

(6) See footnote 5 on page 230.

$$\leqslant \frac{1}{\Lambda^k} \sup_{-\infty<\mu<\infty} \frac{1}{(1+|\mu|)^k} \int_{\mu}^{\mu+1} (\Lambda+|\nu|)^k \left|\tau_n\left(\frac{\nu}{\Lambda}\right)\right| d\nu$$

$$\leqslant C \frac{1}{\Lambda^k} \sup_{-\infty<\mu<\infty} \int_{\mu}^{\mu+1} \left|\tau_\nu\left(\frac{\nu}{\Lambda}\right)\right| d\nu \leqslant \frac{CM}{\Lambda^k}. \tag{2.7}$$

We will show that the Bochner transform (of order $k+2$) of the function $\rho_n(\mu/\Lambda)$ is $(k+2)$-equivalent to zero on the interval $(-1, 1)$. Indeed, if $\rho_n(\mu)$ is a finite trigonometrical sum, it is obvious because equivalence to zero in some interval of the Bochner transform for a trigonometrical polynomial is equivalent to the fact that this interval is free of Fourier exponents of the polynomial. In the general case, we approximate the function $\rho_n(\mu)$ by trigonometrical polynomials (see subsection 2).

Using estimates (2.5) and (2.7), we obtain

$$\sup_{-\infty<\mu<\infty} \frac{1}{(1+|\mu|)^k} \left| A_n + \int_0^{\mu} \rho_n\left(\frac{\nu}{\Lambda}\right) d\nu \right|$$

$$\leqslant C \sup_{-\infty<\mu<\infty} \frac{1}{(1+|\mu|)^k} \int_{\mu}^{\mu+1} \left|\rho_n\left(\frac{\nu}{\Lambda}\right)\right| d\nu \leqslant \frac{CM}{\Lambda^k}.$$

Replacing ν/Λ in the first integral by ν, we find that

$$\sup_{-\infty<\mu<\infty} \frac{1}{(1+|\mu|)^k} \left| A_n + \Lambda \int_0^{\frac{\mu}{\Lambda}} \rho_n(\nu)\, d\nu \right| \leqslant \frac{C\cdot M}{\Lambda^k}.$$

Substituting μ for μ/Λ, we obtain

$$\sup_{-\infty<\mu<\infty} \frac{1}{(1+|\mu|)^k} \left| A_n + \Lambda \int_0^{\mu} \rho_n(\nu)\, d\nu \right| \leqslant \frac{C\cdot M}{\Lambda^k}.$$

Dividing by Λ, we find from the last inequality that

$$\sup_{-\infty<\mu<\infty} \frac{1}{(1+|\mu|)^k} \left| \frac{A_n}{\Lambda} + \int_0^{\mu} \rho_n(\nu)\, d\nu \right| \leqslant \frac{C\cdot M}{\Lambda^{k+1}}.$$

Letting $n \to \infty$ in this inequality for fixed μ, we obtain inequality (2.1) by virtue of the second assertion of Lemma 1.1.

REMARK 1. Inequality (2.1) allows obvious iteration.

REMARK 2. Let $f(\mu) = \mu \cos \Lambda\mu$, where $0 < \Lambda < 1$. It is possible to show by direct calculation that the third-order Bochner transform of the function $f(\mu)$ on the interval $(-\Lambda, \Lambda)$ is 3-equivalent to zero (that is, it is a polynomial of second degree) (see [3], §28.7). Further, we have

$$\int_0^\lambda \nu \cos \Lambda\nu \, d\nu = \frac{\Lambda\lambda \sin \Lambda\lambda + \cos \Lambda\lambda - 1}{\Lambda^2}.$$

This equality shows that at least with respect to the order of Λ the inequality (2.1) is precise.

§3. Proof of the Tauberian theorem

THEOREM 3.1. *Let the function* $\sigma(\mu)$ *satisfy condition* (1.1). *Assume that for any positive number* ϵ *it is possible to find another positive number* $\delta = \delta(\epsilon)$ *and a function* $\sigma_\delta(\mu)$ *whose Bochner-Stieltjes transform on the interval* $(-\delta, \delta)$ *is* $(k+2)$*-equivalent to zero and which satisfies the inequality*

$$\sup_\mu \frac{1}{(1+|\mu|)^k} \overset{\mu+1}{\underset{\mu}{V}} \{\sigma(\mu) - \sigma_\delta(\mu)\} < \varepsilon. \tag{3.1}$$

Under these conditions, when $|\mu| \to \infty$

$$\sigma(\mu) = o(|\mu|^{k+1}). \tag{3.2}$$

PROOF. Assume $\rho_\delta(\mu) = \sigma(\mu) - \sigma_\delta(\mu)$. On the basis of estimate (3.1), when $|\mu| \to \infty$

$$|\rho_\delta(\mu) = O(\varepsilon|\mu|^{k+1}).$$

Further, by the same estimate (3.1), the function $\sigma_\delta(\mu)$ satisfies condition (1.1) (with constant $M + \epsilon$). From inequality (2.1) it follows ($|\mu| \to \infty$) that

$$|\sigma_\delta(\mu)| = O\left(\frac{1}{\delta^{k+1}}|\mu|^k\right).$$

Therefore when $|\mu| \to \infty$

$$|\sigma(\mu)| \leqslant |\sigma_\delta(\mu)| + |\rho_\delta(\mu)| = O\left(\frac{1}{\delta^{k+1}}|\mu|^k\right) + O(\varepsilon|\mu|^{k+1}),$$

and hence

$$\left|\frac{\sigma(\mu)}{\mu^{k+1}}\right| = O\left(\frac{1}{\delta^{k+1}|\mu|}\right) + O(\varepsilon).$$

Because the number δ does not depend on μ, and ϵ is arbitrary, it follows from the last inequality that

$$\lim_{|\mu|\to\infty} \frac{1}{\mu^{k+1}} \sigma(\mu) = 0,$$

which is what was required to be proved.

REMARK. Assume that an infinite set of functions $\sigma(\mu)$ exist satisfying condition (1.1). Denote this set of functions by the letter Σ.

If for all $\sigma \in \Sigma$ the constant M in (1.1) can be taken as one and the same, and if the number $\delta = \delta(\epsilon)$ can be chosen the same for all functions $\sigma \in \Sigma$, then the estimate (3.2) is fulfilled uniformly; that is, for any positive number η it is possible to find a positive number $N = N(\eta)$, depending on η but independent of the function $\sigma \in \Sigma$, such that if $|\mu| > N$, then for all $\sigma \in \Sigma$ the following inequality is fulfilled:

$$|\sigma(\mu)| < \eta |\mu|^{k+1}.$$

§4. Estimation of the remainder in the theorem of the preceding section

Assume, as in the previous section, that the function $\sigma(\mu)$ satisfies condition (1.1). We use the following notation: the function $\sigma(\mu) + C_0$ will be denoted by $\sigma^{-[0]}(\mu)$, where C_0 is an arbitrary constant. The indefinite integral of $\sigma^{-[0]}(\mu)$ will be denoted by $\sigma^{-[1]}(\mu)$; that is, the function

$$\int_0^\mu \sigma(\nu)\, d\nu + C_0\mu + C_1,$$

where C_1 is an arbitrary constant; and so forth.

THEOREM 4.1. *Let the function* $\sigma(\mu)$ *fulfill condition* (1.1). *Assume that for any positive* $\epsilon \leqslant 1$ *it is possible to find a function* $\sigma_\epsilon(\mu)$ *which has a Bochner-Stieltjes transform* (*of order* $k+2$) *on the interval* $(-\epsilon, \epsilon)$ *which is* $(k+2)$-*equivalent to zero* (*that is, it is a* $(k+1)$*th degree polynomial*) *and which satisfies the inequality*

$$\sup_{-\infty<\mu<\infty} \frac{1}{(1+|\mu|)^k} \overset{\mu+1}{\underset{\mu}{V}} \{\sigma(\mu) - \sigma_\varepsilon(\mu)\} \leqslant C \cdot \varepsilon^{(r+1-\alpha)(k+1)}, \tag{4.1}$$

where the constants C, $0 < \alpha < 1$ *and* $r \geqslant 0$ (*an integer*) *are independent of* ϵ. *Then among the functions* $\sigma^{-[r-1]}(\mu)$ *there is a function* $\sigma_0^{-[r-1]}(\mu)$ *satisfying the inequality*

$$\sup_{-\infty<\mu<\infty} \frac{1}{(1+|\mu|)^{k+\alpha}} \int_0^\mu \sigma_0^{-[r-1]}(\nu)\, d\nu \leqslant C_1, \tag{4.2}$$

where the constant C_1 *depends on the constant* C *from estimate* (4.1), *the constant* M *from estimate* (1.1), *and the number* k, *and does not depend on* μ.

When $\alpha = 0$ *the following estimate is valid*([7])

$$\sup_{-\infty<\mu<\infty} \frac{1}{(1+|\mu|)^{k} \ln(1+|\mu|)} \int_0^\mu \sigma_0^{-[r-1]}(\nu)\, d\nu \leqslant C_1. \tag{4.3}$$

PROOF. Assume

$$\varepsilon_p = \frac{1}{2^p} \quad (p = 0, 1, 2, \ldots), \qquad \sigma_p(\mu) = \sigma_{\varepsilon_p}(\mu);$$

$$\tau_p(\mu) = \sigma_p(\mu) - \sigma_{p-1}(\mu), \qquad \tau_0(\mu) = \sigma_0(\mu)$$

and consider the infinite series of functionals

$$\sigma_0(\mu) + [\sigma_1(\mu) - \sigma_0(\mu)] + \cdots = \tau_0(\mu) + \tau_1(\mu) + \cdots = \lim_{p\to\infty} \sigma_p(\mu), \tag{4.4}$$

It is easily seen that at each point of continuity of the function $\sigma(\mu)$ the series (4.4) converges to $\sigma(\mu) + C$.

Consider the function

$$\tau_p(\mu) = \sigma_p(\mu) - \sigma_{p-1}(\mu) \quad (p = 1, 2, \ldots).$$

Its Bochner-Stieltjes transform (of order $k + 2$) on the interval $(-2^{-p}, 2^{-p})$ is $(k+2)$-equivalent to zero. Because

$$\begin{aligned}\sup_{-\infty<\mu<\infty} \frac{1}{(1+|\mu|)^{k}} \overset{\mu+1}{\underset{\mu}{V}} \{\tau_p(\mu)\} &\leqslant \sup_{-\infty<\mu<\infty} \frac{1}{(1+|\mu|)^{k}} \overset{\mu+1}{\underset{\mu}{V}} \{\sigma(\mu) - \sigma_p(\mu)\} \\ &+ \sup_{-\infty<\mu<\infty} \frac{1}{(1+|\mu|)^{k}} \overset{\mu+1}{\underset{\mu}{V}} \{\sigma(\mu) - \sigma_{p-1}(\mu)\} \\ &\leqslant \frac{C}{2^{p(r+1-\alpha)(k+1)}} + \frac{C}{2^{p(r+1-\alpha)(k+1)}} = \frac{C}{2^{p(r+1-\alpha)(k+1)}},\end{aligned}$$

([7]) When $r = 0$ we assume, by definition,

$$\int_0^\mu \sigma_0^{-[1]}(\nu)\, d\nu = \sigma(\mu) + C.$$

Therefore the estimates (4.2) and (4.3) take the forms

$$\sup_{-\infty<\mu<\infty} \frac{1}{(1+|\mu|)^{k+\alpha}} |\sigma(\mu)| \leqslant C_1, \qquad \sup_{-\infty<\mu<\infty} \frac{1}{(1+|\mu|)^{k} \ln(1+|\mu|)} |\sigma(\mu)| \leqslant C_1.$$

on the basis of inequality (2.1) applied to $\tau_p(\mu)$ $r-1$ times we obtain

$$\sup_{-\infty<\mu<\infty} \frac{1}{(1+|\mu|)^k} |\tau_p^{-[r-1]}(\mu)| \leqslant \frac{C}{2^{p(1-\alpha)(k+1)}}. \tag{4.5}$$

Therefore the series

$$\tau_0^{-[r-1]}(\mu) + \sum_{p=1}^{\infty} \tau_p^{-[r-1]}(\mu) = g(\mu) \tag{4.6}$$

converges absolutely and uniformly on each finite interval. We will show that

$$g(\mu) = \sigma_0^{-[r-1]}.$$

In fact, if the series (4.6) is differentiated $r-1$ times, then we obtain the uniformly convergent series (4.4), the sum of which is almost everywhere equal to $\sigma(\mu) + C$.

It remains to estimate

$$\int_0^{\mu} \sigma_0^{-[r-1]}(\nu)\, d\nu = \int_0^{\mu} g(\nu)\, d\nu.$$

We choose an arbitrary real number μ_0 and define a positive integer m from the condition

$$2^{(m-1)(k+1)} \leqslant (1+|\mu_0|) < 2^{m(k+1)}.$$

Let

$$\int_0^{\mu_0} g(\nu)\, d\nu = \int_0^{\mu_0} \tau_0^{-[r-1]}(\nu)\, d\nu + \sum_{p=1}^{m-1} \int_0^{\mu_0} \tau_p^{-[r-1]}(\nu)\, d\nu$$
$$+ \sum_{p=m}^{\infty} \int_0^{\mu_0} \tau_p^{-[r-1]}(\nu)\, d\nu = P(\mu_0) + Q(\mu_0) + R(\mu_0).$$

By virtue of inequality (4.5) we have

$$|R(\mu_0)| \leqslant \sum_{p=m}^{\infty} C \int_0^{|\mu_0|} (1+|\nu|)^k \frac{1}{2^{p(1-\alpha)(k+1)}}\, d\nu$$
$$\leqslant C(1+|\mu_0|)^{k+1} \sum_{p=m}^{\infty} \frac{1}{2^{p(1-\alpha)(k+1)}} \leqslant C(1+|\mu_0|)^{k+1} \frac{1}{2^{m(1-\alpha)(k+1)}}$$
$$\leqslant C(1+|\mu_0|)^{k+1}(1+|\mu_0|)^{\alpha-1} = C(1+|\mu_0|)^{k+\alpha}. \tag{4.7}$$

In order to estimate $Q(\mu_0)$, we again use inequality (2.1). We obtain

$$\left|\int_0^{\mu_0} \tau_p^{-[r-1]}(\nu)\,d\nu\right| \leqslant C\cdot 2^{p\alpha(k+1)}(1+|\mu_0|)^k.$$

Therefore

$$Q(\mu_0)| \leqslant C(1+|\mu_0|)^k \sum_{p=1}^{m-1} 2^{p\alpha(k+1)} \leqslant C(1+|\mu_0|)^k\, 2^{m\alpha(k+1)} \leqslant C(1+|\mu_0|)^{k+\alpha}. \tag{4.8}$$

It remains to estimate $P(\mu_0)$. Because

$$\tau_0(\mu) = \sigma_0(\mu) = \sigma(\mu) - \{\sigma(\mu) - \sigma_0(\mu)\},$$

it follows that

$$\sup_{-\infty<\mu<\infty} \frac{1}{(1+|\mu|)^k} \overset{\mu+1}{\underset{\mu}{V}} \{\tau_0(\mu)\} \leqslant \sup_{-\infty<\mu<\infty} \frac{1}{(1+|\mu|)^k} \overset{\mu+1}{\underset{\mu}{V}} \{\sigma(\mu)\}$$
$$+ \sup_{-\infty<\mu<\infty} \frac{1}{(1+|\mu|)^k} \overset{\mu+1}{\underset{\mu}{V}} \{\sigma(\mu) - \sigma_0(\mu)\} \leqslant C.$$

On the other hand, the Bochner-Stieltjes transform of the function $\tau_0(\mu)$ on the interval $(-1, 1)$ is $(k+2)$-equivalent to zero. Therefore, on the basis of inequality (2.1) (applied r times),

$$|P(\mu_0)| = \left|\int_0^{\mu_0} \tau_0^{-[r-1]}(\nu)\,d\nu\right| \leqslant C(1+|\mu_0|)^k. \tag{4.9}$$

From estimates (4.7), (4.8) and (4.9), estimate (4.2) follows. If $\alpha = 0$, then the estimates for $P(\mu_0)$ and $R(\mu_0)$ remain in force. For $Q(\mu_0)$ we obtain the estimate

$$|Q(\mu_0)| \leqslant C(1+|\mu_0|)^k \sum_{p=1}^{m-1} 2^0 = C(1+|\mu_0|)^k (m-1)$$
$$= C(1+|\mu_0|)^k \cdot \ln(1+|\mu_0|).$$

Therefore estimate (4.3) is valid.

REMARK 1. When $r = 0$, $g(\mu)$ should be understood as the sum of the series (4.4).

REMARK 2. Consider the infinite set of functions $\sigma(\mu)$, which we denote by Σ. If for all $\sigma \in \Sigma$ it is possible to find one and the same constant M in inequality (1.1), and if the constant C in (4.1) can be chosen independent of the function $\sigma \in \Sigma$, then estimates (4.2) and (4.3) are fulfilled uniformly for all $\sigma \in \Sigma$; that is, the constant C in (4.2) does not depend on $\sigma \in \Sigma$.

For the estimation of the iterated integrals, there is a considerably more precise inequality than that which was obtained by means of iterating inequality (2.1). Namely, we have the following result:

Let the function $\sigma(\mu)$ *satisfy the conditions of Theorem* 2.1. *Denote by* l *an arbitrary natural number, and choose a polynomial* $P_l(\mu)$ *of degree* l *such that the Bochner transform of the function*

$$h_l(\mu) = \frac{1}{l!}\int_0^\mu (\mu - \nu)^l \, d\sigma(\nu) + P_l(\mu)$$

on the interval $(-\Lambda, \Lambda)$ *is* $(k+2)$*-equivalent to zero. Then the following inequality is valid*:

$$|h_l(\mu)| < \frac{CM}{\Lambda^{k+1+l}} (1 + |\mu|)^k. \tag{4.10}$$

Inequality (4.10) permits considerably more precision in Theorem 4.1.

Received 8/AUG/52

BIBLIOGRAPHY

1. B. M. Levitan, *On the asymptotic behavior of the spectral function of a self-adjoint second order differential equation*, Izv. Akad. Nauk SSSR Ser. Mat. **16** (1952), 325–352; English transl., Amer. Math. Soc. Transl. (2) **101** (1973), 192–221. MR **15**, 315.

2. ———, *Remark on a theorem of V. A. Marčenko*, Trudy Moskov. Mat. Obšč. **1** (1952), 421–422; English transl., Amer. Math. Soc. Transl. (2) **101** (1973), 105–106. MR **15**, 315.

3. S. Bochner, *Vorlesungen über Fouriersche Integrale*, Akademische Verlagsgesellschaft, Leipzig, 1932; English transl., Ann. of Math. Studies, no. 42, Princeton Univ. Press, Princeton, N. J., 1959. MR **21** #5851.

4. H. Bohr, *Un théorème général sur l'intégration d'un polynôme trigonométrique*, C. R. Acad. Sci. Paris **200** (1935), 1276–1277.

5. N. I. Ahiezer, *On the polynomials of B. M. Levitan*, C. R. (Dokl.) Acad. Sci. URSS **54** (1946), 3–5. MR **8**, 323.

Translated by:
A. S. Rehm

Amer. Math. Soc. Transl.
(2) Vol. 101, 1973

ON THE DISCONTINUITY OF SOLUTIONS OF QUASILINEAR EQUATIONS*

B. L. ROŽDESTVENSKIĬ

The characteristic feature of the solutions of systems of quasilinear equations distinguishing them from the solutions of linear equations is the development of a discontinuity in the solution even with continuous initial data. However, not every system of quasilinear equations of hyperbolic type can develop a discontinuity in the solution if the initial data are continuous. There exist quasilinear equations of hyperbolic type which do not lead to the development of a discontinuity in the solution for any continuous initial data. In connection with these, there arises the problem of identifying the class of systems of quasilinear equations not leading to discontinuities in the solution. Reference [1] was devoted to this problem. Here this problem is solved for a broader class of systems of quasilinear equations, and the method used is quite general and does not depend on the specific form of the system of quasilinear equations.

1. We consider the system of quasilinear equations

$$\frac{\partial u_i}{\partial t} + \sum_{j=1}^{2} a_{ij} \frac{\partial u_j}{\partial x} = b_i \quad (i = 1, 2). \tag{1}$$

We assume that, in the domain of the variables t, x, u_1 and u_2 considered, system (1) is hyperbolic and the coefficients a_{ij} and b_i are continuous and twice continuously differentiable functions of t, x, u_1 and u_2.

From the assumption of hyperbolicity of the system of equations (1) it follows that the roots $\bar{\xi}_1 = \bar{\xi}_1(u_1, u_2, t, x)$ and $\bar{\xi}_2 = \bar{\xi}_2(u_1, u_2, t, x)$ of the equation

$$\begin{vmatrix} a_{11} - \bar{\xi} & a_{12} \\ a_{21} & a_{22} - \bar{\xi} \end{vmatrix} = 0 \tag{2}$$

---

*Translation of Mat. Sb. 47 (89) (1959), 485–494.

AMS (MOS) subject classifications (1970). Primary 35B30, 35L45, 35L60; Secondary 35L65.

are real and distinct for all points of the (t, x, u_1, u_2)-domain under consideration. Thus $\bar{\xi}_1 \neq \bar{\xi}_2$; for definiteness, we assume that

$$\bar{\xi}_1(u_1, u_2, t, x) > \bar{\xi}_2(u_1, u_2, t, x).$$

The system of equations (1) can be reduced to the form

$$\sum_{j=1}^{2} \bar{a}_{ij}\left(\frac{\partial u_j}{\partial t} + \bar{\xi}_i \frac{\partial u_j}{\partial x}\right) = \bar{b}_i \quad (i = 1, 2). \tag{3}$$

Letting the Riemann invariants $r = r(u_1, u_2, t, x)$ and $s = s(u_1, u_2, t, x)$ be the integrals of the equations

$$\begin{aligned} \bar{a}_{11} du_1 + \bar{a}_{12} du_2 &= 0, \\ \bar{a}_{21} du_1 + \bar{a}_{22} du_2 &= 0, \end{aligned} \tag{4}$$

we write the system of equations (3) in the form

$$\begin{aligned} \frac{\partial r}{\partial t} + \xi_1(r, s, t, x)\frac{\partial r}{\partial x} &= f_r(r, s, t, x), \\ \frac{\partial s}{\partial t} + \xi_2(r, s, t, x)\frac{\partial s}{\partial x} &= f_s(r, s, t, x). \end{aligned} \tag{5}$$

We observe that in equations (4) the variables t and x are considered as parameters; ξ_1, ξ_2, f_1 and f_2 are functions of their arguments from $\bar{b}_1, \bar{b}_2, \bar{\xi}_1$ and $\bar{\xi}_2$, while $\xi_1 > \xi_2$.

2. In the consideration of discontinuous solutions of systems of quasilinear equations, the possibility of reducing them to the form

$$\frac{\partial \varphi}{\partial t} + \frac{\partial \psi}{\partial x} = f,$$

where ϕ, ψ and f are functions of the dependent and independent variables, is essential. If a system of quasilinear equations has such a correlation, this is called a *conservation law of the system of quasilinear equations.*

If the system has n independent conservation laws, where n is the number of unknowns, then it is called *conservative*; moreover, the independence of the conservation laws implies that

$$\frac{\partial[\varphi_1; \varphi_2; \ldots; \varphi_n]}{\partial[u_1; u_2; \ldots; u_n]} \neq 0.$$

The question of the possibility of representing a system of quasilinear equations in the form of conservation laws has been considered in [2]; there it was shown that every system of two quasilinear equations is always conservative.

To find the conservation laws of system (5), we write the equations

$$\frac{\partial\psi(r,s,t,x)}{\partial r}=\xi_1(r,s,t,x)\frac{\partial\varphi(r,s,t,x)}{\partial r},$$
$$\frac{\partial\psi(r,s,t,x)}{\partial s}=\xi_2(r,s,t,x)\frac{\partial\varphi(r,s,t,x)}{\partial s}, \tag{6}$$
$$f=f_r\frac{\partial\varphi}{\partial r}+f_s\frac{\partial\varphi}{\partial s}+\varphi_t'+\psi_x'.$$

The system of equations (6) has an infinite number of solutions. Selecting any two independent conservation laws of system (5), we write it in the form

$$\frac{\partial\varphi_i(r,s,t,x)}{\partial t}+\frac{\partial\psi_i(r,s,t,x)}{\partial x}=f_i(r,s,t,x) \quad (i=1,2), \tag{7}$$

in which $\partial[\phi_1;\phi_2]/\partial[r;s]\neq 0$.

To consider discontinuous solutions of the system of equations (7), the concept of the generalized solution of the Cauchy problem is introduced. A *generalized solution of the Cauchy problem for the system of quasilinear equations* (7) is defined to be, generally speaking, discontinuous functions $r=r(t,x)$ and $s=s(t,x)$ which assume the given values on the initial curve and satisfy the relations

$$\oint_C \varphi_i(r,s,t,x)\,dx-\psi_i(r,s,t,x)\,dt=\iint_G f_i(r,s,t,x)\,dt\,dx \quad (i=1,2), \tag{8}$$

where G is any bounded region of the plane of the independent variables t and x, bounded by the closed contour C lying in the region of influence of the initial data.

Assuming the presence of a discontinuity in the solution, lying on the piecewise smooth curve $x=x(t)$, i.e. the existence of left and right limits of the values of the functions $r(t,x)$ and $s(t,x)$ at points of the line $x=x(t)$ (which we denote by r_1, s_1 and r_0, s_0), we obtain from equation (8) the relation

$$[D\varphi_i(r,s,t,x)-\psi_i(r,s,t,x)]_{x=x(t)}=0 \quad (i=1,2), \tag{9}$$

where $D=x'(t)$ and $[g(r,s,t,x)]$ denotes the difference between the left and right limits of the values of the function g at points on the line of discontinuity,

$$[g(r,s,t,x)]\big|_{x=x(t)}=g(r_0,s_0,t,x(t))-g(r_1,s_1,t,x(t)). \tag{10}$$

Equation (9) connects the velocity of movement of the line of discontinuity with the limiting values of the solutions at the discontinuity. In hydrodynamics, analogous relations bear the name Hugoniot's conditions.

It is known that condition (9) is insufficient to insure the uniqueness of a discontinuous solution of the system (7). However, we shall not be concerned here with the question of uniqueness of solutions.

We shall assume that at some point (t, x) the values r_0 and s_0 are fixed in equation (9). We shall consider the dependence of the velocity of the discontinuity front D at this point on the values r_1 and s_1. We show that, if the quantity $(r_1 - r_0)^2 + (s_1 - s_0)^2$ tends to zero, then D converges to $\xi_1(r_0, s_0, t, x)$ or $\xi_2(r_0, s_0, t, x)$.

Let $r_1 = r_0 + \Delta r$ and $s_1 = s_0 + \Delta s$. Then the relation (9) may be rewritten in the form

$$[D\varphi'_{ir}(r_0 + \theta_1\Delta r, s_0 + \theta_1\Delta s, t, x) - \psi'_{ir}(r_0 + \theta_2\Delta r, s_0 + \theta_2\Delta s, t,x)]\,\Delta r$$
$$+ [D\varphi'_{is}(r_0 + \theta_1\Delta r, s_0 + \theta_1\Delta s, t, x) - \psi'_{is}(r_0 + \theta_2\Delta r, s_0 + \theta_2\Delta s, t, x)]\,\Delta s = 0,$$
$$0 < \theta_1 < 1, \quad 0 < \theta_2 < 1. \tag{11}$$

For system (11) to have a nontrivial solution it is necessary that the determinant of the system vanish. Passing to the limit as $\Delta r \to 0$ and $\Delta s \to 0$, we obtain

$$\begin{vmatrix} D\dfrac{\partial\varphi_1}{\partial r} - \dfrac{\partial\psi_1}{\partial r} & D\dfrac{\partial\varphi_1}{\partial s} - \dfrac{\partial\psi_1}{\partial s} \\ D\dfrac{\partial\varphi_2}{\partial r} - \dfrac{\partial\psi_2}{\partial r} & D\dfrac{\partial\varphi_2}{\partial s} - \dfrac{\partial\psi_2}{\partial s} \end{vmatrix} = 0. \tag{12}$$

Rewriting condition (12) in the form

$$(D - \xi_1)(D - \xi_2)\frac{\partial[\varphi_1; \varphi_2]}{\partial[r; s]} = 0,$$

we have, on account of the independence of the laws of conservation (7), that the limiting value of the velocity of the discontinuity D can have only two values

$$D_1 = \xi_1(r_0, s_0, t, x) \text{ and } D_2 = \xi_2(r_0, s_0, t, x),$$

as we were required to prove.

We observe that the limiting value of the propagation velocity of the smaller discontinuity, when the magnitude of the discontinuity tends to zero, coincides with the propagation velocity of a weak discontinuity (discontinuity of derivatives). The property of weak discontinuities of a solution being propagated along characteristics can also be preserved for finite discontinuities in some cases.

3. DEFINITION. Characteristics of the family $dx/dt = \xi_1(r, s, t, x)$ we shall call *contact characteristics* in the region of variation of the dependent and independent variables under consideration if they can be a line of strong discontinuity of the solutions.

It may be noted that it is possible to have a strong discontinuity of the solution for which the slope of the characteristic $dx/dt = \xi_1(r, s, t, x)$ is uniquely determined at the points of the line of discontinuity and coincides with the value of the propagation velocity of the discontinuity. Under this condition, for any values r_0 and s_0 there exist values r_1 and s_1 satisfying the relation (9), in which $D_1 = \xi_1$ and $(r_1 - r_0)^2 + (s_1 - s_0)^2 \neq 0$.

In this way, if $dx/dt = \xi_1(r, s, t, x)$ is a contact characteristic, then we have the simultaneous equations

$$\begin{aligned} &[\xi_1(r, s, t, x)\varphi_i(r, s, t, x) - \psi_i(r, s, t, x)] = 0 \quad (i = 1, 2), \\ &[\xi_1(r, s, t, x)] = 0, \\ &[r]^2 + [s]^2 \neq 0, \\ &\frac{dx}{dt} = \xi_1(r, s, t, x). \end{aligned} \tag{13}$$

Necessary and sufficient conditions for the characteristic $\xi_1(r, s, t, x)$ (i.e. $dx/dt = \xi_1(r, s, t, x)$) to be a contact characteristic are easily obtained.(1)

For the compatibility of equations (13), it is necessary that

$$\left(\xi_1 \frac{\partial\varphi_i}{\partial r} - \frac{\partial\psi_i}{\partial r}\right) dr + \left(\xi_1 \frac{\partial\varphi_i}{\partial s} - \frac{\partial\psi_i}{\partial s}\right) ds = 0 \quad (i = 1, 2),$$

$$\frac{\partial\xi_1}{\partial r} dr + \frac{\partial\xi_1}{\partial s} ds = 0,$$

from which we obtain

$$\frac{\partial\xi_1(r, s, t, x)}{\partial r} = 0. \tag{14}$$

We show that condition (14) is also a sufficient condition for ξ_1 to be a contact characteristic. From equations (6) we have

$$\psi_i(r, s, t, x) = \xi_1(s, t, x)\varphi_i(r, s, t, x) + F_i(s, t, x), \tag{15}$$

and in this case equations (13) are always consistent.

---

(1) The question of under what conditions a system of quasilinear equations can have a contact characteristic is also considered in [3].

THEOREM. *The property that the system of quasilinear equations* (1) *have a contact characteristic is preserved under every nondegenerate transformation of the dependent and independent variables; i. e. this property is an intrinsic invariant property of a system of quasilinear equations. (Here, of course, we must consider only transformations not taking the system out of the class of quasilinear equations.)*

PROOF. We consider first the transformation of dependent variables. Let

$$\begin{aligned} \bar{u}_1 &= g_1(u_1, u_2, t, x), \\ \bar{u}_2 &= g_2(u_1, u_2, t, x) \end{aligned} \tag{16}$$

be new unknown functions. Since the invariants r and s are preserved under such transformations, a contact characteristic remains contact in the new variables also.

We consider a transformation of the independent variables not transforming out of the class of quasilinear equations. This transformation is accomplished with the help of the functional relations

$$\begin{aligned} \frac{\partial}{\partial t} &= A_1(r, s, t, x)\frac{\partial}{\partial t'} - A_2(r, s, t, x)\frac{\partial}{\partial x'}, \\ \frac{\partial}{\partial x} &= -B_1(r, s, t, x)\frac{\partial}{\partial t'} + B_2(r, s, t, x)\frac{\partial}{\partial x'}. \end{aligned} \tag{17}$$

Here t' and x' are new variables, and any function of the variables t and x can be set under the symbol of differentiation.

In order that the transformation (17) be pointwise nondegenerate, it is necessary to fulfill the conditions

$$\begin{aligned} \frac{\partial B_1(r, s, t, x)}{\partial t} + \frac{\partial A_1(r, s, t, x)}{\partial x} &= 0, \\ \frac{\partial B_2(r, s, t, x)}{\partial t} + \frac{\partial A_2(r, s, t, x)}{\partial x} &= 0 \end{aligned} \tag{18}$$

and

$$\Delta = \begin{vmatrix} A_1 & A_2 \\ B_1 & B_2 \end{vmatrix} \neq 0. \tag{19}$$

In the relations (17), (18) and (19), it is intended that r and s be functions of the variables t and x, and the symbols of differentiation in equation (18) denote total partial derivatives.

Since the solution $r = r(t, x)$, $s = s(t, x)$ is not known in advance, the relation (18) must result from the original system of equations (5). Thus, if the system of equations (5) can be written in the form

$$\frac{\partial \widetilde{\varphi}_i (r, s, t, x)}{\partial t} + \frac{\partial \widetilde{\psi}_i (r, s, t, x)}{\partial x} = 0 \quad (i = 1, 2), \tag{20}$$

then, assuming $A_i = \widetilde{\psi}_i$ and $B_i = \widetilde{\phi}_i$, we can effect the change of independent variables (17). We note that the transformation (17) is applied in hydrodynamics to pass from Euler variables to Lagrange variables and vice versa. In the particular case when A_i and B_i do not depend on r and s, the transformation (17) under the conditions (18) and (19) is an ordinary transformation of variables, $t = t(t', x')$, $x = x(t', x')$.

Thus we carry out the transformation (17) in the system of equations (5). Assuming that condition (19) is fulfilled and condition (18) results from the system of equations (5), we obtain a new system of equations:

$$\frac{\partial}{\partial t'}\left[\frac{A_1}{\Delta}\varphi_i - \frac{B_1}{\Delta}\psi_i\right] + \frac{\partial}{\partial x'}\left[\frac{B_2}{\Delta}\psi_i - \frac{A_2}{\Delta}\varphi_i\right] = \frac{f_i}{\Delta} \quad (i = 1, 2). \tag{21}$$

It is easily seen that if the line $dx/dt = \xi_i = \xi_i(r, s, t, x)$ is a characteristic for the system of equations (5), then the line

$$\frac{dx'}{\partial t'} = \frac{B_2\xi_i - A_2}{A_1 - B_1\xi_i} = \xi_i' = \xi_i'(r, s, t', x') \tag{22}$$

is a characteristic of the system of equations (21) corresponding to the characteristic ξ_i in the old variables.

To prove the invariance of the contact property of characteristics with respect to transformation of independent variables we must show, therefore, that if the conditions

$$\begin{gathered}[\xi_1\varphi_i - \psi_i] = 0 \quad (i = 1, 2), \\ [\xi_1] = 0\end{gathered} \tag{23}$$

are satisfied, then so are the relations

$$\begin{gathered}[\xi_1'\varphi_i' - \psi_i'] = 0 \quad (i = 1, 2), \\ [\xi_1'] = 0,\end{gathered} \tag{24}$$

where

$$\varphi_i' = \frac{A_1}{\Delta}\varphi_i - \frac{B_1}{\Delta}\varphi_i, \qquad (i = 1,2). \qquad (25)$$
$$\psi_i' = \frac{B_2}{\Delta}\psi_i - \frac{A_2}{\Delta}\varphi_i$$

Using (22) and verifying condition (24), we obtain the equations $[\xi_i B_i - A_1] = 0$ and $[\xi_1 B_2 - A_2] \doteq 0$, which evidently are satisfied by virtue of equations (23). Thereby the invariance of the contact property with respect to transformations of the independent variables is proved.

4. DEFINITION. We shall call the systems of quasilinear equations (1), (5) *weakly nonlinear* in the region of variation of the variables r, s, t and x if each of the characteristics $dx/dt = \xi_1$ and $dx/dt = \xi_2$ is a contact characteristic in this region; in the opposite case, the systems of equations (1), (5) will be called *strongly nonlinear.*

From the proof above, it follows that this classification is invariant with respect to transformations of variables.

As we have shown, the weakly nonlinear system of equations (5) can be written in the form

$$\frac{\partial r}{\partial t} + \xi_1(s, t, x)\frac{\partial r}{\partial x} = f_1(r, s, t, x),$$
$$\frac{\partial s}{\partial t} + \xi_2(r, t, x)\frac{\partial s}{\partial x} = f_2(r, s, t, x). \qquad (26)$$

We find the conservation laws for the system of equations (26). Solving equations (6), corresponding to (26), we obtain

$$\varphi(r, s, t, x) = \frac{F_1(s, t, x) + F_2(r, t, x)}{\xi_2(r, t, x) - \xi_1(s, t, x)},$$
$$\psi(r, s, t, x) = \frac{\xi_2(r, t, x)F_1(s, t, x) + \xi_1(s, t, x)F_2(r, t, x)}{\xi_2(r, t, x) - \xi_1(s, t, x)}, \qquad (27)$$

where F_1 and F_2 are arbitrary functions of their arguments. Furthermore, it is easily seen that for the weakly nonlinear system of equations (26) the direction of the line of discontinuity, on which only one of the invariants r or s is discontinuous, coincides with the direction of the characteristic ξ_1 or ξ_2.

As in the case of a linear system of equations, such a line of discontinuity satisfies all conservation laws resulting from a weakly nonlinear system of equations.

For the generalized solution of the Cauchy problem for the weakly nonlinear system of equations (26) we require that, on the line of discontinuity, only one of the invariants r or s be discontinuous, or, what is the same thing, that this line of discontinuity satisfy all conservation laws resulting from the system of equations (26).

The following is true.

THEOREM. *In order that the system of quasilinear equations* (1) *or* (5) *not lead to the development of a discontinuity in the solution for arbitrary continuous initial data given on a smooth initial curve, it is necessary and sufficient that this system be weakly nonlinear.*

PROOF. NECESSITY. We show that if the system of equations (5) is not weakly nonlinear, then it is possible to choose continuous initial data on the straight line $t = t_0$ such that a discontinuity develops in the solution.

If system (5) is not weakly nonlinear, then $(\partial\xi_1/\partial r)^2 + (\partial\xi_2/\partial s)^2 \not\equiv 0$ in the region considered. Suppose, for instance, $\partial\xi_1/\partial r \not\equiv 0$ and there exists a region of the variables r, s, t, x where $\partial\xi_1/\partial r > 0$.

Now choose numbers $t_0, x_0, s_0, r_0, \Delta x, \Delta r, \Delta s$ and Δt such that the condition

$$\frac{\partial\xi_1(r, s, t, x)}{\partial r} > \varepsilon > 0 \tag{28}$$

is satisfied for

$$|x - x_0| \leqslant \Delta x, \quad |r - r_0| \leqslant \Delta r, \quad |t - t_0| \leqslant \Delta t, \quad |s - s_0| \leqslant \Delta s. \tag{29}$$

We denote the region of variation of the variables r, s, t, x given by the conditions (29) by G_0, and introduce the notation

$$\begin{aligned} M_1 &= \max_{G_0} |f_1(r, s, t, x)|, \\ M_2 &= \max_{G_0} |f_2(r, s, t, x)|, \\ M_3 &= \max_{G_0} \left|\frac{\partial\xi_1(r, s, t, x)}{\partial s}\right|, \\ M_4 &= \max_{G_0} \left|\frac{\partial\xi_1(r, s, t, x)}{\partial x}\right|, \\ M_5 &= \max_{G_0} |\xi_1(r, s, t, x)|. \end{aligned} \tag{30}$$

On the initial straight line $t = t_0$ we give the initial values $s(x) = s_0$, and we consider two points (t_0, x_1^0) and (t_0, x_2^0) on the initial curve. On the interval $[x_1^0, x_2^0]$ of the initial straight line we give initial values of the function $r(x)$ as any continuous function taking the given values at the end points of the interval; that is, let $r(x_1^0) = r_0 + \Delta r/2$ and $r(x_2^0) = r_0 - \Delta r/2$. We show that if the value $x_2^0 - x_1^0$ $(x_2^0 > x_1^0)$ is chosen sufficiently small, then a discontinuity appears in the solution $r(t, x)$ for $t \leqslant t_0 + \Delta t$. In fact, let $r_1(t)$ be the value of the function $r(t, x)$ on the characteristic $dx_1/dt = \xi_1(r_1, s_1, t, x_1)$ passing through the point (t_0, x_1^0), and $r_2(t)$ the value of $r(t, x)$ on the characteristic $dx_2/dt = \xi_1(r_2, s_2, t, x_2)$ passing through the point (t_0, x_2^0). The values r_1, s_1, x_1 and r_2, s_2, x_2 are connected by the relations

$$
\begin{gathered}
r_1(t) = r_0 + \frac{\Delta r}{2} + \int_{t_0}^{t} f_1(r_1, s_1, t, x_1)\, dt \geqslant r_0 + \frac{\Delta r}{2} - M_1(t - t_0), \\
x_1(t) = x_1^0 + \int_{t}^{t_0} \xi_1(r_1, s_1, t, x_1)\, dt, \\
s_0 - M_2(t - t_0) \leqslant s_1(t) \leqslant s_0 + M_2(t - t_0)
\end{gathered}
\tag{31}
$$

in the region G_0. Similar relations also hold for r_2, s_2 and x_2 in the region G_0.

$$
\begin{gathered}
r_2(t) = r_0 - \frac{\Delta r}{2} + \int_{t_0}^{t} f_1(r_2, s_2, t, x_2)\, dt \leqslant r_0 - \frac{\Delta r}{2} + M_1(t - t_0), \\
x_2(t) = x_2^0 + \int_{t}^{t_0} \xi_1(r_2, s_2, t, x_2)\, dt, \\
s_0 - M_2(t - t_0) \leqslant s_2(t) \leqslant s_0 + M_2(t - t_0).
\end{gathered}
\tag{32}
$$

We form the difference $R(t) = x_2(t) - x_1(t)$, $R(t_0) = x_2^0 - x_1^0 > 0$, and show that it is possible to choose the values $t = t_0 + \Delta t_0$, x_1^0 and x_2^0 so that $R(t_0 + \Delta t_0) < 0$, i.e. so that r-characteristics intersect and a discontinuity occurs in the solution. For this it is sufficient to satisfy the conditions

$$
\begin{aligned}
&\frac{\varepsilon}{2} \Delta r \Delta t_0 - M \Delta t_0^2 > x_2^0 - x_1^0, \\
&M = \varepsilon M_1 + M_2 M_3 + M_4 M_5, \\
&x_2^0 + M_5 \Delta t_0 < x_0 + \Delta x, \\
&x_1^0 - M_5 \Delta t_0 > x_0 - \Delta x, \\
&2M_1 \Delta t_0 < \Delta r, \\
&2M_2 \Delta t_0 < \Delta s, \\
&\Delta t_0 \leqslant \Delta t,
\end{aligned}
\tag{33}
$$

which, of course, can always be accomplished by a suitable choice of the points x_1^0 and x_2^0 and the magnitude of Δt_0. Therefore the necessity is proved.

SUFFICIENCY. We must show that, for a weakly nonlinear system of equations, a discontinuity does not develop if the initial data are continuous. Without loss of generality, we shall assume that the initial data are given on the straight line $t = t_0$, since, if this does not happen, we can perform a transformation of the independent variables changing the initial curve into a straight line; and as we do this the system of equations, by virtue of the invariance of the contact property, remains weakly nonlinear.

We assume, further, that for a weakly nonlinear system of equations there exists a generalized solution satisfying all conservation laws resulting from it. Such a generalized solution is characterized by the fact that only one invariant is discontinuous on the line of discontinuity.

Thus, suppose there exists a generalized solution to the Cauchy problem for the weakly nonlinear system of equations (26), taking continuous initial values on the straight line $t = t_0$, having a finite number of lines of discontinuity, and satisfying all the conservation laws (27). We show that this generalized solution is continuous.

Suppose at a point (x_1, t_1) $(t_1 > t_0)$ a discontinuity develops, and $x = x(t)$ is the line of discontinuity passing through (x_1, t_1). We show that the solution $r(t, x)$, $s(t, x)$ is continuous in a neighborhood of the line $x = x(t)$. In fact, let us assume, for instance, that r is discontinuous on this line, while s is continuous by virtue of the condition of compliance with all conservation laws. Then the r-characteristics

$$\frac{dx}{dt} = \xi_1 [s(t, x), t, x] \tag{34}$$

are determined both on the same line $x = x(t)$ and in a neighborhood of the line of discontinuity. Given the continuous function $s = s(t, x)$, we construct the family of characteristics (34) passing through the neighborhood of the point (t_1, x_1). By virtue of the continuous dependence of the solutions of the system of ordinary differential equations

$$\begin{aligned} \frac{dx}{dt} &= \xi_1 [s(t, x), t, x], \\ \frac{dr}{dt} &= f_1 [r, s(t, x), t, x] \end{aligned} \tag{35}$$

on initial data, we conclude that the function $r = r(t, x)$ is continuous on the line $x = x(t)$, i.e. we get a contradiction. Here we have assumed that in the neighborhood of a line of discontinuity there is no singular point of the differential equation (34).

Thus the assumption of the existence of a discontinuity in the generalized solution to the Cauchy problem for a weakly nonlinear system of equations taking continuous initial values leads to a contradiction; the theorem is proved.

The theorem shows that the basic property of quasilinear equations—development of a discontinuity in the solution—is inherent only in strongly nonlinear systems of quasilinear equations.

In conclusion the author expresses thanks to A. N. Tihonov for discussing the question and for valuable comments.

Received 23/JULY/57

BIBLIOGRAPHY

1. N. N. Janenko, *On discontinuities of solutions of quasilinear equations,* Uspehi Mat. Nauk **10** (1955), no. 2 (64), 195–202. (Russian) MR **17**, 490.

2. B. L. Roždestvenskiĭ, *Systems of quasilinear equations,* Dokl. Akad. Nauk SSSR **115** (1957), 454–457; English transl., Amer. Math. Soc. Transl. (2) **42** (1964), 13–17. MR **20** #175.

3. P. D. Lax, *The initial value problem for nonlinear hyperbolic equations in two independent variables,* Contributions to the Theory of Partial Differential Equations, Ann. of Math. Studies, no. 33, Princeton Univ. Press, Princeton, N. J., 1954, pp. 211–229. MR **16**, 828.

Translated by:
M. Balachandra